Y0-BXQ-630

TRANSITION METAL ORGANOMETALLICS IN ORGANIC SYNTHESIS

Volume I

This is Volume 33 of
ORGANIC CHEMISTRY
A series of monographs
Editors: ALFRED T. BLOMQUIST and HARRY H. WASSERMAN

A complete list of the books in this series appears at the end of the volume.

TRANSITION METAL ORGANOMETALLICS IN ORGANIC SYNTHESIS

Volume I

EDITED BY

Howard Alper
Department of Chemistry
University of Ottawa
Ottawa, Ontario, Canada

ACADEMIC PRESS New York San Francisco London 1976

A Subsidiary of Harcourt Brace Jovanovich, Publishers

ACADEMIC PRESS, INC.
111 Fifth Avenue, New York, New York 10003

United Kingdom Edition published by
ACADEMIC PRESS, INC. (LONDON) LTD.
24/28 Oval Road, London NW1

Library of Congress Cataloging in Publication Data

Main entry under title:

Transition metal organometallics in organic synthesis.

(Organic chemistry series)
Includes bibliographies and index.
1. Chemistry, Organic–Synthesis. 2. Organometallic compounds. 3. Transition metal compounds.
I. Alper, Howard.
QD262.T7 547'.2 75-40604
ISBN 0-12-053101-1

PRINTED IN THE UNITED STATES OF AMERICA

CONTENTS

LIST OF CONTRIBUTORS

Numbers in parentheses indicate the pages on which the authors' contributions begin.

Arthur J. Birch (1), Research School of Chemistry, Australian National University, Canberra, Australia

Charles P. Casey (189), Department of Chemistry, University of Wisconsin, Madison, Wisconsin

Ian D. Jenkins* (1), Research School of Chemistry, Australian National University, Canberra, Australia

R. Noyori (83), Department of Chemistry, Nagoya University, Nagoya, Japan

* Present address: School of Science, Griffith University, Nathan, Brisbane, Queensland, Australia

PREFACE

Transition metal organometallic chemistry has been one of the most active areas of chemical research for the past twenty-five years. A significant part of this research has been concerned with the use of transition metal organometallics in organic synthesis. This two-volume work reviews the literature in this area with particular emphasis on the most effective synthetic transformations.

In Volume I, the versatility of olefin complexes in organic synthesis is amply demonstrated in Chapter 1 by Birch and Jenkins. Noyori has summarized the extensive work on coupling reactions of σ-bonded and π-allyl complexes. In the final chapter of Volume I, Casey considers both achievements and possible areas of use of metal–carbene complexes.

In Volume II, which will be published in the near future, the following topics will be covered: Insertion Reactions of Synthetic Utility (H. Alper); Arene Complexes in Organic Synthesis (G. Jaouen and R. Dabard); Alkyne Complexes and Cluster Compounds (D. Seyferth and M. O. Nestle); Reduction and Oxidation Processes (H. Alper).

An important consideration in discussing stoichiometric reactions is the ease of complexation and decomplexation of the ligands in question. The catalytic area is examined in less detail than the stoichiometric area since reviews of catalytic applications of transition metal organometallics exist in the literature.

This book will be a useful reference for synthetic organic and organometallic chemists and for inorganic chemists who wish to become acquainted with the applications of these organometallic complexes as reagents and catalysts. It should also prove useful to graduate students, either as a reference or as a text for a specialized course in synthesis.

I am grateful to the contributors of this volume and to the staff of Academic Press for their splendid cooperation.

Howard Alper

1

TRANSITION METAL COMPLEXES OF OLEFINIC COMPOUNDS

ARTHUR J. BIRCH and IAN D. JENKINS

I. INTRODUCTION

An indication of the importance of a number of transition metals (Fe, Co, Mo, Zn, Cu, Mn, Cr, and V) in organic synthesis is their known presence in the active centers of enzymes. They act as redox, "super-acid," and alkyl-transfer catalysts. Although π complexes of O_2 and N_2 are known, there have been no reported examples of biological olefin complexes. A tempting speculation is that complexing with $Fe(II)L_3$, for example, of an aromatic system could labilize this complex toward nucleophilic attack or result in "bond fixation" as a complexed diene, baring vulnerable unsaturation. A number of aspects of aromatic metabolism could be so explained. A subject which is no longer speculative is the use of transition metal–olefin complexes in organic synthesis, and the aim of the present chapter is to highlight present and potential uses in this area. An exhaustive review of the literature is not intended, but rather an introduction to the types of transformations that are possible using organometallic intermediates with an emphasis on the principles and reaction mechanisms involved. For chemical syntheses illustrative examples are most instructive, so reaction conditions and yields will be given wherever possible.

Several general principles are to be noted initially. The metal atom and necessary associated structures can act by bringing reacting centers into proximity, by stabilizing otherwise high-energy transition states, by activating one or more reactants in specific ways, or by stabilizing otherwise unobtainable products in reactions. There may also be important stereochemical implications. Another general point is that to be practically useful, the metal must either be relatively cheap or be involved in a catalytic manner. Because of the considerable literature on catalytic processes, this area, although covered here,* is given a less detailed treatment than that involving stoichiometric reactions. More attention has also been given to reactions involving readily available metals. Finally, in assessing potential utility to the synthetic organic chemist, the metal atom must be readily removed if it is to be used as "scaffolding" to assist in the construction of an organic molecule.

* A monograph on the use of noble metal catalysts in organic synthesis is available (Rylander, 1973). Catalytic hydrogenation has been reviewed in detail recently (Birch and Williamson, 1975) and will not be treated here.

II. GENERAL CONSIDERATIONS IN ORGANIC SYNTHESIS. PREPARATION OF STARTING MATERIALS

What effect can a transition metal have on double bonds? According to the metal and the circumstances, it can activate, deactivate, or protect the double bonds for electrophilic or nucleophilic attack, resolve geometric or optical isomers, direct attack stereospecifically, and aromatize or dearomatize appropriate systems. This is a list of achievements often difficult or impossible to reach by standard organic-type reactions alone.

Another important question to the organic chemist is the ease of preparing and handling transition metal π complexes. Usually there are no unfamiliar handling properties, although there is the usual emphasis on the beneficial use of pure solvents and the exclusion of oxygen. Even these precautions are often not necessary.

Since tricarbonyliron complexes are quoted extensively, some general points should be noted to provide some practical knowledge on the subject. Diene complexes are readily prepared from: (a) iron pentacarbonyl, $Fe(CO)_5$, by refluxing the diene with an excess of the reagent in dibutyl ether under nitrogen for 10–30 hr (Cais and Maoz, 1966); (b) triiron dodecacarbonyl, $Fe_3(CO)_{12}$ (King and Stone, 1961; McFarlane and Wilkinson, 1966), by refluxing the diene with an excess of the reagent in a suitable solvent such as benzene or dimethoxyethane; and (c) diiron enneacarbonyl, $Fe_2(CO)_9$ (Braye and Hübel, 1966), by stirring the diene itself with the solid reagent, by refluxing the mixture in pentane, or by treating the diene with the burgundy-colored complex formed when $Fe_2(CO)_9$ dissolves in refluxing acetone under nitrogen. This latter procedure (A. J. Birch and I. D. Jenkins, unpublished work) is particularly mild and can be used for forming complexes of some dienes at room temperature. Another mild method that is sometimes useful involves the transfer of the $Fe(CO)_3$ group from the tricarbonyliron complex of benzylideneacetone (Howell *et al.*, 1972).

Many of the resulting complexes can be chromatographed, distilled, or crystallized, and standard methods such as thin layer (tlc) and gas-liquid (glc) chromatography are frequently applicable. Mass spectrometry and nmr spectra are very useful for defining structures. For key references to the characterization of organometallic compounds by ir, nmr, mass spectrometry, and glc, see Tsutsui (1971), Maddox *et al.* (1965), Bruce (1968), Lewis and Johnson (1968), and Guiochon and Pommier (1973).

Removal of the $Fe(CO)_3$ group is normally achieved by treating the complex with an oxidizing agent such as ceric ammonium nitrate, ferric chloride, or cupric chloride, usually at between 0° and room temperature, in a solvent such as aqueous acetone or ethanol. In some cases, trimethylamine *N*-oxide can be used (Shvo and Hazum, 1974). The nature of the oxidizing agent is important; chromic acid, for example, can oxidize a secondary alcohol to the ketone before removing the iron.

Caution may be required in oxidation reactions. Ceric salts, for example, can alter or destroy products. Ferric salts seem to be more active in 1 *N* hydrochloric acid solution (P. Schudel, personal communication), but the acidic conditions may affect products such as enol ethers. Temperatures also seem to be important in some cases. For example, a substituted cyclohexadiene complex was unaffected by $CuCl_2$ at $-20°$, but it was converted into the free conjugated diene (mainly) at 0° and into the unconjugated diene and substituted benzene derivative at $+20°$ (A. J. Birch and C. Sell, unpublished). In other cases, e.g., arylmanganese derivatives, although interesting synthetic reactions occur, the metal is not so readily removed. This general area is clearly one for further research if useful application is to be made in organic synthesis.

Before considering the subject systematically, a few illustrative examples are given. Iron carbonyl complexes of monoolefins of the type **I** have considerable potential in organic synthesis (Section III,A,1).

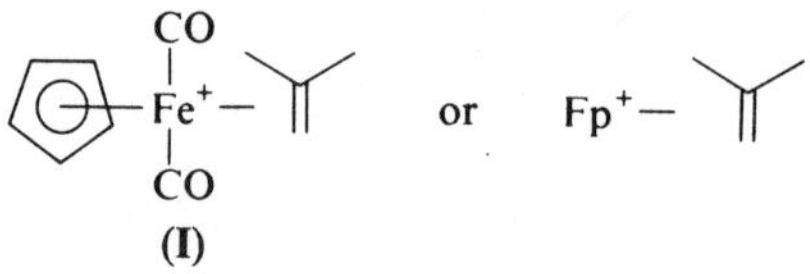

CO, Fe^+, CO or Fp^+—

(I)

The starting materials for these complexes are dicarbonyl(cyclopentadienyl)-iron chloride (Fischer and Moser, 1970) and the dicarbonylcyclopentadienyl-iron anion [as the sodium salt (Green and Nagy. 1963)], both obtainable from tetracarbonyl (di-π-cyclopentadienyl)diiron which is available commercially. The reaction is carried out in tetrahydrofuran (THF). Equation (1) demonstrates the ease of manipulation of these substances in the preparation of an olefin complex from an epoxide. The tetrafluoroborate salt precipitates from ether. This reaction is also useful as a means of reducing expoxides stereospecifically to olefins with retention of configuration. The olefin is readily liberated from the complex by treatment with sodium iodide in acetone at room temperature for 15 min (Giering *et al.*, 1972).

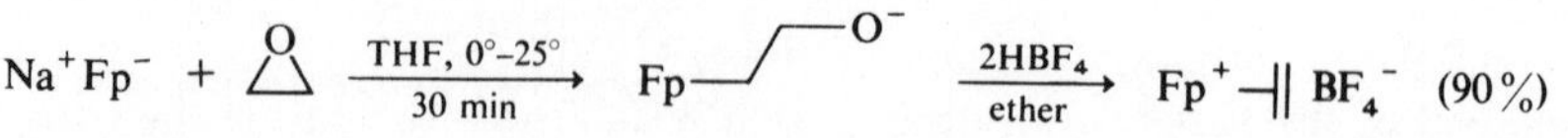

$$Na^+Fp^- + \text{(ethylene oxide)} \xrightarrow[\text{30 min}]{\text{THF, 0°–25°}} Fp\text{—CH}_2\text{CH}_2\text{O}^- \xrightarrow[\text{ether}]{2HBF_4} Fp^+\text{—}\|\ BF_4^- \quad (90\%) \qquad (1)$$

Some olefin complexes of this type are readily prepared by an exchange reaction with the isobutylene complex (**I**). The olefin is heated briefly with **I** in a chlorocarbon solvent at 60°, isobutylene being liberated (Giering and Rosenblum, 1971).

Triene complexes such as (benzene)tricarbonylchromium are readily prepared from the aromatic compound and chromium hexacarbonyl (Rausch, 1974), from $(NH_3)_3Cr(CO)_3$ (Moser and Rausch, 1974), or from $(MeCN)_3$-$Cr(CO)_3$ (Me = methyl) (Knox *et al.*, 1972). Another efficient method is to heat a readily available dihydroanisole with chromium hexacarbonyl; the benzene derivative results in high yield by loss of methanol (Birch *et al.*, 1965, 1966). Similar reactions can be performed with molybdenum or tungsten carbonyls. The $Cr(CO)_3$, $Mo(CO)_3$, or $W(CO)_3$ can be removed oxidatively with iodine in ether at 0°—25° (Semmelhack *et al.*, 1975), with ferric chloride in ethanol (Trahanovsky and Baumann, 1974), by displacement with phosphine ligands (Mathews *et al.*, 1959), in some cases by nitrogen ligands such as pyridine (Kutney *et al.*, 1974), or simply by bubbling air through a solution of the complex in ether (Birch *et al.*, 1965, 1966). This also represents a practical procedure for indirect removal of an OR group from a phenol ether.

Before discussing some applications of organometallics in organic synthesis, it is necessary to consider briefly, coordination number and bonding in these complexes.

A. The Concept of Coordinative Unsaturation

The 16- and 18-electron rule (recently reviewed by Tolman, 1972) is best explained by a few examples. Iron has 26 electrons and has a "d^8" configuration in the Fe(0) state. It can achieve the next inert gas (Kr) electronic configuration of 36 electrons (i.e., 18 outer electrons) by coordinating to the equivalent of five 2-electron donors as, for example, in $Fe(CO)_5$ and $(Ph_3P)_2Fe(CO)_3$. The iron atom in these compounds is said to be "coordinatively saturated." The mechanism by which $Fe(CO)_5$ reacts with an olefin is by initial dissociation of a molecule of CO to give the 16-electron ("coordinatively unsaturated") $Fe(CO)_4$ species, which can then coordinate with the olefin to give once again an 18-electron species, (olefin)$Fe(CO)_4$. Oxidation of the iron to Fe(II) results in a d^6 configuration, so that six 2-electron donors are now required to achieve the inert gas configuration, e.g., in ferrocene and $(C_5H_5)Fe^+(C_6H_6)$. [Ferrocene could also be regarded as an Fe(0) species

Ferrocene

coordinated to two 5-electron donors.] In σ-bonded species such as $(C_5H_5)Fe(CO)_2CH_2CH_3$, the iron is probably best regarded as Fe(II) with a 6-electron $C_5H_5^-$ donor, a 2-electron ethyl (Et) anion donor and two 2-electron CO donors, but it could also be treated (for electron bookkeeping purposes) as an Fe(0) species with a 5-electron C_5H_5 radical donor, a 1-electron ethyl radical donor, and two 2-electron CO donors.

Similarly, Cr and Mo (Group VI transition metals) also have a d^6 configuration and therefore give rise to compounds such as $Cr(CO)_6$ and $(CO)_3Cr(C_6H_6)$.

While some transition metal complexes are more stable with an 18-electron configuration than a 16-electron configuration, others prefer the latter, for example, complexes of Ti(IV), Zr(IV), Pt(II), and Pd(II). Some Pt(0) and Pd(0) complexes show similar stabilities with either configuration. Thus, in solution, $Pd(PPh_3)_4$ is in equilibrium with $Pd(PPh_3)_3$ (Hartley, 1973).

Reaction pathways usually involve only 16- and 18-electron species or intermediates. Complexes of 18 electrons undergo ligand dissociation, reductive elimination, insertion, and oxidative coupling, whereas 16-electron complexes undergo ligand association, oxidative addition, reversal of insertion type, and reductive coupling reactions. These two mechanisms lead to different types of synthetically useful reactions.

B. Bonding

In monoolefin complexes such as Zeise's salt, $K^+[Cl_3Pt(C_2H_4)]^-$, and $(Ph_3P)_2Pt(C_2H_4)$ and $(CO)_4Fe(EtO\text{–}CH{=}CH_2)$, the olefinic carbon atoms are both bonded to the metal atom in the form of a "π bond" (**II**) in which the π electrons of the olefin overlap with orbitals of suitable symmetry on the metal atom, while at the same time there is "back-donation" of electron density from filled *d* orbitals of suitable symmetry on the metal into low-lying antibonding orbitals of the olefin (Dewar, 1951; Chatt and Duncanson, 1953). The dual character of the metal–olefin bond has important consequences, as the two components are synergically related. As one component increases, it facilitates an increase in the other, thus tending to keep the metal–olefin bond essentially electroneutral. In very simple terms, the π bonding which would result in effective charge separation is neutralized electronically by the back-bonding from the metal.

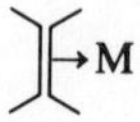

M (**II**)

M (**III**)

Depending on the metal and its oxidation state, there may be a considerable contribution from the σ-bonded form (**III**). For example, in $K[PtCl_3(C_2H_4)]$ there is a low barrier to rotation about the metal–olefin bond (implying a bonding structure as in **II**), while in $Pt(PPh_3)_2(C_2H_4)$, the rotation barrier is too high to be observed by nmr (implying a bonding structure more like **III**) (Hartley, 1972). Similarly, with diolefin complexes, the metal atom is bonded to the four olefinic carbon atoms and usually lies between the two extremes, **IV** and **V**.

M
(IV)

M
(V)

In the vinylnaphthalene–tricarbonyliron complex (**VI**), for example, the C-1–C-2, C-2–C-11, and C-11–C-12 bond lengths are all approximately equal (Davis and Pettit, 1970).

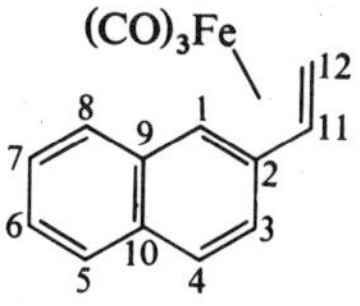

$(CO)_3Fe$
(VI)

Approximate bond lengths

1,2 = 2,11 = 11,12 = 1.41 Å

2,3 = 4,10 = 9,1 = 1.46 Å

3,4 = 1.31 Å

10,5 = 5,6 = 6,7 = 7,8 = 8,9 = 1.38 Å

The tricarbonyl(cyclohexadienyl)iron salt (**VII**) has five carbon atoms bonded to the iron (cf. ferrocene).

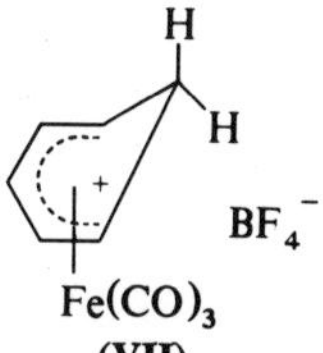

$Fe(CO)_3$ BF_4^-
(VII)

III. C–C BOND FORMATION

A. Nucleophilic Attack on Coordinated Double Bond Systems

1. Monoolefins

Simple olefins are very inert to attack by nucleophiles. However, an olefin coordinated to an electrophile, as in the bromium or mercurinium ions (**VIII**, X = Br^+ or Hg^{2+}), is well known to be readily attacked by nucleophiles, and the same is true when X = Fp^+, Pd^{2+}, or Pt^{2+}. Thus, **IX** reacts with a nucleophile Nu^-, under mild conditions to give the σ-bonded iron species **X**.

or

(VIII)

The iron is readily removed from such systems by hydrochloric acid [Eq. (2)]. Bromine cleaves the Fe–C σ bond selectively in the presence of an olefin, and with inversion of configuration, gives the corresponding bromo derivative [Eq. (3)]. Oxidation results in "CO insertion" and the corresponding carboxy derivatives are formed in good yield [Eq. (4)]. The olefin from complexes such as **IX** is readily liberated on treatment with sodium iodide in acetone at room temperature (RT) for 15 min.

Fp^+ (IX) + Nu^- → Fp–CH₂CH₂–Nu (X) —HCl, CH_2Cl_2, RT→ H–CH₂CH₂–Nu (2)

—Br_2, −78%→ Br–CH₂CH₂–Nu (3)

—Ce(IV), MeOH, RT→ MeO_2C–CH₂CH₂–Nu (4)

—$HgCl_2$→ ClHg–CH₂CH₂–Nu (5)

Typical examples of C–C bond-forming reactions that are possible using olefins coordinated through iron are given in Eqs. (6)–(8) (Rosan *et al.*, 1973; Rosenblum, 1974; Nicholas and Rosan, 1975).

^+Fp + N (enamine) —0°→ Fp (iminium, N+) (80%) (6)

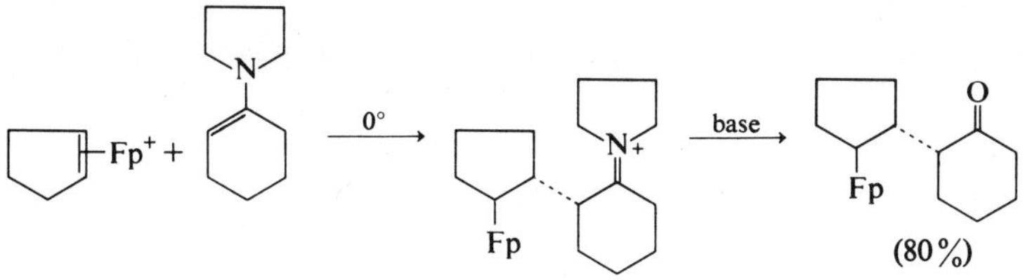

0° base (80%) (7)

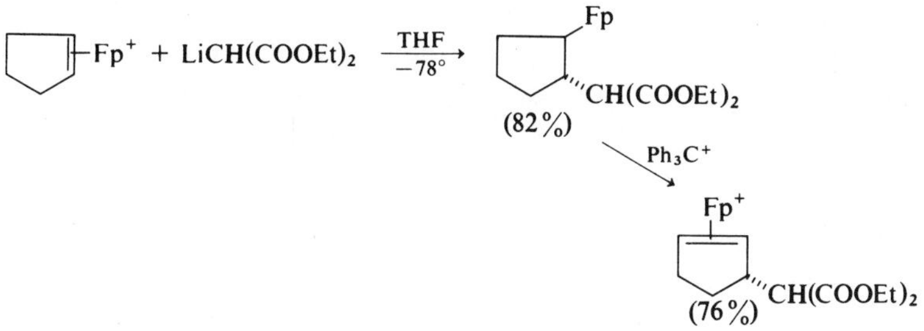

THF −78° (82%) Ph_3C^+ (76%) (8)

Use of strongly anionic nucleophiles such as alkylmagnesium halides and lithium alkyls leads to reductive dimerization or displacement of the olefin. It is quite possible however that alkylation of olefin–iron cations (**IX**) would be feasible with alkyl derivatives of zinc, cadmium, or tin reagents. Alkylation of dienyliron cations with dialkylzinc and dialkylcadmium reagents has recently been demonstrated (Section III,A,4). Alkylation of olefin–iron cations can be achieved under mild conditions with σ-bonded allyliron compounds [Eqs. (9) and (10), (Rosan *et al.*, 1973)].

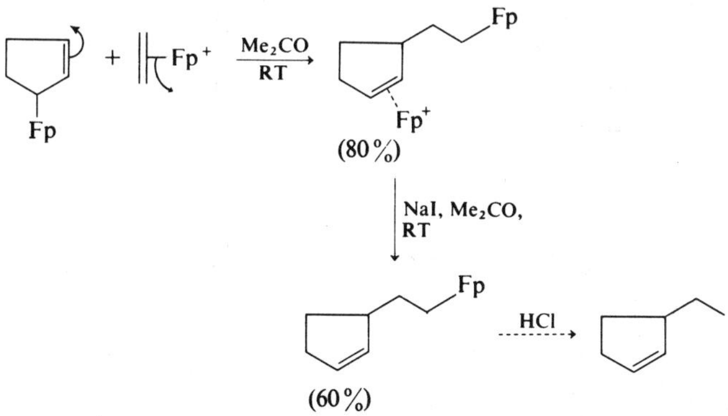

Me_2CO, RT (80%) NaI, Me_2CO, RT (60%) HCl (9)

Fp + Fp$^+$ → Fp ... Fp + Fp$^+$ ... Fp

NaI NaI

Br ← Br_2, −78° Fp ... Fp

40%

(10)

Olefins can be arylated or alkylated at room temperature in the presence of Pd(II) salts and air or moisture to give yields varying from a few percent to near quantitative. Carbon–carbon bond formation usually takes place at the less alkyl-substituted carbon atom of the double bond (Heck, 1968, 1969). Pd(II) salts catalyze nucleophilic attack on olefins more readily than do Pt(II) salts and both cis and trans addition are observed (Hartley, 1969).

The main differences between using an iron cation and Pd(II) to activate an olefin are: (a) Pd(II) reactions are often catalytic processes: (b) the C–Pd σ bond that results from nucleophilic attack on the coordinated olefin is reactive and undergoes either decomposition or further transformations, usually with reduction of the Pd species to Pd(0). There are numerous examples of arylation reactions of Pd(II)-coordinated olefins, but only a few simple alkylation reactions. Pd-catalyzed reactions are usually more important for forming C–X bonds where X is an electronegative element (Section IV). Some examples of typical Pd-catalyzed C–C bond-forming reactions are given in Eqs. (11)–(15) (Heck, 1968, 1969), Eq. (16) (Heck, 1972), and Eq. (17) (Brown, 1974) (see also S. Tsuji, 1969; J. Tsuji, 1972; Bird, 1972, Henry, 1973).

$$\text{PhCH=CH}_2 + \text{Me}_4\text{Sn} \xrightarrow[\text{RT, 24 hr}]{\text{Li}_2\text{PdCl}_4\text{, MeOH}} \textit{trans}\text{-PhCH=CHMe} \quad (95\%) \tag{11}$$

$$\text{MeOOCCH=CH}_2 + \text{Me}_4\text{Sn} \xrightarrow[\text{RT, 24 hr}]{\text{Li}_2\text{PdCl}_4\text{, MeOH}} \textit{trans}\text{-MeOOCCH=CHMe} \quad (57\%) \tag{12}$$

$$\text{PhCH=CH}_2 + p\text{-MeC}_6\text{H}_4\text{HgCl} \xrightarrow[\text{RT, 24 hr}]{\text{Li}_2\text{PdCl}_4\text{, MeOH}} \textit{trans}\text{-PhCH=CH(C}_6\text{H}_4\text{-Me-}p\text{)} \quad (47\%) \tag{13}$$

$$C_2H_4 + ClHgCOOMe \xrightarrow[\text{RT, 24 hr}]{Li_2PdCl_4,\ AcOH} CH_2{=}CH{-}COOMe \quad (14)$$

(Ac=acetyl) (50%)

$$C_2H_4 + PhHgCl \xrightarrow[AcOH,\ H_2O]{Li_2PdCl_4,\ CuCl_2,\ LiCl} PhCH_2CH_2Cl \quad (15)$$

(70%)

$$CH_2{=}CHOAc + PhHgOAc \xrightarrow[\text{RT, 16 hr}]{Pd(OAc)_2} \textit{trans}\text{-}PhCH{=}CHOAc$$

(74%)

$$+ \textit{cis}\text{-}PhCH{=}CHOAc + PhCH{=}CH_2 \quad (16)$$

(17%) (5%)

Hydrocyanation

$$MeCH{=}CH_2 + HCN \xrightarrow{Pd[(PhO)_3P]_4} MeCH_2CH_2CN + MeCH(CN)Me \quad (17)$$

(75%) (25%)

The mechanism of these transformations differs from that for nucleophilic attack on iron-coordinated olefins in that the nucleophile appears to coordinate first to the Pd and then to migrate to the coordinated olefin (Scheme 1).

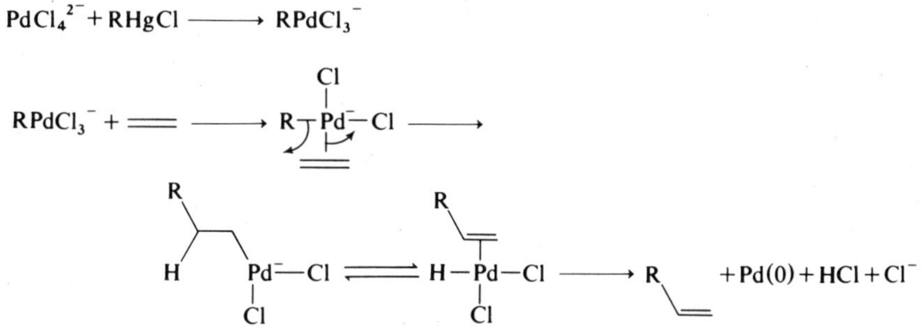

$$PdCl_4^{2-} + RHgCl \longrightarrow RPdCl_3^-$$

$$RPdCl_3^- + {=} \longrightarrow R{-}Pd^-(Cl)_2(\eta^2\text{-olefin}) \longrightarrow$$

$$R(H)CH{-}CH_2{-}Pd^-Cl_2 \rightleftharpoons H{-}Pd(Cl)_2(RCH{=}CH_2) \longrightarrow RCH{=}CH_2 + Pd(0) + HCl + Cl^-$$

Scheme 1

A related C–C bond-forming reaction [Eq. (18)], involving coupling of two vinyl groups to give (stereoselectively) a 2,4-diene, promises to be a synthetically useful reaction (Vedejs and Weeks, 1974). Vinylic mercurials are

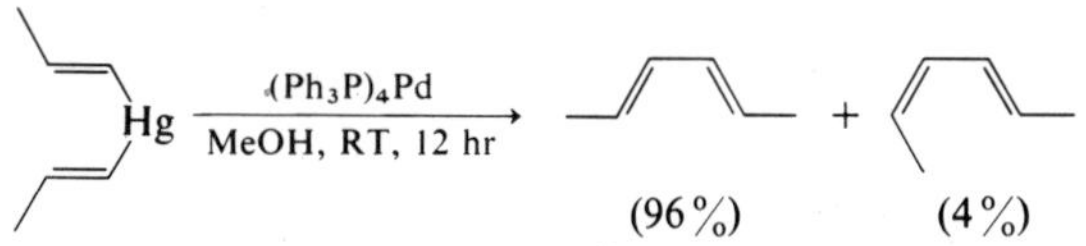

$$\text{(dipropenyl)Hg} \xrightarrow[\text{MeOH, RT, 12 hr}]{(Ph_3P)_4Pd} \text{2,4-hexadiene (96\%)} + \text{isomer (4\%)} \quad (18)$$

readily available from alkynes via a hydroboration/mercuration procedure. The reaction presumably takes place by a double oxidative addition of the Hg–C bonds to the Pd(0), with Hg extrusion, followed by Pd extrusion [Eq. (19)]. Addition of the C–Cl bond across an olefin in the presence of a

Hg Pd(Ph$_3$P)$_2$ ⟶ Pd Hg ⟶ Pd (19)

cobalt catalyst may involve oxidative addition followed by a similar $\pi \rightarrow \sigma$ rearrangement mechanism [Eq. (20) (Mori and Tsuji, 1972)].

$$+ \mathrm{Cl_3CCOOMe} \xrightarrow[150°,\ 16\ \mathrm{hr}]{\mathrm{Co_2(CO)_8}} \quad \mathrm{CCl_2COOMe},\ \mathrm{Cl}\ (74\%) \tag{20}$$

2. Olefinic (Allyl) Carbonium Ions

An Fe(0) atom coordinated to an olefin stabilizes a neighboring (conjugated) carbonium ion **XI**.

Fe(CO)$_4$ ≡ [Fe(CO)$_4$]$^+$

(XI) **(XII)**

This stabilization is presumably due to electron donation by the Fe(0), perhaps with significant contributions from an allyl-Fe(II) species, **XII**. Fe(0)-stabilized carbonium ions have been used very little in organic synthesis, but their potential value is illustrated in Eqs. (21) and (22) (Whitesides *et al.*, 1973). The closely related Fe(0)-stabilized dienyl carbonium ions

$$\mathrm{Fe(CO)_3} \xrightarrow[(2)\ \mathrm{DBF_4/CO}]{(1)\ \mathrm{CF_3COOD}} \mathrm{CD_3}\ \mathrm{CD_2}\ \mathrm{Fe(CO)_4} \tag{21}$$

specifically labeled isoprenyl units

Fe(CO)$_4$ + MeCO$\bar{\mathrm{C}}$HCOOMe

(1) THF, RT, Few minutes
(2) saponification
(3) decarboxylation

COMe + MeCO (22)

(68%) (17%)

(Section III,A,4) have proved extremely useful as intermediates in organic synthesis so there is good precedent for predicting reactions such as Eqs. (23)* and (24).

* Reaction (23) has recently been accomplished (A. J. Pearson, personal communication).

$$\text{R-allyl}^{+}\text{Fe(CO)}_4 \xrightarrow[\text{(2) air}]{\text{(1) } R_2'Cd} \text{R-CH=CH-CH}_2\text{-R}' \tag{23}$$

$$\text{R-allyl}^{+}\text{Fe(CO)}_4 \xrightarrow[\text{(2) air}]{\text{(1) Zn/Cu}} \text{CH}_2\text{=CH-CHR-CHR-CH=CH}_2 \tag{24}$$

Fe(0)-stabilized carbonium ions are normally prepared by protonation of Fe(0)–diene complexes with HBF_4 in the presence of carbon monoxide [Eq. (21)] or by protonation of diene–tetracarbonyliron complexes [Eq. (25)] (Gibson and Vonnahme, 1974).

$$\text{diene} \xrightarrow{Fe_2(CO)_9} \text{diene-Fe(CO)}_4 \xrightarrow{HBF_4} \text{allyl}^{+}\text{Fe(CO)}_4\ BF_4^- \ (84\%) \tag{25}$$

Preparation might be possible directly from the olefin, $Fe_3(CO)_{12}$, and trityl fluoroborate.

3. Dienes

Nucleophilic attack on dienes coordinated to Pd(II) is facile [Eq. (26)] (Takahashi and Tsuji, 1968).

$$\text{(1,5-cyclooctadiene)PdCl}_2 \xrightarrow[\text{RT, Et}_2\text{O, 15 hr}]{CH_2(COOEt)_2,\ Na_2CO_3} \textbf{(XIII)} \tag{26}$$

(XIII): $[\text{CH(COOEt)}_2\text{-cyclooctenyl-Pd-Cl}]_2$

(XIII) $\xrightarrow[\text{RT, 2 hr}]{NaBH_4,\ EtOH,\ CH_2Cl_2}$ cyclooctyl-$CH(COOEt)_2$ (75%)

(XIII) $\xrightarrow{NaCH_2SMe}$ $[\text{C(COOEt)}_2\text{-cyclooctenyl-Pd-Cl}]_2$ $\longrightarrow$ bicyclo[6.1.0]non-4-ene-9,9-$(COOEt)_2$ (50%)

4. Dienyl Carbonium Ions

Stabilized dienyl cations of the type **XIV** or **XV** are potentially very useful as intermediates in organic synthesis.

$Fe(CO)_3$ **(XIV)** $\quad$ $+$—$Fe(CO)_3$ **(XV)**

As the tetrafluoroborate salts, they are stable crystalline solids, and are readily obtainable from suitable tricarbonyldieneiron derivatives, for example, by hydride abstraction with trityl fluoroborate in methylene chloride followed by precipitation of the salt with ether. The more common methods of synthesis are illustrated in Eqs. (27) (Fischer and Fischer, 1960), (28)–(30) (Birch *et al.*, 1968; Birch and Haas, 1971), (31) (Mahler and Pettit, 1963), and (32) (Dauben and Bertelli, 1961).

$Fe(CO)_3$ $\xrightarrow[\text{RT, 30 min}]{Ph_3C^+BF_4^-,\ CH_2Cl_2}$ $Fe(CO)_3$ (+) BF_4^- (27)

H_3C $Fe(CO)_3$ $\xrightarrow[\text{RT, 30 min}]{Ph_3C^+BF_4^-,\ CH_2Cl_2}$ H_3C $Fe(CO)_3$ (+) (56%) + H_3C $Fe(CO)_3$ (+) (38%) (28)

H_3CO $Fe(CO)_3$ $\xrightarrow[\text{RT, 30 min}]{Ph_3C^+BF_4^-,\ CH_2Cl_2}$ H_3CO $Fe(CO)_3$ (+) (90%) $+ BF_4^-$ (29)

H_3CO $Fe(CO)_3$ $\xrightarrow[NH_4PF_6]{H_2SO_4}$ $Fe(CO)_3$ (+) (70–80%) PF_6^- (30)

CH_2OH $\xrightarrow[\text{Et-cyclohexane}]{Fe(CO)_5,\ 36\ hr}$ CH_2OH $Fe(CO)_3$ (45%) $\xrightarrow[Et_2O]{HCl,\ Ac_2O}$ ClO_4^- $Fe(CO)_3$ (80%) (31)

$Fe(CO)_3$ $\xrightarrow{HBF_4}$ BF_4^- $Fe(CO)_3$ (90%) (32)

Stabilized dienyl cations are good alkylating agents and undergo a range of C–C bond-forming reactions with suitable nucleophiles under mild conditions. Nucleophilic attack usually takes place on the face of the molecule opposite the tricarbonyliron group. Which carbon in the system is attacked is determined by both steric and electronic factors in a way at present not completely understood (see also Section IX). Some typical C–C bond-forming reactions using dienyl carbonium salts and nucleophilic reagents such as ketones, enamines, electron-rich π systems, and carbanions are illustrated in Eqs. (33) (Birch *et al.*, 1968), (34) (Birch *et al.*, 1973*a*; Ireland *et al.*, 1974), (35) (Birch *et al.*, 1973a), (36) (Birch *et al.*, 1973b), (37) and (38) (Kane-Maguire and Mansfield, 1973), (39) (Pelter *et al.*, 1974), and (40) (Mansfield *et al.*, 1964).

MeO $Fe(CO)_3$ $\xrightarrow[RT]{NaCN,\ H_2O}$ MeO $Fe(CO)_3$ CN (58%) $\xrightarrow{\text{Pd/C, toluene, reflux}}$ MeO CN (66%) (33)

OMe

Fe(CO)$_3$ + N

OMe

Fe(CO)$_3$

O

(88%)

CrO_3, RT, few minutes (34)

O

O

(69%)

Fe(CO)$_3$ + O $\xrightarrow[\text{reflux, 7 hr}]{\text{EtOH}}$ (CO)$_3$Fe O

(70%)

$FeCl_2$, EtOH, HCl, RT, 3 hr (35)

O

(75%)

MeO Fe(CO)$_3$ + O O $\xrightarrow[\text{reflux, 2 hr}]{\text{EtOH}}$ MeO (CO)$_3$Fe O O

(95%)

$\xrightarrow[\text{reflux, 2 hr}]{MnO_2,\ C_6H_6}$ MeO (CO)$_3$Fe H O H O (36)

(75%)

$Fe(CO)_3$ ... + ... $\xrightarrow[CH_2Cl_2]{H_2O \text{ or}}$ ... N H ... Me ... $Fe(CO)_3$ ... N H ... Me (37)

$Fe(CO)_3$ ... + (Fe) ⟶ ... Fe ... $Fe(CO)_3$ (38)

MeO ... $Fe(CO)_3$ + $R'_3\bar{B}{-}C{\equiv}C{-}R^2$ ⟶ MeO ... $Fe(CO)_3$ ... $C{=}C(R')BR'_2$ ... R^2

(XVI)

(39)

(XVI) $\xrightarrow{Me_3NO}$ MeO ... $Fe(CO)_3$ ... $CH{-}COR'$ ... R^2 (48%) $\xrightarrow{Me_3NO}$ MeO ... $CH{-}COR'$ ... R^2 (46%)

(XVI) $\xrightarrow{H^+}$ MeO ... $Fe(CO)_3$ ... $C{=}CHR'$ ... R^2 (77%) $\xrightarrow{Ce(IV)}$ O ... $C{=}CHR'$ ... R^2 (72%)

$Fe(CO)_3$ ... + ... OMe ... OMe ⟶ MeO ... $Fe(CO)_3$ ... OMe ... OMe (40)

Although alkylmagnesium halides lead to reductive dimerization of the dienyl cations, synthetically useful alkylations and arylations have recently been achieved using zinc and cadmium reagents [Eqs. (41) and (42) (Birch and Pearson, 1975)].

$Fe(CO)_3$ + Ph_2Cd —THF, 0°, 5 min→ $(CO)_3Fe$ Ph (83%)

$Fe(CO)_3$ + $(PhCH_2)_2Cd$ —THF, 0°, 5 min→ $(CO)_3Fe$ CH_2Ph (72%) (41)

$Fe(CO)_3$ + Cd —THF, 0°, 5 min→ $(CO)_3Fe$ (82%)

$Fe(CO)_3$ + Cd —THF, 0°, 5min→ $Fe(CO)_3$ (52%) (42)

The use of reactions such as Eq. (42) in asymmetric synthesis is discussed in Section IX. Similar alkylations of acyclic dienyl cations with zinc and cadmium reagents have also been achieved (Birch and Pearson, 1975).

Another useful C–C bond-forming reaction employing dienyl cation reagents is reductive dimerization with zinc or, better, Zn/Cu couple. [Eqs. (43) (A. J. Birch and B. Chauncy, unpublished work; Hashmi *et al.*, 1967) and (44) (Mahler and Pettit, 1963)].

$Fe(CO)_3$ —Zn/Cu, THF, RT, 16 hr→ $(CO)_3Fe$ $Fe(CO)_3$ (70%) (43)

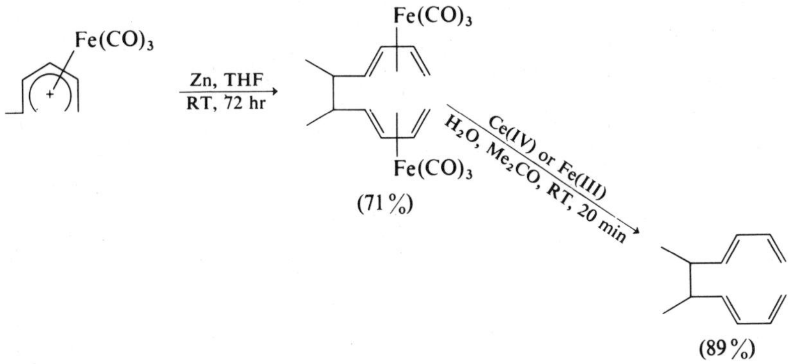

$Fe(CO)_3$ Zn, THF; RT, 72 hr → $Fe(CO)_3$ $Fe(CO)_3$ (71%) → Ce(IV) or Fe(III); H_2O, Me_2CO, RT, 20 min → (89%) (44)

These reactions may take place via a radical mechanism (Scheme 2).

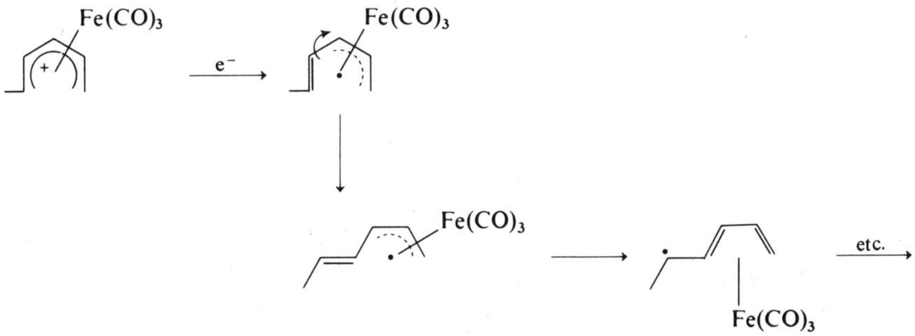

$Fe(CO)_3$ e^- → $Fe(CO)_3$ → $Fe(CO)_3$ → $Fe(CO)_3$ etc. →

Scheme 2

A similar dimerization reaction takes place with strongly anionic reagents and alkylmagnesium halides. In the presence of strong bases, an iron(0) species can apparently act as a reducing agent for the dienyl cation. Basic alumina, however, results in a nonreductive dimerization reaction with **XVII** to give a mixture of diastereoisomeric pentaenes (Anderson *et al.*, 1973). This reaction presumably involves an initial elimination to give a triene which undergoes electrophilic attack by the dienyl cation [Eq. (45)].

$Fe(CO)_3$ $Fe(CO)_3$

(XVII) Al_2O_3, 60 hr (45)

$Fe(CO)_3$ $Fe(CO)_3$ $Fe(CO)_3$ $Fe(CO)_3$

(60%)

5. Trienes and Arenes

Although aromatic compounds are discussed in detail elsewhere in the text, they will be treated briefly here for the sake of completeness.

Nucleophilic attack on trienes, especially aromatic compounds, is a particularly difficult reaction. Direct alkylation of benzene by lithium reagents, for example, occurs very slowly at 165° (Dixon and Fishman, 1963).

Coordination of the arene to Fe(II), tricarbonylmanganese(I), or tricarbonylchromium(0), however, renders the aromatic ring susceptible to nucleophilic attack. Some typical C–C bond-forming reactions on coordinated trienes and arenes are illustrated in Eqs. (46) (Semmelhack *et al.*, 1975) (47) (Helling and Braitsch, 1970), (48) (Card and Trahanovsky, 1973), and (49)–(51) (Walker and Mawby, 1973a,b,c). Borohydride reduces arene metal cations to π-cyclohexadienyl complexes (Jones *et al.*, 1962), and some nucleophiles add (Helling and Cash, 1974).

$Cr(CO)_3$ + Bu

$Cr(CO)_3$ + Bu^tLi $\xrightarrow[(2)\ I_2,\ 25°]{(1)\ THF,\ 0°,\ 30\ min}$ $PhBu^t$ (97%)

+ Li(S-dithiane) $\xrightarrow[(2)\ I_2,\ 25°]{(1)\ THF,\ 0°,\ 30\ min}$ Ph-(dithiane) (93%) (46)

+ Li–C_6H_4–Me $\xrightarrow[(2)\ I_2,\ 25°]{(1)\ THF,\ 0°,\ 30\ min}$ Ph–C_6H_4–Me (71%)

$[(\text{Mesitylene})_2\text{Fe}](\text{PF}_6)_2 + 2\text{PhLi} \xrightarrow[-70°\text{–RT}]{\text{THF}}$ Ph, H, Fe, H, Ph (57%) $\xrightarrow{\text{KMnO}_4}$ Ph (95%) (47)

Et Cr(CO)_3 $\xrightarrow[(2)\ \text{Ce(IV)},\ H_2O,\ \text{MeCN}]{(1)\ \text{Bu}^t\text{Li, THF, pentane, }-10°,\ 25\text{ min}}$ Et (49%) + Et Bu^t (32%) + Et Bu^t (9%) (48)

$^+\text{Mn(CO)}_3$ $\xrightarrow[H_2O]{CN^-}$ CN H Mn(CO)_3 (50–80%) $\xrightarrow[4\text{ days}]{\text{Ce(IV)},\ H_2O}$ PhCN (80%) (49)

$^+\text{Mn(CO)}_3$ $\xrightarrow[\Delta,\ 2\text{ hr}]{\text{KCN},\ H_2O}$ CN H Mn(CO)_3 (80%) (50)

$^+\text{Mn (CO)}_3$ $\xrightarrow[0°,\ 1\text{ hr}]{\text{MeLi},\ Et_2O}$ CH_3 Mn(CO)_3 (87%) (51)

Removal of the Mn group in these two reactions [Eqs. (50) and (51)] might be difficult however (see also Section IV, A,5).

The mechanism of these reactions probably usually involves direct nucleophilic attack on the carbon system of the coordinated triene, although initial attack on coordinated carbon monoxide or on the metal, followed by rearrangement, may occur in some cases. Nucleophilic displacement of halogen from aryl halides is facilitated by coordination of the ring to the tricarbonylchromium group [Eq. (52)], but the reaction mechanism is more complicated than simple nucleophilic attack at the carbon carrying the chlorine atom, followed by elimination of chloride ion. Attack takes place at both ortho and meta positions in the aryl ring (Semmelhack and Hall, 1974) [Eq. (48)].

$$(C_6H_5Cl)Cr(CO)_3 + Me_2\bar{C}CN \xrightarrow[(2)\ I_2,\ Et_2O,\ 0^\circ]{(1)\ 25^\circ,\ 20\ hr} C_6H_5CMe_2CN\ (78\%) \tag{52}$$

6. Trienyl Carbonium Ions

Cations such as **XVIII** undergo nucleophilic addition reactions as shown in Eq. (53), but many nucleophiles (e.g., PhLi, NaOAc) bring about a facile reductive dimerization. The same dimeric product is obtained with zinc dust [Eq. (54) (Munro and Pauson, 1961)].

$$C_7H_8 \xrightarrow[diglyme]{Cr(CO)_6} (C_7H_8)Cr(CO)_3\ (64\%) \xrightarrow[CH_2Cl_2]{Ph_3C^+} [(C_7H_7)Cr(CO)_3]^+\ (97\%)\ \textbf{(XVIII)} \xrightarrow{NaCMe(COOEt)_2,\ RT} (C_7H_7\text{-}MeC(COOEt)_2)Cr(CO)_3\ (79\%) \tag{53}$$

XVIII $\xrightarrow[\text{THF, RT}]{\text{Zn}}$ [bis-$Cr(CO)_3$ complex] $\xrightarrow{\text{DET}}$ free ligand (54)

$Cr(CO)_3$ $Cr(CO)_3$

(58 %)

(DET = Diethylenetriamine)

B. Electrophilic Attack on Coordinated Double Bonds

1. Dienes, Trienes, and Tetraenes

The reactivity of olefins toward electrophilic reagents is modified considerably on coordination to a transition metal. Cycloctatetraene, for example, is usually polymerized by electrophilic reagents. However, tricarbonyl(cyclooctatetraene)iron undergoes a facile formylation in reasonable yield. The coordinated formyl compound **XIX** can be used in further synthetic transformations, as shown below, and the tricarbonyliron group readily removed by oxidation with ceric ion [Eq. (55) (Johnson *et al.*, 1969, 1971)].

$Fe(CO)_3$ $\xrightarrow[\text{4°, 30 min}]{\text{POCl}_3\text{, DMF}}$ CHO, $Fe(CO)_3$ (60 %) (**XIX**) $\xrightarrow[\text{EtOH, RT}]{\text{Ce(IV), H}_2\text{O}}$ **CHO** (80 %) (55)

(DMF = Dimethylformamide)

(XIX) $\xrightarrow[\text{0°, 5 min}]{\text{NaBH}_4\text{, EtOH}}$ CH_2OH, $Fe(CO)_3$ (93 %) $\xrightarrow[\text{Et}_2\text{O, 20°}]{\text{HPF}_6\text{, H}_2\text{O}}$ +, $Fe(CO)_3$ (89 %)

MeOH, **RT,** 20 min

CH_2OMe (90 %) $\xleftarrow{\text{Ce(IV)}}$ CH_2OMe, $(CO)_3Fe$ (89 %)

Presumably the iron atom facilitates electrophilic attack by stabilizing an intermediate carbonium ion **XX** (Scheme 3). This could either lose a proton as observed in the formylation reaction or under strongly acidic conditions rearrange into the bicyclic carbonium ion **XXI** by a type of cycloaddition process. The carbonium ion **XXI** is the major product isolated (28% as the PF_6^- salt) under Friedel–Crafts acetylation conditions ($E = CH_3CO^+$) and is readily converted into the bicyclooctadiene **XXII** (Johnson *et al.*, 1971). Such a mechanism suggests that in order to achieve normal electrophilic substitution reactions with tricarbonyl(cyclooctatetraene)iron, the conditions should be as neutral as possible. Thus, acetylation might be best carried out with acetic anhydride in pyridine or with only a trace of fluoroboric acid as a catalyst. Tricarbonyl(cycloheptatriene)iron undergoes analogous electrophilic substitutions with high yields of products (Johnson *et al.*, 1972).

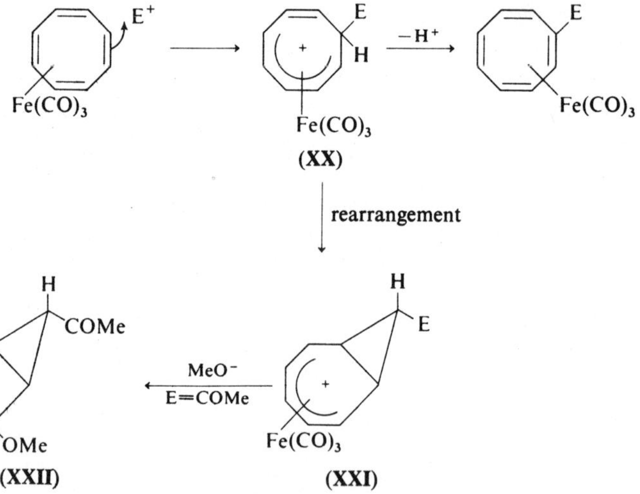

E^+ Fe(CO)$_3$ → E H Fe(CO)$_3$ (XX) $\xrightarrow{-H^+}$ E Fe(CO)$_3$

(XX) ↓ rearrangement

H COMe OMe (XXII) $\xleftarrow[E=COMe]{MeO^-}$ H E Fe(CO)$_3$ (XXI)

Scheme 3

Another reaction that becomes possible by employing the tricarbonyliron group as a protecting/activating group is electrophilic ring substitution of heptafulvenes [Eq. (56) (Johnson *et al.*, 1972)]. Normally, heptafulvenes undergo electrophilic attack at the 8-position to give a tropylium ion.

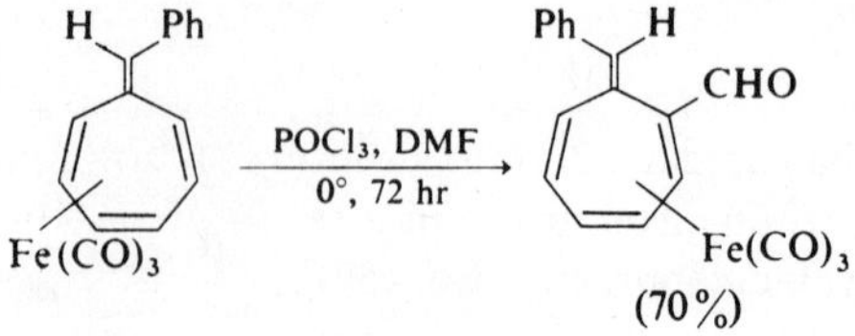

H Ph Fe(CO)$_3$ $\xrightarrow[0°,\ 72\ hr]{POCl_3,\ DMF}$ Ph H CHO Fe(CO)$_3$ (70%) (56)

The same principle has been used by Gill *et al.* (1969) to prepare substituted azepines [Eq. (57)]. Introduction of substituents into the azepine nucleus was not previously possible.

COOEt, N, $Fe(CO)_3$ —[Ac_2O, HBF_4; 0°]→ COOEt, N, COMe, $Fe(CO)_3$ (42%) (57)

A metal atom can lead to significant changes in the reaction pattern toward electrophilic reagents. For example, tricarbonyliron derivatives of cycloheptatriene and cyclooctatetraene undergo facile, unprecedented, 1,3-cycloaddition reactions with electrophiles such as hexafluoroacetone and tetracyanoethylene. Removal of the tricarbonyliron group can lead to useful products [Eq. (58) (Green *et al.*, 1973)].

$Fe(CO)_3$ —[$(CF_3)_2CO$; RT, 2 days]→ $(CF_3)_2C$—O, +, $Fe(CO)_3$ → CF_3, F_3C—O, $Fe(CO)_3$ (69%) (58)

—[CO, 100 atm; 60°, 24 hr]→ CF_3, F_3C—O, O (84%)

Paquette (1974) has used a 1,3-cycloaddition reaction followed by oxidative removal of the tricarbonyliron group in a ready preparation of triquinacene derivatives from cyclooctatetraene [Eq. (59)].

$$\text{(COT)Fe(CO)}_3 \xrightarrow[\text{C}_6\text{H}_6,\ \text{RT}]{\text{TCNE}} \text{TCNE adduct–Fe(CO)}_3\ (96\%) \xrightarrow{\text{Ce (IV)}} \text{tetracyano product}\ (100\%) \xrightarrow{\text{readily converted}} \text{HOOC product} \tag{59}$$

(TCNE = tetracyanoethylene)

Relatively little work has been done with simple coordinated dienes and electrophilic reagents in organic synthesis. Butadiene, as its tricarbonyliron derivative, undergoes Friedel–Crafts acylation much more readily than does benzene ($K_{rel} \simeq 3800$). The product arises from attack by the acyl cation on the same side of the molecule as the iron atom [Eq. (60) (Greaves *et al.*, 1969, 1974; Graf and Lillya, 1972)] and at an outer position only.

$$\text{(butadiene)Fe(CO)}_3 \xrightarrow[\text{CH}_2\text{Cl}_2]{\text{MeCOCl, AlCl}_3} (\text{CO})_3\text{Fe}\leftarrow\text{O cation}\ \text{AlCl}_4^-\ (86\%) \xrightarrow{\text{dicyclohexylethylamine}} \text{Fe(CO)}_3\text{-diene-COMe} \xrightarrow{\text{H}^+\ \text{or OH}^-} \text{Fe(CO)}_3\text{-diene-COMe} \tag{60}$$

Similarly, norbornadiene coordinated to tricarbonyliron is formylated under fairly mild conditions [Eq. (61) (Graf and Lillya, 1973)].

$$\text{(norbornadiene)Fe(CO)}_3 \xrightarrow[\text{(2) hydrolysis}]{\text{(1) Cl}_2\text{CHOMe, TiCl}_4,\ \text{CH}_2\text{Cl}_2,\ 0^\circ,\ 30\ \text{min}} \text{CHO-(norbornadiene)Fe(CO)}_3\ (27\%) \tag{61}$$

It is possible that the acylation of dienes coordinated to tricarbonyliron might be improved as a synthetic reaction by carrying out the reaction in the presence of carbon monoxide. This would facilitate formation of a more stable intermediate **XXIII**. Treatment with base, followed by removal of the tricarbonyliron group, should provide a ready synthesis of 2,4-dienones [Eq. (62)]. However, acylations of cyclohexadieneiron derivatives have so far not given favorable results.

$$\text{diene-}Fe(CO)_3 \xrightarrow[CO]{R\overset{+}{C}O} ? \ \text{(allyl)}^{+}\text{-COR}\cdot Fe(CO)_4 \ \textbf{(XXIII)} \longrightarrow \ \longrightarrow \ \text{dienyl-COR} \qquad (62)$$

A tetracarbonyl(allyl)iron cation analogous to **XXIII** is formed from allene tetracarbonyliron derivatives by acylation. This cation is readily converted to a cross-conjugated dienone derivative [Eq. (63) (Gibson *et al.*, 1971)].

$$\text{Me}_2C{=}C{=}C\text{Me}_2\cdot Fe(CO)_4 \xrightarrow[AlCl_3]{RCOCl} \text{(COR-allyl)}^{+} Fe(CO)_4 \xrightarrow[Me_2CO]{\text{warm}} \text{(COR-diene)}\ Fe(CO)_3\ (60\%) \qquad (63)$$

2. Ferrocene and Tricarbonyl(cyclobutadiene)iron

There is a vast literature on the chemistry of ferrocene, but very little use has been made of this substance in organic synthesis. As ferrocene is readily available (Jolly, 1968) and undergoes a great variety of reactions such as acylation, alkylation, sulfonation, and metalation (Perevalova and Nikitina, 1972), a simple means of removing the iron at the end of the reaction sequence would provide a ready route to substituted cyclopentane derivatives. Some typical acylation reactions [which take place 3.3×10^6 times faster than with benzene (Rosenblum *et al.*, 1963b) are illustrated in Eqs. (64) (Hauser and Lindsay, 1957), (65) (Rosenblum and Woodward, 1958), and (66) (Rosenblum *et al.*, 1963a).

$$\text{Fc} \xrightarrow[\text{(2) RT, 4 hr; (3) NaOAc, } H_2O]{\text{(1) } Ac_2O,\ CH_2Cl_2,\ BF_3,\ 0^\circ} (C_5H_4COMe)Fe(C_5H_5)\ (90\%) \qquad (64)$$

$$\text{(Fe)} \xrightarrow[\text{CH}_2\text{Cl}_2,\ \text{RT, 2 hr}]{\text{MeCOCl, AlCl}_3} (\text{MeCOC}_5\text{H}_4)\text{Fe}(\text{C}_5\text{H}_4\text{COMe}) \quad (76\%) \tag{65}$$

$$\text{(Fe)} \xrightarrow[\text{(2) NaOAc, H}_2\text{O}]{\text{(1) PhN(Me)CHO, POCl}_3,\ 65^\circ,\ 2\ \text{hr}} (\text{OHCC}_5\text{H}_4)\text{Fe}(\text{C}_5\text{H}_5) \quad (80\%) \tag{66}$$

A useful intermediate in ferrocene chemistry is the stable salt, α-ferrocenylcarbonium tetrafluoroborate, readily obtained from α-ferrocenylcarbinols [Eq. (67) (Allenmark, 1974)].

$$\text{(Fe)}\text{—C—OH} \xrightarrow[\text{Et}_2\text{O}]{\text{HBF}_4} \text{(Fe)}\text{—C}^+\ \text{BF}_4^- \xrightarrow{\text{R}_2\text{NH}} \text{(Fe)}\text{—C—NR}_2 \quad (90\text{–}98\%\ \text{overall}) \tag{67}$$

Removal of the iron from ferrocene has been accomplished by brief treatment (5 min) with lithium in ethylamine to give reasonable yields (~70%) of cyclopentadiene (Trifan and Nicholas, 1957). More recently, a hydrogenation procedure for ferrocenes employing Pd/C as catalyst under mild conditions (1 atm hydrogen, room temperature) but in acid solution (aqueous perchloric acid in acetic acid) has been reported as a synthetic route to substituted cyclopentanes. No yields were given (Van Meurs *et al.*, 1975).

Tricarbonyl(cyclobutadiene)iron is also readily acylated as shown in Eq. (68) (Fitzpatrick *et al.*, 1965), thus providing easy access to substituted cyclobutadienes by removal of the tricarbonyliron group with ceric ion (Section VI,A). Other cyclobutadiene derivatives have been prepared (Roberts *et al.*, 1969).

$$(\text{C}_4\text{H}_4)\text{Fe(CO)}_3 \xrightarrow[\text{CS}_2,\ 20^\circ,\ 45\ \text{min}]{\text{MeCOCl, AlCl}_3} (\text{MeCOC}_4\text{H}_3)\text{Fe(CO)}_3 \quad (60\%) \tag{68}$$

C. Carbonylation of Olefins

1. Hydroformylation. The Oxo Reaction

An extensive literature can be found for these reactions, so only a brief treatment in the way of illustrative examples will be attempted here. The interested reader may refer to the reviews by Paulik (1972), Orchin and Rupilius (1972), Rosenthal (1968), Falbe (1971), and Markó (1974).

Hydroformylation of olefins [Eq. (69)] requires the presence of a transition metal catalyst. It is less successful with conjugated dienes than with noncon-

$$RCH{=}CH_2 + CO + H_2 \xrightarrow{\text{catalyst}} RCH_2CH_2CHO \qquad (69)$$

jugated dienes or simple olefins. α,β-Unsaturated aldehydes are hydrogenated under hydroformylation conditions, but α,β-unsaturated esters are hydroformylated in good yield. The reaction is usually carried out under conditions of homogeneous catalysis, although supported (i.e., heterogeneous) catalysts have many advantages—low volatility, long life, and fewer losses of expensive catalyst (Pittman and Hanes, 1974). In general, hydroformylation using traditional cobalt carbonyl catalysts requires fairly vigorous conditions (for example, 150° and 250 atm). Much milder conditions are possible, however, using rhodium catalysts or stoichiometric reagents. For example, hydridocarbonyltris(triphenylphosphine)rhodium(I), $RhH(CO)(PPh_3)_3$, is an efficient catalyst for hydroformylation of alkenes at 25° and 1 atm (Brown and Wilkinson, 1970). The rate of hydroformylation depends on the structure of the olefin and on the metal catalyst (Wender *et al.*, 1956; Whyman, 1974). Use of an asymmetric catalyst leads to asymmetric induction in the hydroformylation reaction (Section IX). Some examples of hydroformylation reactions are given in Eqs. (70) (Rosenthal and Koch, 1965), (71) (Chini *et al.*, 1972), (72) (Pino and Botteghi, 1973), and (73) (Adkins and Krsek, 1949).

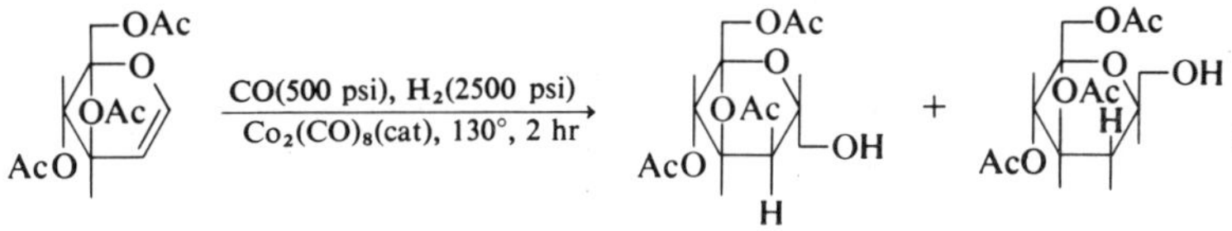

OAc, O, OAc, AcO $\xrightarrow[Co_2(CO)_8\text{(cat), 130°, 2 hr}]{CO\text{(500 psi)}, H_2\text{(2500 psi)}}$ OAc, O, OAc, AcO, OH, H + OAc, O, OH, OAc, H, AcO (70)

(almost quantitative, ratio 1:1)

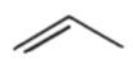

$$+ Rh_4(CO)_{12} \xrightarrow[\text{Toluene}]{H_2\text{(1 atm), RT}} Me_2CHCHO + MeCH_2CH_2CHO \qquad (71)$$

(90–95%, ratio 1:1)

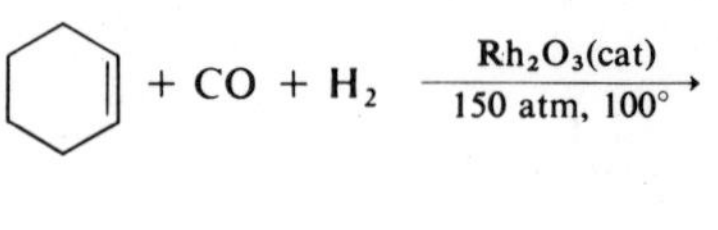

$$\text{cyclohexene} + CO + H_2 \xrightarrow[150\text{ atm}, 100°]{Rh_2O_3(\text{cat})} \text{cyclohexane-CHO} \quad (95\%) \tag{72}$$

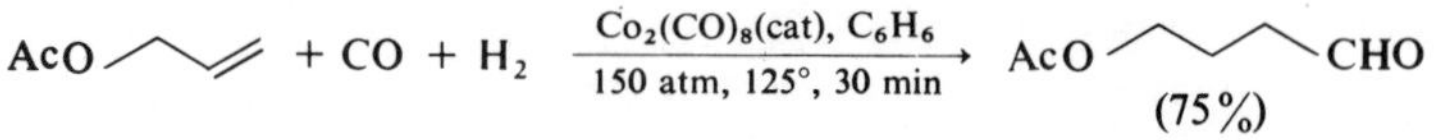

$$AcO\text{–}CH_2CH=CH_2 + CO + H_2 \xrightarrow[150\text{ atm}, 125°, 30\text{ min}]{Co_2(CO)_8(\text{cat}), C_6H_6} AcO(CH_2)_3CHO \quad (75\%) \tag{73}$$

The general mechanistic principles involved in these reactions are illustrated in Scheme 4.

$$HCo(CO)_4 \rightleftharpoons HCo(CO)_3 + CO \xrightleftharpoons{RCH=CH_2} RCH{=}CH_2 \cdots H{-}Co(CO)_3$$

$$RCH{=}CH_2 \cdots H{-}Co(CO)_3 \rightleftharpoons RCH_2CH_2{-}Co(CO)_3 \xrightleftharpoons{CO} RCH_2CH_2Co(CO)_4 \rightleftharpoons RCH_2CH_2COCo(CO)_3$$

$$RCH_2CH_2COCo(CO)_3 \xrightleftharpoons{CO} RCH_2CH_2COCo(CO)_4 \xrightarrow{H_2} RCH_2CH_2CHO + HCo(CO)_4$$

Scheme 4

2. Reactions Related to Hydroformylation

Treatment of olefins with carbon monoxide and a transition metal catalyst in the presence of alcohols or amines can lead to useful synthetic reactions. Some examples are given in Eqs. (74) (Iqbal, 1971; Markó and Bakos, 1974), (75), (Falbe and Korte, 1965), (76) (Bittler *et al.*, 1968), (77) (Bott, 1973), and (78) (Stille and James, 1975).

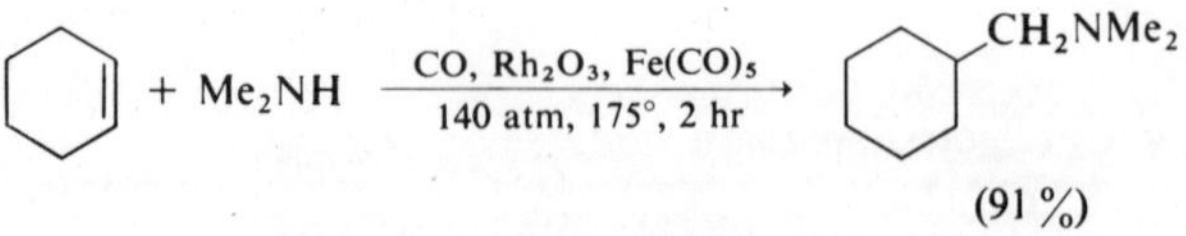

$$\text{cyclohexene} + Me_2NH \xrightarrow[140\text{ atm}, 175°, 2\text{ hr}]{CO, Rh_2O_3, Fe(CO)_5} \text{cyclohexyl-}CH_2NMe_2 \quad (91\%) \tag{74}$$

$$CH_2{=}CHCH_2NHMe \xrightarrow[120°\text{–}300°, 100\text{–}300\text{ atm}]{CO, Co_2(CO)_8(\text{cat})} \text{N-methyl-2-pyrrolidone (N–Me)} \quad (78\%) \tag{75}$$

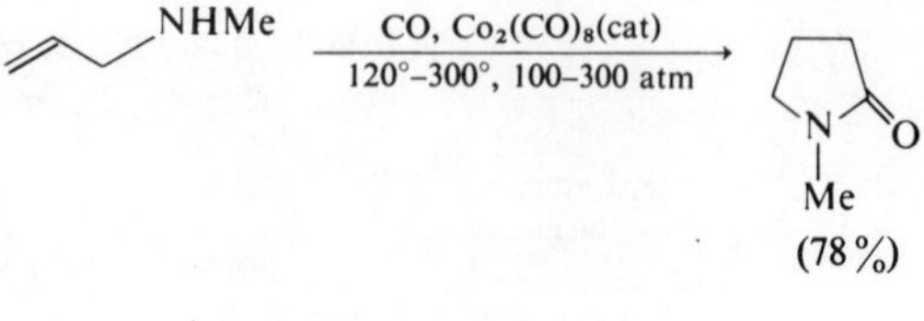

$$\text{CH}_3\text{CH}{=}\text{CHCH}_3 + \text{EtOH} \xrightarrow[300\text{–}700\ \text{atm},\ 60^\circ\text{–}100^\circ]{\text{CO},\ (\text{Ph}_3\text{P})_2\text{PdCl}_2(\text{cat})} \text{CH}_3\text{CH}_2\text{CH}(\text{CH}_3)\text{COOEt}\ (95\%) \tag{76}$$

$$\text{Cyclohexene} + \text{PhCOCOPh} \xrightarrow[20\ \text{atm},\ 120^\circ,\ 11\ \text{hr}]{\text{CO},\ \text{H}_2,\ \text{Rh}_2\text{O}_3(\text{cat})} \text{C}_6\text{H}_{11}\text{C}(\text{=O})\text{O–CHPh(COPh)}\ (80\%) \tag{77}$$

$$\text{Cycloocta-1,5-diene} \xrightarrow{\text{PdCl}_2} (\text{C}_8\text{H}_{12})\text{PdCl}_2 \xrightarrow[\text{Na}_2\text{CO}_3]{\text{H}_2\text{O}} [(\text{HO-C}_8\text{H}_{12})\text{PdCl}]_2\ (96\%) \xrightarrow{\text{CO},\ \text{H}_2\text{O},\ \text{NaOAc}}$$

$$\text{HO-C}_8\text{H}_{12}\text{-C(O)-Pd(CO)Cl} \longrightarrow \beta\text{-lactone}\ (78\%) \xrightarrow[\text{Na}_2\text{CO}_3]{\text{MeOH}} \text{HO-C}_8\text{H}_{12}\text{-COOMe}\ (94\%) \tag{78}$$

A recently reported reaction that appears to have potential in organic synthesis is "hydrozirconation" (Hart and Schwartz, 1974). The reagents involved are inexpensive; they are easy to prepare, require only moderate care in handling, and give high yields of products. Hydrozirconation converts olefins via π-bonded Zr–H intermediates to σ-bonded zirconium derivatives in which the zirconium atom appears in the least hindered position in the molecule. The σ-bonded zirconium compound undergoes coordination to carbon monoxide, followed by alkyl migration, resulting in a facile hydroformylation reaction [Eq. (79) (Bertelo and Schwartz, 1975)].

$$\text{CH}_3\text{CH}_2\text{CH}{=}\text{CHCH}_2\text{CH}_3 \xrightarrow[\text{RT, few hours}]{\text{Cp}_2\text{Zr(H)Cl},\ \text{C}_6\text{H}_6} \text{CH}_3(\text{CH}_2)_5\text{Zr(Cp)}_2\text{Cl} \xrightarrow[\text{RT, few hours}]{\text{CO(20 psi)},\ \text{C}_6\text{H}_6}$$

$$\text{CH}_3(\text{CH}_2)_5\text{C(=O)Zr(Cp)}_2\text{Cl} \xrightarrow{\text{dilute HCl}} \text{CH}_3(\text{CH}_2)_5\text{CHO}\ (99\%) \tag{79}$$

3. Miscellaneous Carbonylation Reactions

Various tricarbonyliron compounds undergo CO insertion reactions to give cyclic ketones as shown in Eqs. (80) (Johnson *et al.*, 1974), (81) (Mantzaris and Weissberger, 1974), and (82) (Whitesides and Shelley, 1975).

$Fe(CO)_3$ $\xrightarrow[C_6H_6,\ RT]{AlCl_3}$ =O (80)

(48%)

$\xrightarrow[85°,\ 42\ hr]{Fe_2(CO)_9,\ octane}$ (81)

O (67%)

$Fe(CO)_3$ $\xrightarrow[h\nu]{C_2H_4}$ Fe $(CO)_3$ $\xrightarrow[or\ O_2]{Ce(IV)}$ O (82)

(moderate yields)

D. Polymerization, Oligomerization, and Cyclooligomerization of Alkenes

1. Ziegler–Natta Polymerization

There is extensive literature on Ziegler–Natta catalysis and the interested reader is referred to the recent review by Fischer *et al.* (1973), the reviews edited by Ketley (1967), and the summary by Mole and Jeffrey (1972). The following very brief description is intended only to illustrate how the special properties of transition metal–olefin bonds can lead to extremely important C–C bond-forming reactions.

Ziegler and co-workers (1955) were the first to observe that ethylene could be polymerized at room temperature and low pressure (few atmospheres) in the presence of catalysts of the type $TiCl_4/AlEt_3$. Natta (1955, 1956, 1957) showed that such catalysts could be modified to produce stereoregular polymers of high molecular weight.

Polymerization of an alkene such as propylene on the surface of a Ziegler–Natta catalyst produces one asymmetric center each time a monomer unit is added to the growing chain. If all asymmetric centers have the same configuration, the polymer is said to be "isotactic," if the configurations alternate along the chain, the polymer is "syndiotactic," and if they are random, "atactic." The stereoregular polymers produced on Ziegler–Natta catalysts have a much more uniform structure and are of higher molecular weight than polymers produced by traditional free-radical polymerization methods. They consequently have superior and more well-defined macroscopic properties such as high crystallinity, higher density, good mechanical strength, and higher melting point.

The mechanisms of Ziegler–Natta polymerization are not fully understood, but the heterogeneous nature of the catalysts appears to play an important role in controlling the stereochemistry of the polymer. The mechanism suggested in Scheme 5 therefore possibly reflects the process taking place on the (modified) surface of a titanium chloride crystal.

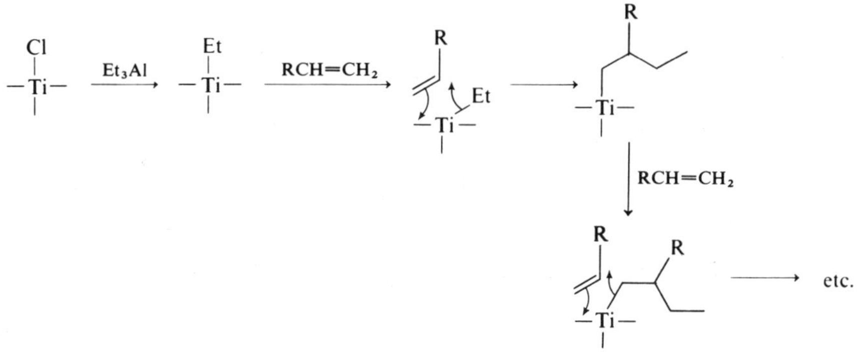

Et_3Al ; $RCH{=}CH_2$; $RCH{=}CH_2$; etc.

Scheme 5

2. Oligomerization and Cyclooligomerization*

In the reactions of olefins with olefins, different transition metal catalysts can give rise to markedly different products (Heimbach, 1973; Buchholz *et al.*, 1972; Kricka and Ledwith, 1974). Certain nickel catalysts, for example, normally trimerize butadiene as shown in Eq. (83) (Bogdanović *et al.*, 1969). These same catalysts, however, in the presence of suitable ligands, will cyclodimerize butadiene to give initially divinylcyclobutane (which can be isolated in 40% yield) and finally 1,5-cyclooctadiene in high yield [Eq. (84) (Brenner *et al.*, 1969)].

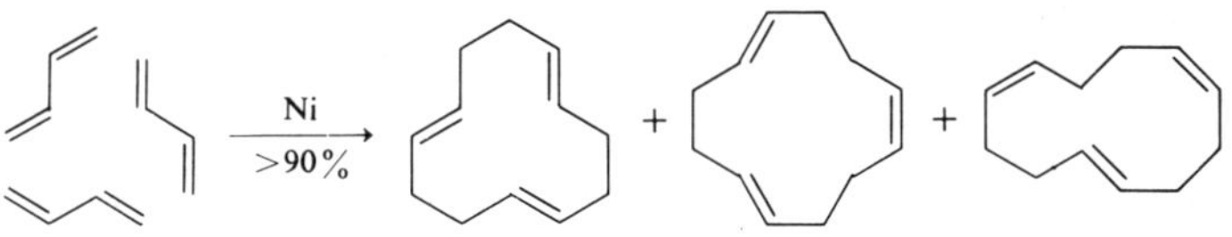

Ni, >90% (83)

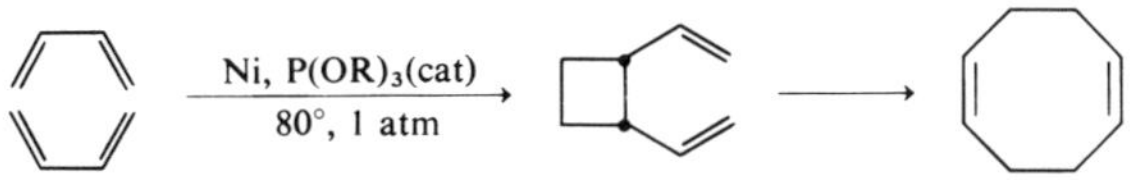

Ni, $P(OR)_3$(cat), 80°, 1 atm (84)

(R = *o*-PrC_6H_4) (97%)

* Also see this volume, Chapter 2 by Noyori, Section III,C.

Other nickel catalyst systems can cyclodimerize butadiene to 2-methylene-vinylcyclopentane (Kiji *et al.*, 1970a, b) or linearly dimerize it to (E,E)-1,3,6-octatriene in high yield [Eq. (85) (Pittman and Smith, 1975)].

$(Ph_3P)_2NiBr_2$, $NaBH_4$ / THF, EtOH, 100°, 24 hr (95%) (85)

Similarly, butadiene and ethylene can react together in the presence of "naked nickel" or nickel–ligand catalytic systems to give *cis,trans*-1,5-cyclodecadiene in yields up to 80% [Eq. (86) (Wilke, 1963; Heimbach and Wilke, 1969)]. Norbornadiene dimerizes or reacts specifically with olefins using Ni(O) (Schrauzer and Glockner, 1964; Schrauzer *et al.*, 1970).

Ni(cat) + (80%) (86)

Butadiene and ethylene will codimerize to give *cis*-1,4-hexadiene in the presence of cobalt (Henrici-Olivé and Olivé, 1972) or rhodium catalysts [Eq. (87) (Alderson *et al.*, 1965)]. The 2,4-hexadiene formed presumably arises by isomerization of the 1,4-diene.

+ $RhCl_3$(cat), EtOH / 50°, 16 hr (67%) (22%) (87)

Some miscellaneous transition metal-catalyzed olefin and olefin addition reactions are given in Eqs. (88) (Singer, 1974), (89) (Binger *et al.*, 1974), (90) (Binger, 1973), (91) (Agnes *et al.*, 1968), and (92) (Bennett *et al.*, 1973; Noyori *et al.*, 1973).

COOMe COOMe + $Ni(acac)_2$, Et_3Al, Ph_3Sb / 80°, 5 hr COOMe COOMe

(acac = acetylacetone) (50%) (88)

$(Ph_3P)_4Pd$(cat) C_6H_6 / 45°, 4 hr (100%) (89)

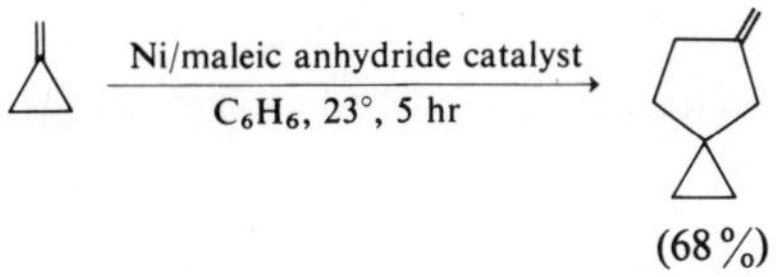

Ni/maleic anhydride catalyst; C_6H_6, 23°, 5 hr (68%) (90)

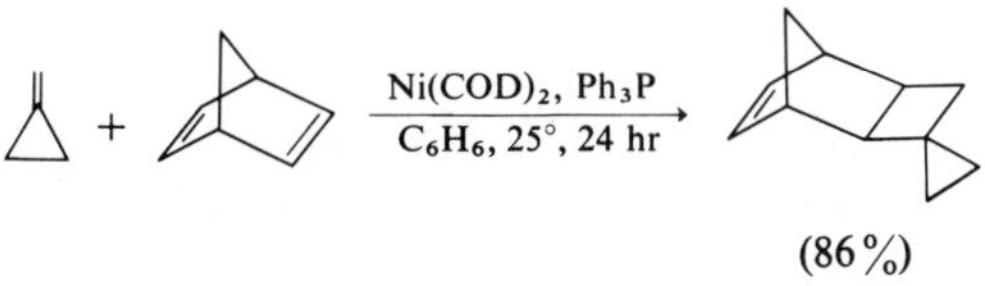

Ni(COD)$_2$, Ph_3P; C_6H_6, 25°, 24 hr (86%) (91)

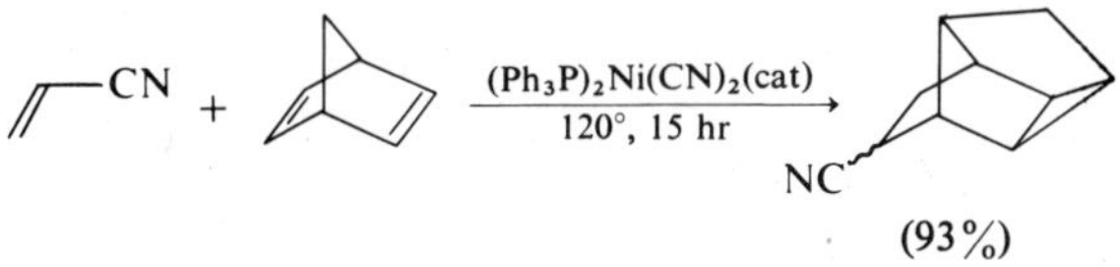

CN + ; $(Ph_3P)_2Ni(CN)_2$(cat); 120°, 15 hr; NC (93%) (92)

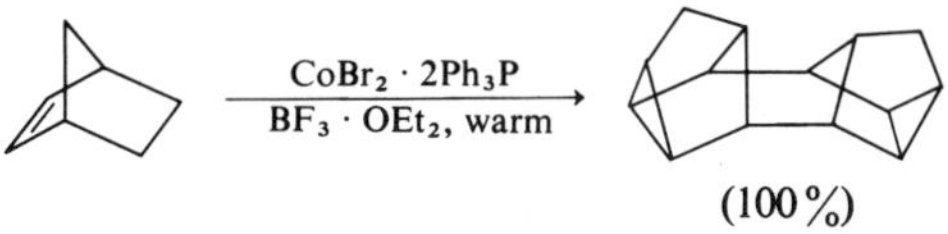

$CoBr_2 \cdot 2Ph_3P$; $BF_3 \cdot OEt_2$, warm (100%) (93)

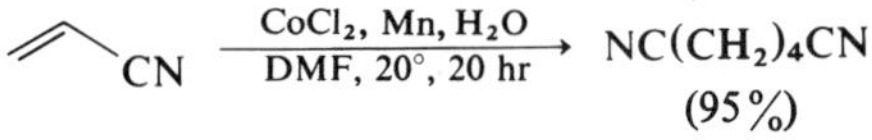

CN; $CoCl_2$, Mn, H_2O; DMF, 20°, 20 hr → $NC(CH_2)_4CN$ (95%) (94)

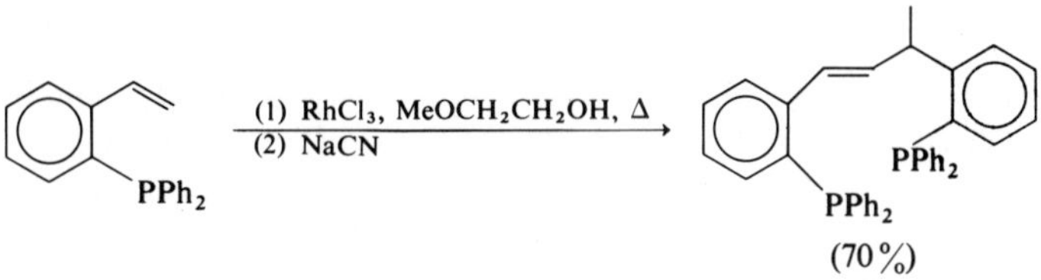

PPh_2; (1) $RhCl_3$, $MeOCH_2CH_2OH$, Δ (2) NaCN; PPh_2 PPh_2 (70%) (95)

Mechanisms illustrating the possible types of processes involved in these reactions are suggested for the codimerization reaction [Eq. (87)] and the dimerization reaction [Eq. (90)] (Schemes 6 and 7).

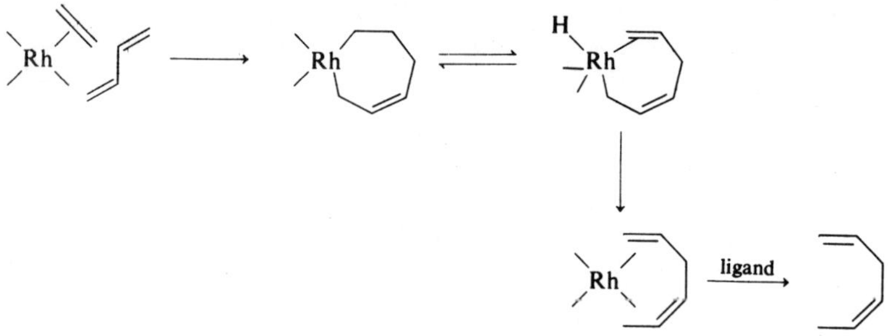

ligand

Scheme 6

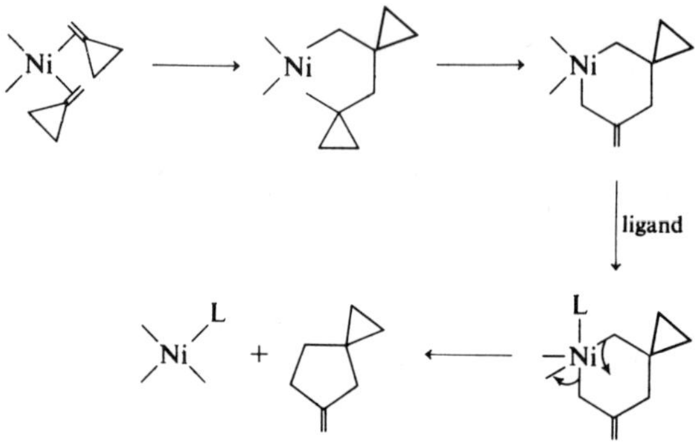

ligand

Scheme 7

IV. C–X BOND FORMATION (X = ELEMENT OTHER THAN CARBON)

A. Nucleophilic Attack on Coordinated Double Bonds

1. Monoolefins. Wacker Process

C–X bond formation by nucleophilic attack on coordinated olefins is in most respects analogous to C–C bond formation (Section IIIA). However, for olefins coordinated to Fp^+, the reaction appears to be limited either by the reversibility of the addition process or by preferential nucleophilic attack at the iron atom or coordinated carbon monoxide. For example, in the reaction of $[C_5H_5Fe(CO)_2C_2H_4]^+$ with azide ion, the cyanato complex $[C_5H_5Fe(CO)(C_2H_4)(NCO)]$ is formed; methoxide ion attacks at coordinated ethylene to give $[C_5H_5Fe(CO)_2(C_2H_4OCH_3)]$, but treatment of this adduct with HCl regenerates the original iron cation (Busetto *et al.*, 1970). Although very little work has been done on these reactions, their use in organic synthesis appears at present to be restricted to nucleophilic attack by carbon or other "soft" nucleophiles such as phosphite or sulfite (see Sections III and IV,A,4) where the reactions are essentially irreversible. Triphenylphosphine, for, example, is alkylated by an olefin coordinated to Fe(II) [Eq. (96) (A. Rosan, unpublished; Rosenblum, 1974)]. The resulting phosphonium salt would be expected to undergo Fe–C bond cleavage on treatment with HCl. C–P bonds are not cleaved under these conditions.

$$\text{olefin–Fp}^+ \xrightarrow{Ph_3P} \text{Fp–CH}_2\text{–CH(R)–}\overset{+}{PPh_3} \xrightarrow{HCl} \text{CH}_3\text{–CH(R)–}\overset{+}{PPh_3} \quad (96)$$

exclusive product obtained

The Pd(II)- and Pt(II)-promoted nucleophilic additions to olefins are important reactions in organic synthesis. Nucleophiles that have been shown to add to coordinated olefins include OH^-, RO^-, AcO^-, Cl^-, amines, and amides, and with the possible exception of OH^-, nucleophilic attack takes place on the side of the double bond remote from the metal. The main difference between Pd(II)- or Pt(II)-promoted and Fe(II)-promoted addition is that nucleophilic attack on Pd(II)- or Pt(II)-coordinated monoolefins results in reduction of the divalent metal to the zero valent state. In most cases, Pd(II) salts promote nucleophilic attack on olefins more readily than Pt(II) salts (Hartley, 1973; Maitlis, 1971). Equations (97) (Stern and Spector, 1961), (98) (Baird, 1966), (99) (Kitching *et al.*, 1966), (100) (Akermark *et al.*, 1974), and (101) (Kohl and Van Helden, 1968; Henry, 1972) illustrate some typical Pd(II)- and Pt(II)-promoted additions to olefins.

$$\text{propene} \xrightarrow[\text{isooctane, RT, 5 days}]{PdCl_2,\ HOAc,\ Na_2HPO_4} \text{CH}_2\text{=C(OAc)CH}_3\ (36\%) \quad (97)$$

$$\text{norbornadiene} \xrightarrow[\text{HOAc, 80°, 72 hr}]{PdCl_2(cat),\ CuCl_2,\ NaOAc} \text{OAc, Cl adduct}\ (80\%) \xrightarrow[\text{DMSO}]{KOBu^t} \text{OH product}\ (70\%) \quad (98)$$

$$\text{2-butene} \xrightarrow[25°]{Pd(OAc)_2,\ HOAc} \text{CH}_2\text{=CH–CH(OAc)–CH}_3\ \text{(high yield)} \quad (99)$$

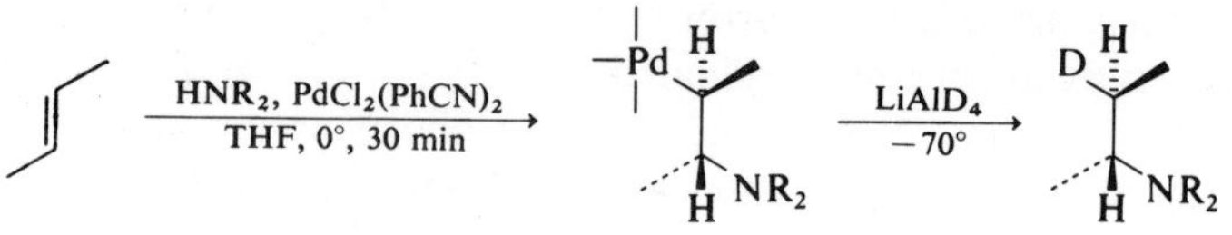

$$\xrightarrow[\text{THF, 0°, 30 min}]{\text{HNR}_2\text{, PdCl}_2\text{(PhCN)}_2} \quad \xrightarrow[-70°]{\text{LiAlD}_4} \tag{100}$$

$$\text{CH}_2{=}\text{CHCH}_2\text{Cl} + \text{NaOAc} \xrightarrow[\text{65°, 1–2 hr}]{\text{PdCl}_2\text{, HOAc}} \text{CH}_2{=}\text{CHCH}_2\text{OAc} \; (94\%) \tag{101}$$

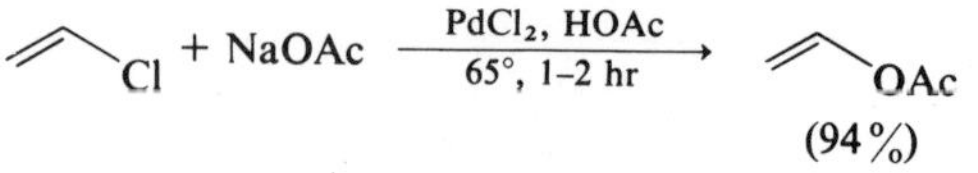

The Pd(II)-catalyzed addition of water to ethylene is the basis of the Wacker process for the manufacture of acetaldehyde. Using a flow reactor, greater than 99% conversion of ethylene to acetaldehyde is achieved (Schmidt *et al.*, 1962; Aguilo, 1967; Hartley, 1969; Henry, 1973). A general mechanism illustrating the principles involved in these Pd-promoted addition/oxidation reactions is given for the conversion of ethylene to acetaldehyde (Scheme 8) The mechanism for these reactions is still not completely understood (Henry, 1973). One possibility that does not appear to have been considered and which would account for the apparent cis addition in the case of the ethylene hydration reaction (trans addition of OH^- is observed with diene complexes) is that in the presence of a large concentration of a nonsterically demanding

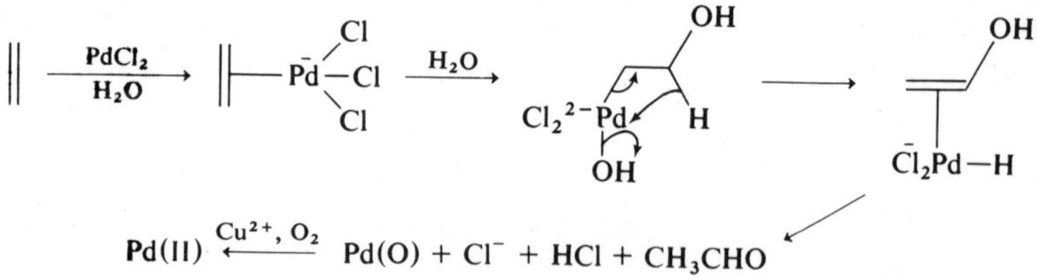

$$\text{Pd(II)} \xleftarrow{\text{Cu}^{2+}\text{, O}_2} \text{Pd(O)} + \text{Cl}^- + \text{HCl} + \text{CH}_3\text{CHO}$$

Scheme 8

nucleophile (i.e., water or OH^-) an 18-electron 5-coordinate Pd(II) species might be formed which could internally transfer a hydroxyl to the coordinated ethylene (Scheme 9). Five-coordinate Pd(II) and Pt(II) species are known

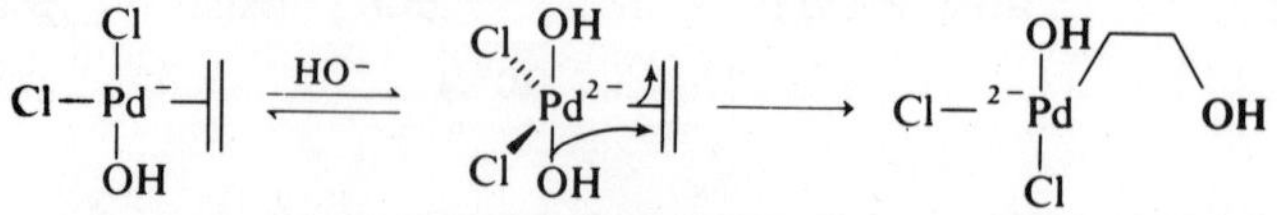

Scheme 9

(Nyholm *et al.*, 1969; Venanzi, 1964). When formation of a 5-coordinate species is not favored, then presumably the more generally observed trans attack by the nucleophile on a 16-electron metal–olefin species takes place [Eq. (102) (Panunzi *et al.*, 1970)].

$$\text{Cl–Pt(Cl)(Et-olefin)} + \text{NHEt}_2 \longrightarrow \text{Cl–Pt(Cl)–CH}_2\text{CH(Et)NEt}_2 \xrightarrow{H^+} \text{Et–CH(CH}_3\text{)–}\overset{+}{\text{N}}\text{HEt}_2\ \text{Cl}^- \quad (102)$$

S-amine *S*-amine *S*-isomer

Another possibility is that there is an initial (cis) transfer of a ligand such as Cl from the metal to the olefin and that this ligand then undergoes a metal-assisted S_N2 displacement by the attacking nucleophile to give the trans product ("ping-pong effect").

Some olefin oxidations related to the Wacker process are given in Eqs. (103) and (104) (Lloyd and Luberoff, 1969).

$$CH_2{=}CH_2 + HOCH_2CH_2OH \xrightarrow[50^\circ,\ 40\ \text{min}]{PdCl_2(cat),\ CuCl_2,\ O_2} CH_3CH\langle OCH_2CH_2O\rangle + CH_3CHO \quad (103)$$

(91%) (9%)

$$\text{cyclohexene} \xrightarrow[EtOH,\ 90^\circ,\ 2\ \text{hr}]{PdCl_2,\ CuCl_2,\ O_2} \text{cyclohexanone} \quad (104)$$

(95%)

Catalysts other then Pd(II) and Pt(II) have also been used in olefin oxidation reactions (Henry, 1974). For example (Dudley *et al.*, 1974), hex-1-ene, hept-1-ene, and oct-1-ene are converted into the corresponding methyl ketones in yields of 70–80% at room temperature and pressure in benzene in the presence of oxygen and $RhCl(PPh_3)_3$ catalyst. There is evidence that oxidations catalyzed by Rh(I) are free-radical processes (Kurkov *et al.*, 1968).

One other important Pt-catalyzed C–X bond-forming reaction is hydrosilylation which is formally analogous to hydrogenation and results in the addition of $H{-}SiR_3$ to an olefin (see Eaborn and Bott, 1968). An interesting example of a transition metal-photocatalyzed hydrosilylation reaction was reported recently. Treatment of butadiene with R_3SiH and uv light (300–380 nm) in the presence of chromium hexacarbonyl at room temperature gave near quantitative yields of *cis*-$MeCH{=}CHCH_2SiR_3$ (Wrighton and Schroeder, 1974).

2. Olefinic (Allyl) Carbonium Ions

C–X bond formation by nucleophilic attack on metal-stabilized allyl cations is analogous to C–C bond formation (Section III,A,2) and is illustrated by Eqs. (105) and (106) (Whitesides *et al.*, 1973).

$Fe(CO)_4$ BF_4^- $\xrightarrow[\text{few minutes}]{PPh_3,\ THF,\ RT}$ $\overset{+}{P}Ph_3$ BF_4^- (74%) (105)

$Fe(CO)_4$ BF_4^- $\xrightarrow{\text{pyridine}}$ $\overset{+}{N}$ BF_4^- (80%) (106)

The Pd(II)-catalyzed exchange of allylic groups as illustrated in Eq. (107) (Atkins *et al.*, 1970) could be regarded as a nucleophilic attack on an incipient allyl carbonium ion, but the reaction mechanism probably involves a simple aminopalladation, followed by dehydroxypalladation (Henry, 1973).

OH + Et_2NH $\xrightarrow[Ph_3P,\ 50^\circ,\ 30\ min]{Pd(acac)_2(cat)}$ NEt_2 (95%) (107)

(acac = acetylacetone)

3. Dienes

C–X bond formation by nucleophilic attack on coordinated dienes is illustrated by the examples in Eqs. (108) (Stille and Fox, 1970), (109) (Stille and Morgan, 1966), (110) (Paiaro *et al.*, 1967), and (111) (Stille and James, 1975).

Pt, Cl, Cl $\xrightarrow[RT,\ 5\ hr]{PhCH_2NH_2,\ CH_2Cl_2}$ $PhCH_2NH$, Pt, Cl, 2 (100%) $\xrightarrow{H_2,\ MeOH}$ $PhCH_2NH$ (74%) (108)

$\xrightarrow[RT,\ 16\ hr]{Na_2PdCl_4,\ MeOH}$ MeO, Pd, Cl, 2 (88%) $\xrightarrow{H_2}$ MeO (78%) (109)

$\xrightarrow[\text{CHCl}_3\text{, RT}]{\text{Na}_2\text{PdCl}_4}$ $\xrightarrow[\text{H}_2]{\text{NaOAc}}$ OAc (110)

$\xrightarrow{\text{PdCl}_2}$ Pd Cl Cl; $\xrightarrow[\text{H}_2\text{O}]{\text{Na}_2\text{CO}_3}$ OH Pd Cl (96%) $\xrightarrow[\text{Me}_2\text{CO}]{\Delta}$ O (70%) (111)

4. Dienyl Carbonium Ions

As described in Section III,A,4, dienyl carbonium ions stabilized by coordination to a tricarbonyliron group are mild alkylating or electrophilic reagents that undergo facile attack by nucleophiles leading to numerous possibilities in organic synthesis. One of the most important recent developments in this field is a mild procedure for phenylating (or arylating) amines. Examples of this novel arylation reaction are given in Eqs. (112) and (113) (Birch and Jenkins, 1975).

$$\text{Fe(CO)}_3 + \text{PhNH}_2 \xrightarrow[75^\circ\text{, 16 hr}]{\text{HOAc}} \text{Ph}_2\text{NH} \quad (90\%) \qquad (112)$$

$$\text{Fe(CO)}_3 \xrightarrow[\text{CH}_2\text{Cl}_2\text{, RT, 16 hr}]{\text{Et}_3\text{O}^+\text{BF}_4^-} \text{OEt, Fe(CO)}_3 \xrightarrow[\text{CH}_2\text{Cl}_2\text{, RT, 2 min}]{\text{cyclohexylamine}} \text{C}_6\text{H}_{11}\text{NHPh} \quad (70\%) \qquad (113)$$

A mechanism for the acid-catalyzed phenylation of arylamines by tricarbonyl(cyclohexadienone)iron is suggested in Scheme 10.

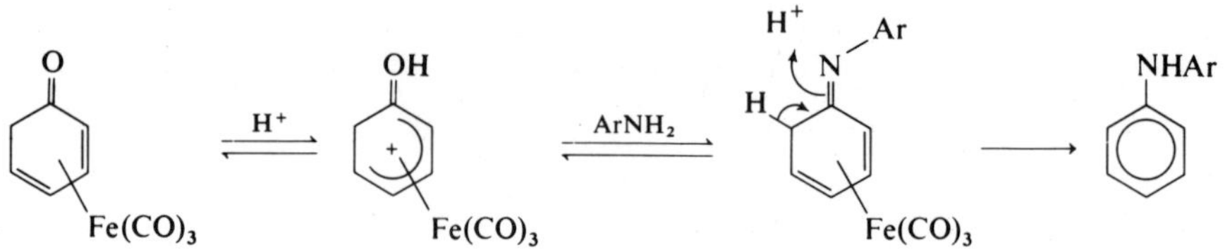

Scheme 10

Amides and carbamates undergo analogous alkylation reactions with dienyl carbonium salts (A. J. Liepa, personal communication) as do purines and pyrimidines (I. D. Jenkins, unpublished). Another useful application of dienyl carbonium salts is in the synthesis of 2,4-dienylphosphonic acid derivatives through a facile Arbusov reaction with phosphites [Eqs. (114) and (115) (Birch *et al.*, 1975)]. This type of compound would be difficult to synthesize by normal procedures.

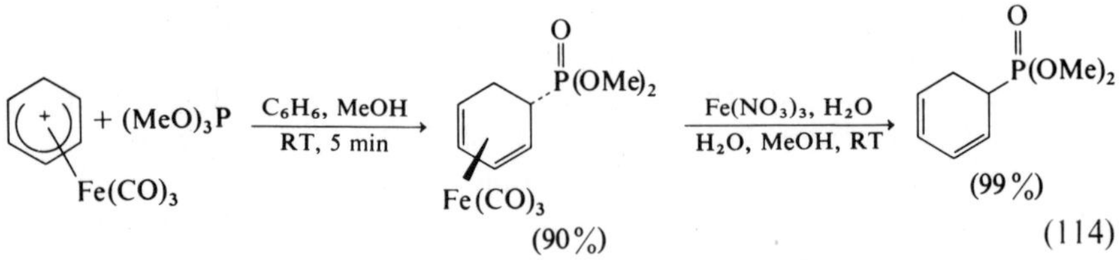

(114)

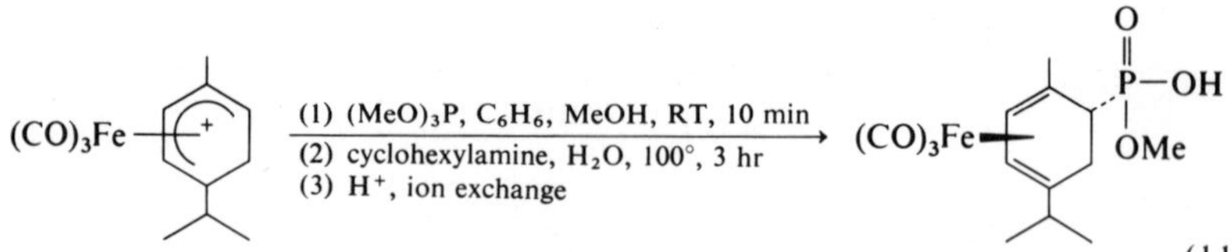

(115)

Examples of other useful C–X bond-forming reactions where the tricarbonyliron group should be readily removed are given in Eqs. (116)–(119) (Birch *et al.*, 1975) and (120) (Evans *et al.*, 1973).

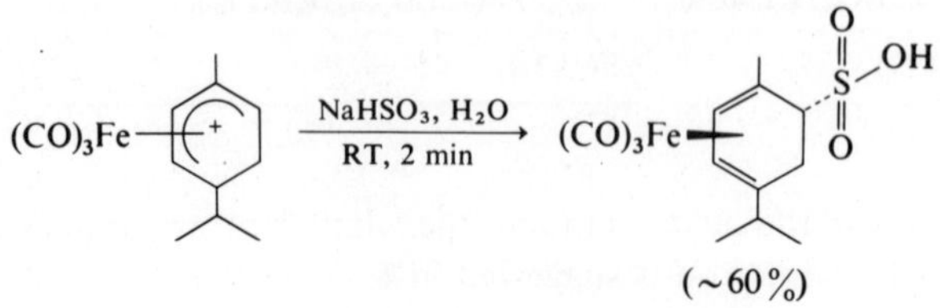

(116)

$$\text{(tricarbonyl)cyclohexadienyl cation-Fe(CO)}_3 \xrightarrow[65°,\ 2\ \text{hr}]{H_2PO_3,\ H_2O} (CO)_3Fe\text{-cyclohexadienyl-P(O)(H)OH} \ (42\%) \xrightarrow[RT,\ 16\ \text{hr}]{HgO,\ C_6H_6,} (CO)_3Fe\text{-cyclohexadienyl-P(O)(OH)}_2 \quad (117)$$

$$Fe(CO)_3\text{-cyclohexadienyl}^+ \xrightarrow[RT,\ 5\ \text{min}]{Na_2S_2O_4,\ H_2O} (CO)_3Fe\text{-}C_6H_7\text{-SO}_2\text{-SO}_2\text{-}C_6H_7\text{-}Fe(CO)_3 \quad (118)$$

$$Fe(CO)_3\text{-cyclohexadienyl}^+ \xrightarrow[RT,\ 1.5,\ \text{hr}]{Bu_3SnD,\ C_6H_6} (CO)_3Fe\text{-}C_6H_6D \ (>90\%) \quad (119)$$

$$Fe(CO)_3\text{-cycloheptadienyl}^+ \xrightarrow[CH_2Cl_2,\ RT]{Ph_3P} Fe(CO)_3\text{-cycloheptadienyl-}P^+Ph_3 \quad (120)$$

Nucleophiles such as OH^-, S^{2-}, N_3^- (Birch *et al.*, 1975), RO^- (Hine *et al.*, 1975), and amines (McArdle and Sherlock, 1973; Maglio and Palumbo, 1974) also add to dienyl carbonium ions, but these reactions tend to be reversible in the presence of acids so that removal of the tricarbonyliron group may be more difficult. This problem should be less pronounced with acyclic than with cyclic dienyl carbonium salts (owing to the greater stability of the latter cations) and should be absent in the case of allylcarbonium salts where the monoolefin–tetracarbonyliron complexes resulting from nucleophilic attack are unstable and lose the $Fe(CO)_4$ group when left in solution (Section IV,A,3). The stereospecificity of nucleophilic attack on the side of the molecule opposite the tricarbonyliron group appears to be a result of kinetic control (see Section IX). One unusual example of nucleophilic attack on a dienyl carbonium salt where the nucleophile does not attack at a terminal position is shown in Eq. (121) (Salzer and Werner, 1975).

$$\text{C}_7\text{H}_8\text{Mo(CO)}_3 \xrightarrow[\text{RT, 15 min}]{\text{HBF}_4\text{, (EtCO)}_2\text{O}} [\text{C}_7\text{H}_9\text{Mo(CO)}_3]^+ \text{BF}_4^- \ (50\%) \xrightarrow[\text{CH}_2\text{Cl}_2\text{, RT}]{\text{PPh}_3} \text{Ph}_3\text{P}^+\text{-cycloheptadiene} \ (46\%) \quad (121)$$

Other examples of nonterminal nucleophilic attack on cyclic dienylium systems are known. The position of attack depends on ring size, the transition metal involved, and the other coordinating ligands. These effects have been discussed (Edwards *et al.*, 1974; Deeming *et al.*, 1974). The reason for nucleophilic attack at the 3-position of the dienylium system in Eq. (121) may be due to initial coordination of triphenylphosphine to the coordinatively unsaturated Mo atom. The resulting tricarbonyl(triphenylphosphine)Mo complex could then undergo transfer of triphenylphosphine from Mo or attack by a second molecule of triphenylphosphine at the 3-position of the dienylium system.

5. Trienes and Arenes

It was shown in Section III,A,5 that trienes and arenes coordinated to Fe(II) Mn(I), and Cr(0) are susceptible to nucleophilic attack under mild conditions, resulting in useful C–C bond-forming reactions. Examples of analogous C–X bond-forming reactions are given in Eqs. (122) (Nicholls and Whiting, 1959), (123) (Bunnett and Hermann, 1971), and (124) (Haque *et al.*, 1971). The penul-

$$\text{C}_6\text{H}_5\text{Cl}\cdot\text{Cr(CO)}_3 \xrightarrow[\text{65°, 24 hr}]{\text{NaOMe/MeOH}} \text{C}_6\text{H}_5\text{OMe}\cdot\text{Cr(CO)}_3 \ (90\%) \quad (122)$$

$$\text{C}_6\text{H}_5\text{F}\cdot\text{Cr(CO)}_3 + \text{HN(CH}_2)_5 \xrightarrow[\text{RT}]{\text{MeCN}} \text{C}_6\text{H}_5\text{N(CH}_2)_5\cdot\text{Cr(CO)}_3 \ (99\%) \quad (123)$$

$$\text{C}_7\text{H}_8\text{Mn(CO)}_3^+ \xrightarrow[\text{RT, 15 min}]{\text{HNMe}_2\text{, H}_2\text{O}} \text{Me}_2\text{N-C}_7\text{H}_8\text{Mn(CO)}_3 \ (87\%) \quad (124)$$

timate reaction [Eq. (123)] constitutes a mild procedure for phenylating amines, although this fact does not appear to have been recognized in the literature. The reaction also works well with *n*-butylamine and pyrrolidine, but the scope and generality of the reaction are not known. The tricarbonylchromium group should be readily and quantitatively removed from these complexes with iodine in ether at 0° (cf. Semmelhack and Hall, 1974). Removal of the manganese tricarbonyl group could prove more difficult. It does not appear to have been investigated in the case of (cycloheptadienylium)manganese complexes, although oxidative removal might conceivably result in hydride abstraction to give 1-dimethylaminocycloheptatriene [cf. Eq. (49)].

6. Trienyl Cations

Tricarbonylchromium complexes of tropylium salts undergo facile attack by nucleophiles such as MeO^- [Eq. (125) (Munro and Pauson, 1961)], but reductive dimerization occurs in many cases (Section III,A,6).

$Cr(CO)_3$ $\xrightarrow[\text{RT, 2 hr}]{\text{NaOMe, MeOH}}$ OMe $Cr(CO)_3$ (88%) (125)

B. Electrophilic Attack on Coordinated Double Bond Systems

1. Dienes, Trienes and Tetraenes

C–C bond formation resulting from electrophilic attack on coordinated double-bond systems is a useful synthetic reaction (Section III,B). The formation of C–X bonds by electrophilic attack of X^+, however, is restricted essentially to H^+ or D^+. Electrophiles such as Br^+ or $NO_2{}^+$ usually lead to oxidative cleavage of the metal–olefin complex, although NBS (*N*-bromosuccinimide) has been reported to act as a hydride acceptor in some cases (Khand *et al.*, 1968, 1969).

C–H or C–D bond formation is usually a stereospecific process in which the electrophile attacks the double-bond system on the same side as the metal atom [Eqs. (126), (127) (Whitesides and Arhart, 1971), (128) (Winstein *et al.*, 1965), (129), (130) (Birch and Williamson, 1973; A. J. Birch, B. J. Chauncy, and D. J. Thompson, unpublished results, 1975 and (131) (Birch *et al.*, 1973b; A. J. Birch, B. J. Chauncy, and D. J. Thompson, unpublished results, 1975)].

$(CO)_3Fe$ —— CF_3COOD, RT → D, D, $(CO)_3Fe$ (126)

Ph, $Fe(CO)_3$ —— CF_3COOD, RT → Ph, D, D, D, $Fe(CO)_3$ (127)

$Mo(CO)_3$ —— D_2SO_4 → H, D, +, $Mo(CO)_3$ (128)

COOMe, $(CO)_3Fe$ —— D_2SO_4, MeOD, 40 hr → COOMe, D, $(CO)_3Fe$ (84%) (129)

OMe, $(CO)_3Fe$ —— (1) D_2SO_4, RT, 15 min (2) Et_2O (3) NH_4PF_6, H_2O → D, $(CO)_3Fe$, +, PF_6^- (80%) (130)

OMe, Me, $Fe(CO)_3$ —— (1) D_2SO_4, RT, 15 min (2) Et_2O (3) NH_4PF_6, H_2O → D, Me, +, $Fe(CO)_3$, PF_6^- (131)

A mechanism for the stereospecific monodeuteration of tricarbonyl(1-carbomethoxycyclohexa-1,3-diene)iron [Eq. (129)] is suggested in Scheme 11. This mechanism is consistent with all the available data on stereospecific deuterations of variously substituted tricarbonyl(cyclohexadiene)iron complexes (A. J. Birch, B. J. Chauncy, and D. J. Thompson, unpublished results, 1975) and with the large D/H (deuterium/hydrogen) isotope effect, i.e., rate-determining protonation at Fe (Whitesides and Nielan, 1975). It should be noted that the principle of microscopic reversibility may not strictly apply in

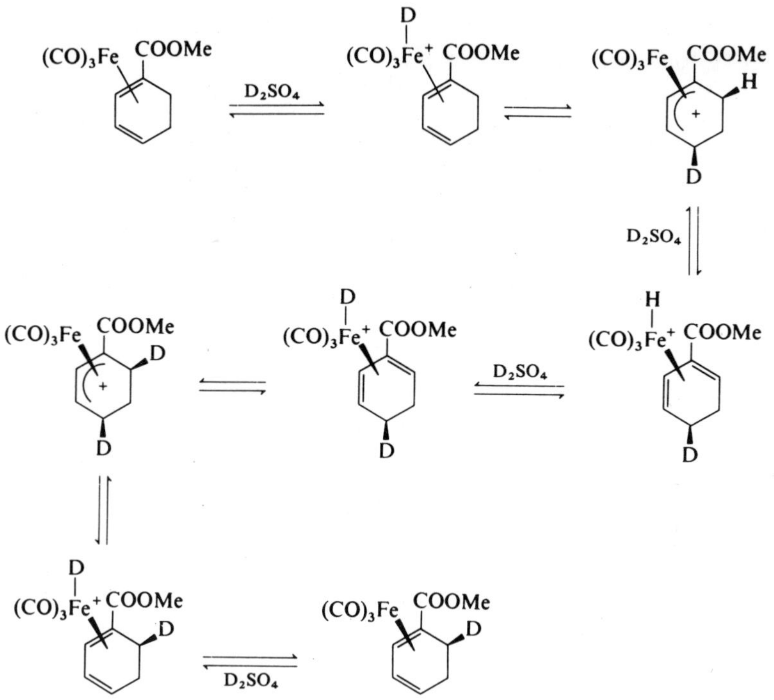

Scheme 11

this type of mechanism as the reverse of *proton* donation from iron(0) to the diene system to give a 16-electron π-allyl iron complex might be *hydride* transfer from the π-allyl system to the iron atom. (The iron would undergo a formal change of oxidation state from Fe(0) to Fe(II) at the π-allyl cation stage, thus ensuring that the metal atom retained most of the positive charge at all stages in the reaction pathway.) These mechanistic considerations offer a ready explanation for the lack of deuterium incorporation into the 5-position of the 1-carbomethoxydiene [Eq. (129)] complex, as dideuteration [cf. Eq. (126)] would necessitate initial proton (deuteron) transfer from iron to the 1-carbon atom and this is not favored for both electronic and steric reasons. If the carbomethoxy group is replaced by the methoxyl group, however, deuteron transfer to the 1-carbon atom is more facile. Elimination of methanol from the resulting π-allyl iron system by a 1,5-elimination leads to the 1-deuteriodienylium system [Eq. (130)]. In contrast to the 1-carbomethoxydiene system, the corresponding 5-carbomethoxycyclohexa-1,3-diene complex undergoes dideuteration to give the 5,6-dideuterio-1-carbomethoxy-1,3-diene complex as predicted by Scheme 11.

One example of a stereospecific deuteration on the side of the molecule opposite the tricarbonyliron group has been observed [Eq. (132) (Hunt *et al.*,

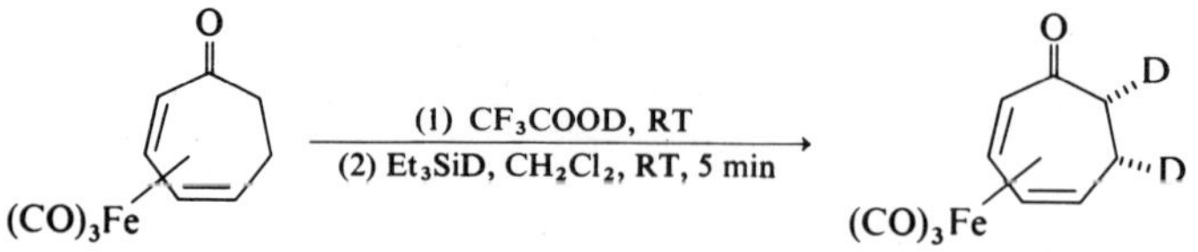

$(CO)_3Fe$ — (1) CF_3COOD, RT; (2) Et_3SiD, CH_2Cl_2, RT, 5 min → $(CO)_3Fe$ (D, D) (132)

$(CO)_3Fe$ (D^+) ⇌ $(CO)_3Fe$ (O–D, D^+) ⇌ $(CO)_3Fe$ (D)

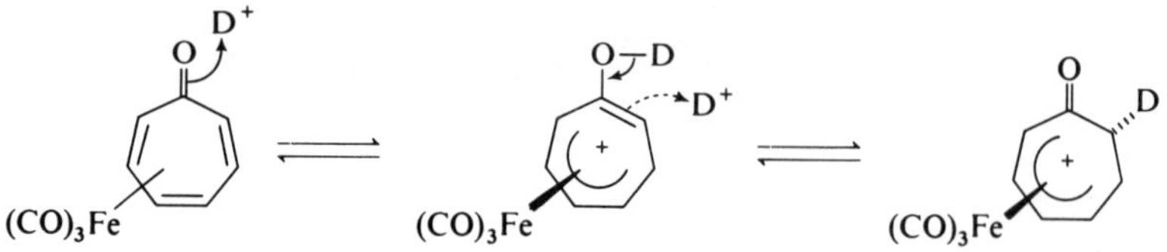

Scheme 12

1972)]. In this case, the tricarbonyliron group almost certainly facilitates initial protonation on oxygen. Protonation of the resulting enol would then be expected to occur on the least hindered side of the molecule, i.e., opposite to the tricarbonyliron group (Scheme 12).

The 1-CO_2Me complex above undergoes base-catalyzed exchange with MeO^-, MeOD to introduce a 5α-D (A. J. Birch, B. J. Chauncy, and D. J. Thompson, unpublished results, 1975). This would be expected on the basis of an anionic intermediate.

2. Ferrocene and Tricarbonyl(cyclobutadiene)iron

The potential use of ferrocene in organic synthesis was discussed in Section III,B2. Examples of typical C–X bond-forming reactions are given in Eqs. (133) (Nesmeyanov *et al.*, 1961), (134) and (135) (Perevalova and Nikitina, 1972).

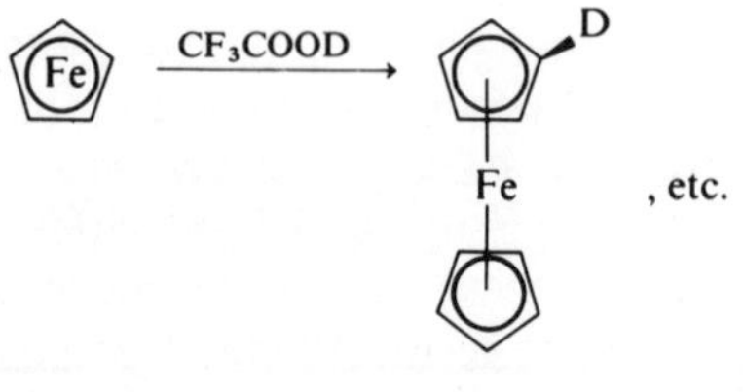

Fe — CF_3COOD → Fe (D), etc. (133)

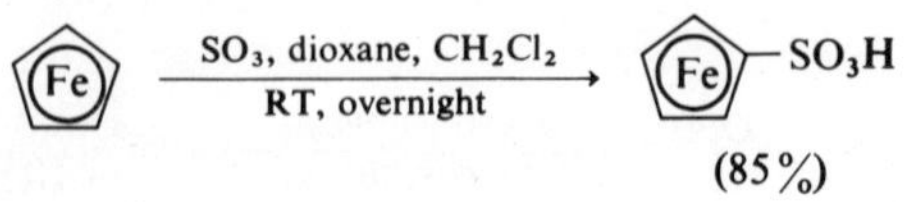

Fe — SO_3, dioxane, CH_2Cl_2; RT, overnight → Fe–SO_3H (85%) (134)

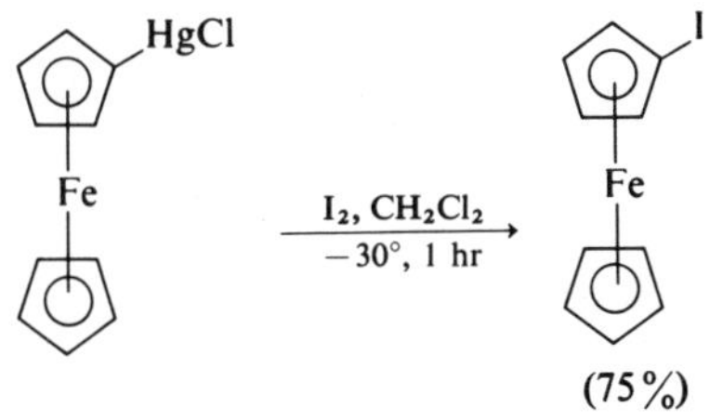

HgCl Fe I_2, CH_2Cl_2, $-30°$, 1 hr I Fe (75%) (135)

Tricarbonyl(cyclobutadiene)iron undergoes similar, facile, electrophilic substitution reactions (Maitlis, 1966; Fitzpatrick *et al.*, 1965).

V. ISOMERIZATION AND REORGANIZATION REACTIONS OF OLEFINS

A. Isomerization of Olefins

1. Monoolefins

Many transition metal complexes catalyze the isomerization of olefins (Manuel, 1962: Cramer and Lindsey, 1966; Davies, 1967; Hubert and Reimlinger, 1970; Maitlis, 1971; Bingham *et al.*, 1974).

These reactions are only of synthetic importance when there is some other functionality in the molecule that either stabilizes a particular olefin isomer or irreversibly blocks the isomerization process. For example, 1-octene is isomerized into a mixture of 2,3- and 4-octenes in the presence of $PdCl_2/HOAc$ (Davies, 1964) while this same catalyst converts allylbenzene mainly into *trans*-β-methylstyrene. [Eq. (136) (Cruickshank and Davies, 1966)].

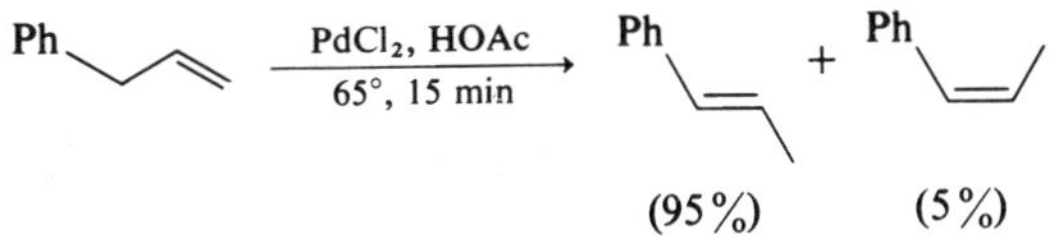

Ph $PdCl_2$, HOAc, 65°, 15 min Ph + Ph (95%) (5%) (136)

One of the most useful olefin isomerizations is the conversion of allyl alcohols into aldehydes and ketones [Eqs. (137) (Sasson and Rempel, 1974) and (138) (Strohmeier and Weigelt, 1975)].

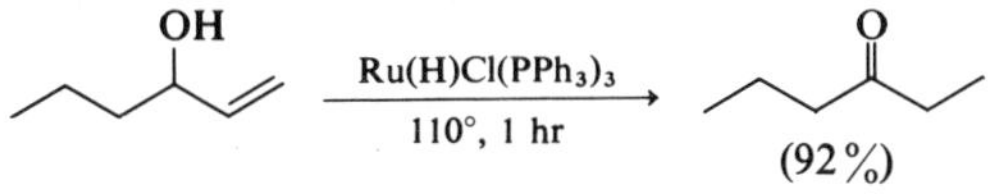

OH $Ru(H)Cl(PPh_3)_3$, 110°, 1 hr O (92%) (137)

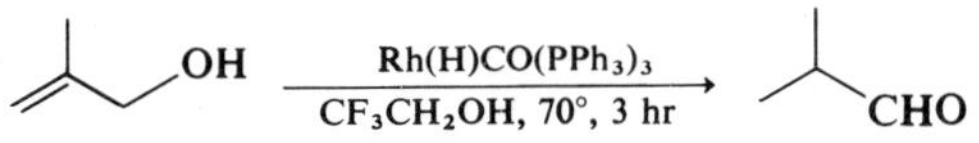

OH $Rh(H)CO(PPh_3)_3$, CF_3CH_2OH, 70°, 3 hr CHO (138)

This isomerization is a basis of using allyl ethers as protecting groups for alcohols [Eq. (139) (Corey and Suggs, 1973)].

$$\mathrm{ROCH_2CH{=}CH_2} \xrightarrow[\text{neutral, aprotic conditions}]{\mathrm{RhCl(PPh_3)_3}} \mathrm{ROCH{=}CHCH_3} \xrightarrow{\text{pH } 2} \underset{(>90\%)}{\mathrm{ROH + CH_3CH_2CHO}} \qquad (139)$$

In the transition metal-catalyzed isomerizations of olefins, above room temperatures are normally employed. Some metal carbonyl-catalyzed isomerizations can be effected at low temperature in the presence of uv light [Eq. (140) (Jolly *et al.*, 1965; Hubert *et al.*, 1972, 1973; Wrighton *et al.*, 1974)]. Presumably photolysis facilitates generation of a coordinatively unsaturated metal–olefin complex thus allowing formation of a π-allyl metal hydride intermediate (Scheme 13) (cf. Alper *et al.*, 1969).

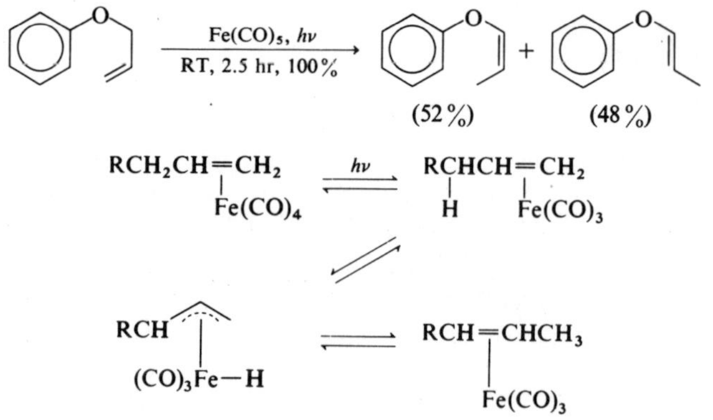

$$\mathrm{PhOCH_2CH{=}CH_2} \xrightarrow[\text{RT, 2.5 hr, 100\%}]{\mathrm{Fe(CO)_5},\ h\nu} \underset{(52\%)}{\mathrm{PhOCH{=}CHCH_3}} + \underset{(48\%)}{\mathrm{PhOCH{=}CHCH_3}}$$

$$\mathrm{RCH_2CH{=}CH_2 \cdot Fe(CO)_4} \underset{}{\overset{h\nu}{\rightleftharpoons}} \mathrm{RCH(H)CH{=}CH_2 \cdot Fe(CO)_3} \rightleftharpoons \mathrm{RCH(\pi\text{-allyl})(CO)_3Fe{-}H} \rightleftharpoons \mathrm{RCH{=}CHCH_3 \cdot Fe(CO)_3} \qquad (140)$$

Scheme 13

Pd(II) complexes appear to effect olefin isomerization, including cis–trans isomerization, under milder conditions than most other complexes (Maitlis, 1971). This may reflect a more facile formation of a π-allyl metal hydride, perhaps via a σ-bonded Pd intermediate as suggested in Scheme 14. Pd(II) is well known to form σ-bonded carbon species and halide ion is necessary for the isomerization to occur to a significant extent (Maitlis, 1971).

2. Dienes

Unconjugated dienes are readily brought into conjugation in the presence of iron carbonyl compounds. Usually, the less sterically crowded of the pos-

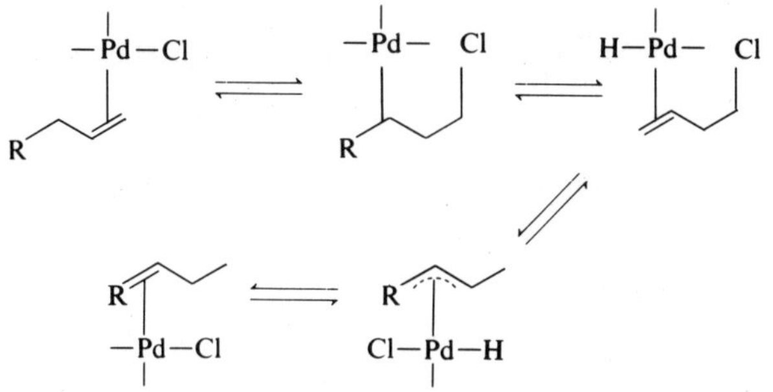

—Pd—Cl R ⇌ —Pd— Cl R ⇌ H—Pd— Cl

⇌

R —Pd—Cl ⇌ R Cl—Pd—H

Scheme 14

sible diene or diene tricarbonyliron complexes is formed [Eqs. (141) (Arnet and Pettit, 1961), (142) (Alper *et al.*, 1969), (143) (Alper and Edward, 1968). and (144) (Corey and Moinet, 1973)].

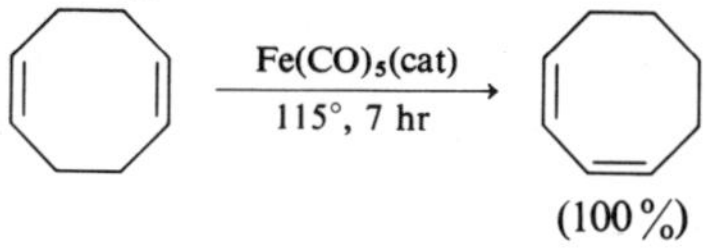

$Fe(CO)_5$(cat), 115°, 7 hr → (100%) (141)

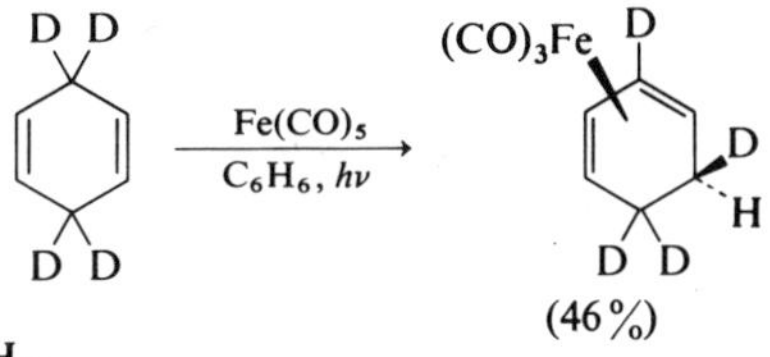

D D, D D → $Fe(CO)_5$, C_6H_6, $h\nu$ → $(CO)_3Fe$ D, D, H, D D (46%) (142)

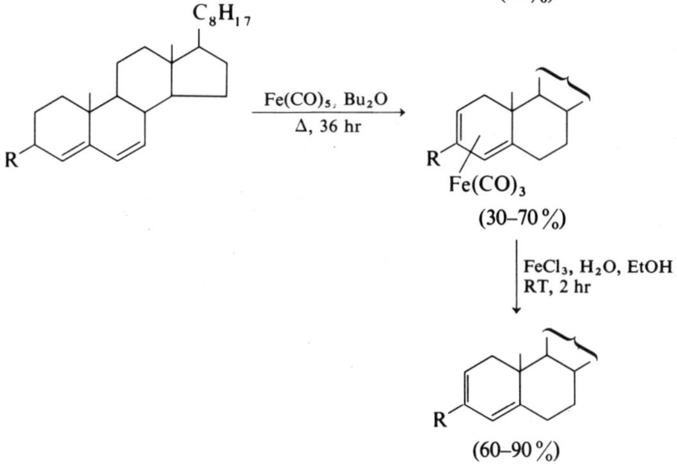

C_8H_{17}, R → $Fe(CO)_5$, Bu_2O, Δ, 36 hr → R, $Fe(CO)_3$ (30–70%)

$FeCl_3$, H_2O, EtOH, RT, 2 hr (143)

R (60–90%)

(OTHP = tetrahydropyranyl ether)

$Fe_3(CO)_{12}$ / DME, 95°, 30 min → (40–60%); CrO_3/pyridine, CH_2Cl_2, −23° → (86%) (144)

Conjugation can also be effected in some cases with Rh(I) [Eq. (145) (Birch and Subba Rao, 1968)], whereas deconjugation is possible in special cases [Eq. (146) (Rinehart and Lasky, 1964)].

1% $RhCl(PPh_3)_3$ / $CHCl_3$, 2 hr → (80%) (145)

$RhCl_3 \cdot 3H_2O$ / EtOH, 50°, 24 hr → (66%); KCN/H_2O → (146)

Isomerization of conjugated dienes is possible by protonation of the corresponding tricarbonyliron complexes [Eqs. (147) (Birch and Williamson, 1973) and (148) (Birch *et al.*, 1975)]. These isomerizations presumably also involve transfer of allylic hydrogen to the transition metal (cf. Scheme 11).

COOMe, $Fe(CO)_3$ or COOMe, $Fe(CO)_3$ — H_2SO_4 / MeOH → COOMe, $Fe(CO)_3$ (~80%) (147)

$$\text{(bicyclic diene)}\,Fe(CO)_3 \xrightarrow{H^+} (CO)_3Fe\,\text{(isomeric bicyclic diene)} \qquad (148)$$

Although substituted conjugated dienes normally give the corresponding complex, unconjugated dienes frequently give mixtures of complexes. In both cases acid-catalyzed equilibration may lead to different isomeric complexes of higher stabilities [e.g., 2-alkylcyclohexadiene complexes are more stable than the l-alkyl ones (A. J. Birch, B. J. Chauncy, and D. J. Thompson, unpublished results; A. J. Birch, C. S. Sell, and I. D. Jenkins, unpublished results). This equilibrium can lead to further synthetic possibilities, not possible with the uncomplexed dienes.

B. Olefin Disproportionation (Exchange of Alkylidene Groups)

Transition metal-catalyzed olefin disproportionation [Eq. (149) (R = alkyl, aryl; R = heteroatom such as Cl deactivates the double bond toward the metathesis reaction)] has been used very little in organic synthesis, but the reaction has considerable potential and has been the subject of several reviews (Banks, 1972; Calderon, 1972; Hughes, 1972; Cardin *et al.*, 1973; Haines and Leigh, 1975).

$$R_2C{=}CR_2 + R_2'C{=}CR_2' \rightleftharpoons 2R_2C{=}CR_2' \qquad (149)$$

Catalysts for the reaction usually consist of a transition metal (often tungsten) and a nontransition metal cocatalyst (R_4Sn, $RAlCl_2$) (cf. Ziegler–Natta polymerization). Rigorous exclusion of air and moisture, and pure olefins and solvents are necessary. The olefin disproportionation reaction attains a thermodynamic equilibrium with respect to both the exchange of alkylidene groups and the cis/trans isomer distribution of the component olefins. The reversible nature of the reaction limits its use in organic synthesis except in those cases where one of the component olefins (for example, ethylene) is readily removed from the reaction mixture. Some examples of alkylidene exchange reactions are shown in Eqs. (150) (Calderon *et al.*, 1968), (151) (Knoche, 1970), and (152)–(154) (Zuech *et al.*, 1970).

$$CD_3CD{=}CDCD_3 + CH_3CH{=}CHCH_3 \xrightarrow[\text{RT, few minutes (50\% conversion)}]{WCl_6,\ EtAlCl_2,\ EtOH} 2CD_3CD{=}CHCH_3 \qquad (150)$$

Et, Me $\xrightarrow[\text{hexane, RT, 15 min}]{WO(OPh)_4,\ EtAlCl_2}$ Et, Me / Me, Et + = ↑ (151)

1 : 1(cis : trans)

(95 %)

Pr $\xrightarrow[Me_3Al_2Cl_3,\ \text{RT, 21 hr}]{(Ph_3P)_2Cl_2(NO)_2Mo,\ PhCl}$ SM + Pr, Pr + = ↑ (152)

(33 %) (61 %)

$\xrightarrow[Me_3Al_2Cl_3,\ \text{RT, 21 hr}]{(Ph_3P)_2Cl_2(NO)_2Mo,\ PhCl}$ SM + + = ↑ (153)

(6 %) (91 %)

+ ‖ $\xrightarrow[0^\circ,\ 4\ \text{hr}]{(Ph_3P)_2Cl_2(NO)_2Mo,\ PhCl,}$ SM + (154)

(83 %) (17 %)

Cyclic olefins readily undergo transalkylidenation with other cyclic olefins to give large ring dienes, but, of course, the product reacts further with other cyclic olefin molecules, leading to a polymeric mixture of macrocyclic rings.

The mechanism of transalkylidenation is not known, but metal–carbene or 3- or 4-carbon–metal heterocycles may be intermediates in the reaction (Cardin *et al.*, 1973). The cocatalyst appears to generate the catalytically active transition metal species, but such activation can also be achieved photolytically in the absence of a cocatalyst [Eq. (155) (Krausz *et al.*, 1975)]. A mechanism for this latter reaction is suggested in Scheme 15. There is good precedent for

Et, Me $\xrightarrow[h\nu,\ \text{RT, (5\% conversion)}]{W(CO)_6,\ CCl_4}$ Et, Et + Me, Me + SM

(2.5 %) (2.5 %) (95 %)

(155)

the intermediacy of metal–carbenes in olefin metathesis reactions (Cardin *et al.*, 1973; Casey and Burkhardt, 1974) and presumably, only an infinitesimal amount of an (unstable) metal–carbene complex is necessary to initiate the reaction.*

* See this volume, Chapter 3 by Casey, Section V,B for further discussion regarding the mechanism.

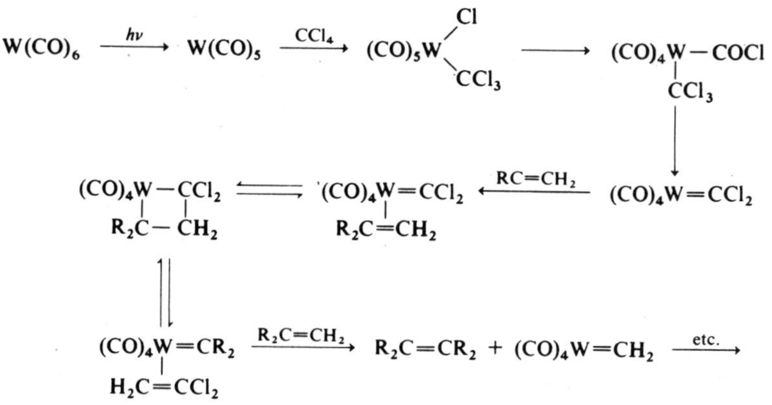

$$W(CO)_6 \xrightarrow{h\nu} W(CO)_5 \xrightarrow{CCl_4} (CO)_5W\begin{smallmatrix}Cl\\ CCl_3\end{smallmatrix} \longrightarrow (CO)_4W(COCl)(CCl_3) \longrightarrow (CO)_4W{=}CCl_2$$

$$(CO)_4W{=}CCl_2 \xrightarrow{RC{=}CH_2} (CO)_4W{=}CCl_2 \cdot R_2C{=}CH_2 \rightleftharpoons \begin{matrix}(CO)_4W-CCl_2\\ |\quad\ \ |\\ R_2C-CH_2\end{matrix} \rightleftharpoons (CO)_4W{=}CR_2 \cdot H_2C{=}CCl_2$$

$$(CO)_4W{=}CR_2 \xrightarrow{R_2C{=}CH_2} R_2C{=}CR_2 + (CO)_4W{=}CH_2 \xrightarrow{etc.}$$

Scheme 15

VI. STABILIZATION OF UNSTABLE INTERMEDIATES

The examples given in this section are intended to illustrate the use of transition metal complexes in stabilizing compounds that cannot normally be isolated. In principle, the reactive intermediate or structure may then be generated, for reaction *in situ* or rapid examination, by removal of the "scaffolding." Sometimes, useful transformations can be carried out on the complex before removal of the metal.

A. Cyclobutadiene

Although cyclobutadiene is stable in a noble gas matrix at 8°K (Schmidt, 1973), this is not a convenient form for use in organic synthesis. The tricarbonyliron complex of cyclobutadiene is a readily prepared, stable, pale yellow liquid [Eq. (156) (Pettit and Henery, 1970)]. "Free" cyclobutadiene (Schmidt,

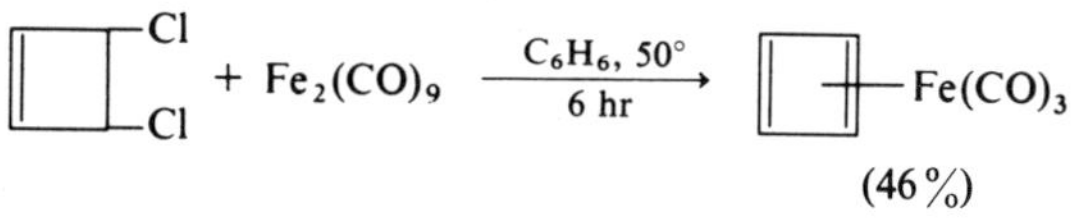

$$\text{3,4-dichlorocyclobutene} + Fe_2(CO)_9 \xrightarrow[6\ hr]{C_6H_6,\ 50^\circ} C_4H_4Fe(CO)_3\ (46\%) \tag{156}$$

1973) can be generated *in situ* by oxidation of the complex with ceric ion (Watts and Pettit, 1966; Sanders *et al.*, 1974). This reaction has been used in the synthesis of a variety of compounds such as Dewar benzenes (Watts *et*, *al.* 1965; Reeves *et al.*, 1969; Meinwald and Mioduski, 1974), cubanes [Eq. (157) (Barborak and Pettit, 1967)], and caged keto sulfones [Eq. (158) (Paquette and Wise, 1967)]. Intramolecular reactions give higher yields of products if the dienophile is positioned close to the incipient diene by coordination to iron [Eq. (159) (Grubbs *et al.*, 1974)].

(157)

(158)

(159)

In a related reaction leading to a synthesis of homopentaprismanone, the cyclobutadiene complex behaves as the dienophile [Eq. (160) (Ward and Pettit, 1970)].

hν, Et$_2$O, 4 hr; (1) $(NH_4)_2Ce(NO_3)_6$ (2) hν, Et$_2$O, 1 hr (3) hydrolysis

Fe(CO)$_3$ (CO)$_2$Fe (10%) (14%)

(160)

In most of its reactions, cyclobutadiene generated from the tricarbonyliron complex behaves as a *diene*, but it sometimes acts as a dienophile (Watts *et al.*, 1966). A cyclobutadiene complex that appears to behave *only* as a *dienophile* has recently been described and should prove as useful as "free" cyclobutadiene in organic synthesis. The complex is generated *in situ* from a readily prepared starting material [Eq. (161) (Sanders and Giering, 1975)]. An example of the difference between these two types of "cyclobutadiene" is provided by the reaction with dimethylfumarate. Cyclobutadiene generated from the tricarbonyliron complex undergoes a facile Diels–Alder reaction, whereas the cyclobutadiene complex generated as in Eq. (161) does not react with dimethylfumarate.

Cl, Fp; $AgPF_6$, CH_2Cl_2; Fp$^+$; cyclopentadiene, −78°; Fp$^+$ (58%)

(161)

Variously substituted tricarbonyliron–cyclobutadiene complexes are readily prepared (Fitzpatrick *et al.*, 1965; Roberts *et al.*, 1969; Agar *et al.*, 1974) and cyclobutadiene complexes with other transition metals are known (Maitlis, 1966), but few of these have been used in organic synthesis.

B. Trimethylenemethane (Trismethylenemethyl)

This elusive compound (recently reviewd by Dowd, 1972) is readily obtained as its pale yellow, liquid, tricarbonyliron complex [Eq. (162) (Emerson *et al.*, 1966)]. Generation of free trimethylenemethane from this complex by removal of the tricarbonyliron group has been achieved by photolysis (Day and Powell,

1968) and by oxidation with ceric ion (Ward and Pettit, 1970), but very little work has been published on the use of trimethylenemethane in organic synthesis. The photolysis work is a little discouraging because photolysis in cyclopentadiene, for example, gave 16 products. One of these (23%) was the expected 1,4-cycloadduct **(XXIV)** of trimethylenemethane and cyclopentadiene.

CH_2Cl / CH_2Cl — $Fe_2(CO)_9$, Et_2O; RT, 12 hr → H_2C, CH_2, CH_2, $Fe(CO)_3$ (30%) (162)

(XXIV)

Another example of the trimethylenemethane type of compound is provided by the tricarbonyliron complex of heptafulvene. This complex undergoes a cycloaddition reaction with methyl acetylenedicarboxylate and the product can be converted into 1,2-dicarbomethyoxyazulene [Eq. (163) (Kerber and Ehntholt, 1973)].

$Fe(CO)_3$ + COOMe / COOMe — (1) xylene, 150°, 8 hr; (2) Pd/C, air, 2 days → COOMe, COOMe (16%) (163)

This latter reaction may not involve a trimethylenemethane species as an intermediate, however.

Clearly, trimethylenemethane would be a synthetically useful species if procedures for its generation *in situ* could be improved.

C. Cyclohexadienone

As its tricarbonyliron complex, the phenol tautomer, cyclohexa-2,4-dienone is a stable, yellow, crystalline solid, that is readily prepared [Eq. (164) (Birch *et al.*, 1968; Birch and Chamberlain, 1973)]. Reports of arylation of amines [Section IV.A,4] and of an alkylzinc reagent [Eq. (165) (Cowles *et al.*, 1972)]

OMe → → OMe $Fe(CO)_3$

Ph_3C^+, CH_2Cl_2
RT, 30 min

OMe + $(CO)_3Fe$ (90%) → O $(CO)_3Fe$ (100%) (164)

using tricarbonyl(cyclohexadienone)iron derivatives indicate the potential use of these substances in organic synthesis as mild arylating agents. Strongly basic nucleophiles must be avoided for two reasons: ready enolization by basic

O $Fe(CO)_3$ —(1) $BrCH_2COOMe$, Zn, C_6H_6, Δ, 4 hr; (2) 2 *N* H_2SO_4→ HO CH_2COOMe $Fe(CO)_3$ (74%)

$Ph_3C^+BF_4^-$,
CH_2Cl_2, RT

CH_2COOMe + $Fe(CO)_3$ (71%) —H_2O, 100°, 3 min→ H—C—COOMe $Fe(CO)_3$ (98%) (165)

Ce(IV), EtOH
H_2O, RT, 1 hr

CH_2COOMe (70%)

reagents leads to loss of iron and formation of phenol, and the ready reducibility of the carbonyl, even by Grignard reagents, may lead to dimeric pinacols.

D. Norbornadiene-7-one

This otherwise unstable compound may be isolated as its tricarbonyliron complex [Eq. (166) (Landesberg and Sieczkowski, 1968, 1969)]. The free ligand appears to be generated *in situ* on photolysis of the complex.

OH → (pyridine·SO_3, DMSO / Et_3N) → O

Fe $(CO)_3$ → Fe $(CO)_3$ (45%) (166)

E. Pentalene

Attempts to prepare the "antiaromatic" compound pentalene have so far failed. Various transition metal complexes of pentalene are, however, easily prepared [for example, ruthenium carbonyl complexes, Eq. (167) (Brookes *et al.*, 1973)] so that pentalene might be generated *in situ* from such complexes and trapped with various reagents (Knox and Stone, 1974).

→ ($Ru(CO)_4(SiMe_3)_2$ / heptane, reflux) → $SiMe_3$, $Ru(CO)_2SiMe_3$

↓ (167)

$Me_3Si-Ru(CO)_2-Ru(CO)_2-SiMe_3$ (~30%) ⇢

F. Tricyclo(4.4.0.0^{2,5})deca-7,9-diene

It has been postulated as an intermediate in the photochemical reaction of benzene with cyclobutene, but it has not been isolated in the free state. Its tricarbonyliron complex, however, is readily isolated as a stable crystalline solid [Eq. (168) (Cotton and Deganello, 1972)] presenting the possibility of generating the free ligand *in situ* by oxidative removal of the tricarbonyliron group.

$\xrightarrow[\text{RT, 90 hr}]{Fe_2(CO)_9,\ Et_2O}$ Fe(CO)$_3$ (23%) $\xrightarrow[\Delta,\ 26\ \text{hr}]{\text{toluene}}$ (CO)$_3$Fe (40%)

(168)

VII. TRANSITION METALS AS PROTECTING, ACTIVATING, AND DIRECTING GROUPS

In Section VI, the use of transition metals to stabilize otherwise unstable molecules was demonstrated. Transition metals can also be used in the reverse sense to destabilize otherwise relatively stable systems [e.g., the palladium(II) chloride derivative of Dewar(hexamethyl)benzene, decomposes quantitatively at 33° in 20 min into hexamethylbenzene and palladous chloride, whereas the free ligand has a half-life of 105 hr at 120° (Koser and St. Cyr, 1974)].

Many symmetry-disallowed reactions are catalyzed by transition metals and their compounds: valence isomerizations of strained cyclic systems, disproportionation and hydrogenation of alkenes, and cyclooligomerizations. The reactions may proceed by multistep pathways involving transition metal π- and σ-bonded species rather than by changes in symmetry requirements. Some such reactions are useful for organic preparations since they provide routes to several unique hydrocarbons (Labunskaya *et al.*, 1974) [see also Eqs. (58) and (59)]. An example of a $[2\pi + 6\pi]$-cycloaddition reaction is given in Eq. (169) (Davis *et al.*, 1974). The iron atom appears to serve two main purposes: (1) it binds to one of the π bonds of the triene, rendering this ligand "diene-like," and (2) it alters the entropy factor by holding in proximity the acetylene which still has a remaining π bond.

Fe(CO)$_3$ $\xrightarrow{h\nu,\ -78°,\ \text{THF}}$ Fe-THF (CO)$_3$ $\xrightarrow[\text{RT}]{Ph_2C_2}$ Fe(CO)$_3$ · Ph–C≡C–Ph → Ph, Ph (25%)

(169)

It is to be noted, however, that simple dienes coordinated to tricarbonyliron do not undergo Diels–Alder reactions nor do they undergo catalytic hydrogenation (Pettit *et al.*, 1963). This ability of the tricarbonyliron group to act as a protecting group for *cis*-dienes is illustrated by the conversion of thebaine to *N*-cyanonorthebaine, where in the absence of the tricarbonyliron group extensive rearrangement occurs [Eq. (170) (Birch and Fitton, 1969)], and by the conversion of myrcene into a dihydromyrcene in which only the isopropylidene double bond has been reduced [Eq. (171) (Banthorpe *et al.*, 1973)]. A procedure which would be regarded as even more difficult by usual methods is the hydrogenation of the middle double bond of a conjugated triene, but this has been carried out by complexing the outer bonds [Eq. (172) (Grimme, 1972)].

Thebaine $\xrightarrow[h\nu,\ 80°,\ 12\ hr]{Fe(CO)_5,\ C_6H_6}$ MeO, O, NR, MeO, $Fe(CO)_3$ **(XXV)**

(XXV) (R = Me) $\xrightarrow[RT,\ overnight]{BrCN,\ CCl_4,}$ **(XXV)** (R = CN) $\xrightarrow[Me_2CO]{FeCl_3}$ *N*-cyanonorthebaine (95–100%) (170)

$Fe(CO)_3$ $\xrightarrow[(2)\ MeOH,\ -78°]{(1)\ B_2H_6,\ THF;\ 20°,\ 12\ hr}$ $Fe(CO)_3$, $B(OMe)_2$ (80%) $\xrightarrow{(1)\ HOAc,\ 25°,\ 12\ hr;\ (2)\ Ce(IV),\ H_2O,\ Et_2O,\ RT}$ (43% overall from myrcene) (171)

hexane
(87%)
H_2, Pd/C hexane
(172)
KCN, H_2O
(93%)
(90%)

Another partial hydrogenation reaction which would not be easy to achieve by normal procedures is the quantitative conversion of *trans,trans*-2,4-hexadiene into *cis*-3-hexene by $Cr(CO)_6$ under photolytic conditions, which takes place selectively in the presence of cis,trans and cis,cis isomers [Eq. (173) (Platbrood and Wilputte-Steinert, 1974a,b)]. This 1,4-addition of hydrogen to 1,3-dienes can also be effected using tris(acetonitrile)tricarbonylchromium(0) as the catalyst (Schroeder and Wrighton, 1974).

$Cr(CO)_6$, $h\nu$, H_2
pentane, Me_2CO
$Cr(CO)_4$
(100%)
(173)

Just as the tricarbonyliron group has been used as a protecting group for dienes, the tricarbonylchromium group has been used in a similar fashion to stabilize dihydropyridines which are normally difficult to handle and to isolate. The dihydropyridine is readily liberated from the complex by treatment with pyridine at room temperature (Kutney *et al.*, 1974). Dihydropyridines can also be generated from the corresponding tricarbonyliron complexes, recently synthesized by Alper (1975).

Use of $Cr(CO)_3$ as a blocking group is illustrated by the stereo-selective deuteration of the olefin **XXVI** in which one side of the olefin is shielded by the $Cr(CO)_3$ group [Eq. (174) (Trahanovsky and Baumann, 1974)].

(XXVI) $\xrightarrow[\text{dioxane, }\Delta\text{, 14 min}]{(CO)_3Cr(MeCN)_3}$ $Cr(CO)_3$ complex (44%) $\xrightarrow{D_2\text{, Pd/C, dioxane}}$ D, H, D, H $Cr(CO)_3$ (87%) (174)

The tricarbonylchromium group also acts as a protecting group to inhibit polymerization of styrene under free radical or anionic conditions (Knox *et al.*, 1972).

Because of its electron-withdrawing ability (comparable to that of a p-NO_2 group) the tricarbonylchromium group increases the rates of hydrolysis of benzoic acid esters (Klopman and Calderazzo, 1967) and affects the position of oxazolidine–imino alcohol tautomerism (Alper *et al.*, 1973).

The instability of monoolefin complexes limits the use of this means of protection, although tetracarbonyliron complexes of olefins containing electron-withdrawing groups usually show increased stability (Herberhold, 1974; Kemmitt, 1972) and the derivatives of α,β-unsaturated ketones are reported to lose their normal high reactivity toward nucleophiles (Nesmeyanov *et al.*, 1974). Further work in this area is needed.

VIII. REARRANGEMENTS AND THE SYNTHESIS OF SOME UNUSUAL COMPOUNDS

Some miscellaneous examples illustrate many of the principles developed in the previous sections. Some of the reactions result in rather unexpected products and the subject is a fruitful one for speculation and extrapolation.

Synthesis of homotropone (Holmes and Pettit, 1963)

$Fe(CO)_3$ $\xrightarrow{HBF_4}$ + $Fe(CO)_3$ $\xrightarrow{NaOH}$ OH $Fe(CO)_3$ $\xrightarrow[(2)\ Ce(IV)]{(1)\ CrO_3}$ O (175)

Synthesis of barbaralone (Heil *et al.*, 1974)

$Fe(CO)_3$ — $AlCl_3$, benzene, 10° → $Fe(CO)_3$ (40%) — CO, 120°, 100 atm → (90%) (176)

Synthesis of 9-oxabicyclo[4.2.1]nona-2,4,7-triene (Aumann and Averbeck 1975)

$Fe(CO)_5$, $h\nu$, 6 hr, 25° → $Fe(CO)_3$ → $(CO)_4Fe$, $Fe(CO)_3$ (70%) — Me_3NO, 8 hr, 25° → (48%) (177)

Synthesis of 6-*exo*-chloro-2-*endo*-acetoxybicyclo[3.3.0]octane (Chung and Scott, 1975)

$PdCl_2$ — $Pb(OAc)_4$, HOAc, RT, 20 hr → H, Cl, H, OAc (60%) (178)

Conversion of a cholestadiene to cholestatriene (Alper and Huang, 1973)

$Fe_3(CO)_{12}$, C_6H_6, Δ, 12 hr → $(CO)_3Fe$ complex (72%)

$Ph_3C^+BF_4^-$, CH_2Cl_2, Δ, 72 hr (179)

$(CO)_3Fe$ (68%) — $NaHCO_3$, RT, 10 min → $(CO)_3Fe$ (45%)

Reduction and rearrangement of santonin (Alper and Keung, 1972)

$Fe_2(CO)_9$, C_6H_6, 40°, 4 hr → HO, H, $(CO)_3Fe$, Me (14%) (180)

Tropylium ring contraction (Munro and Pauson, 1961; Pauson *et al.*, 1967)

$Cr(CO)_3$ — Na cyclopentadienide, THF, RT, 16 hr → $Cr(CO)_3$ (46%) (181)

Cyclopentadiene → cyclohexadienylium ring expansion (Herberich and Müller, 1971)

CH_2OTs, Me, $Fe(CO)_5$ — (1) HBF_4, Ac_2O (2) NH_4PF_6 → $(CO)_5Fe$, Me, PF_6^- (80%) (182)

Conversion of cobalticinium ion into azulene (Attridge *et al.*, 1970)

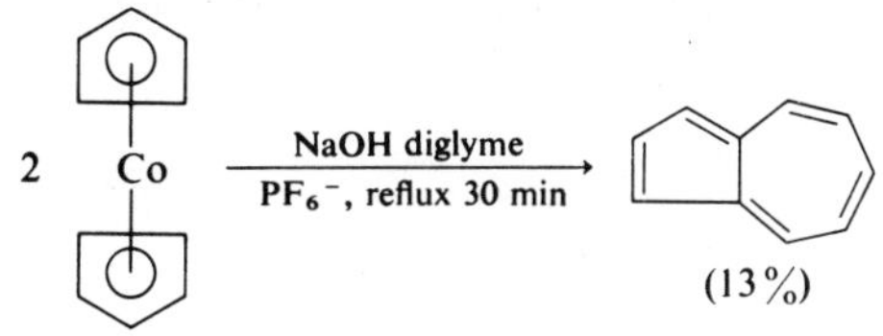

2 Co —NaOH diglyme / PF_6^-, reflux 30 min→ (13%) (183)

As an alternative to the rather unlikely mechanism suggested by the authors for this reaction is Scheme 16. This is based on an initial reductive dimerization of the salt [cf. Eqs. (43) and (54) and is one of several schemes possible in respect of subsequent reactions. The metal acts both as a point of entry and exit of electrons and as a stabilizer of intermediate cations.

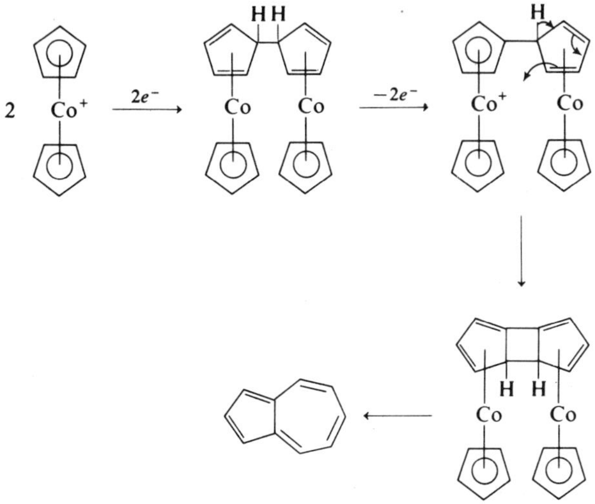

2 Co^+ —$2e^-$→ HH Co Co —$-2e^-$→ H Co^+ Co → H H Co Co →

Scheme 16

IX. STEREOCHEMICAL IMPLICATIONS

Although several examples of steric effects have been mentioned in previous sections, it is worthwhile noting some general implications for the organic chemist. When a metal atom [in the form of a group such as $M(CO)_3$] is attached to a molecule, the number of symmetry elements is decreased. For example, any 1,3-diene possessing only a plane of symmetry passing through both double bonds gives rise to a pair of enantiomeric tricarbonyliron complexes, i.e., the molecule as a whole becomes asymmetric on coordination to a tricarbonyliron group, even though the diene group is not asymmetric [e.g., the $Fe(CO)_3$ complex of α-terpinene (A. J. Birch, D. J. Thompson, and I. D. Jenkins, unpublished work)]. The two faces of the diene, formerly equivalent, are now clearly differentiated, and a reagent can attack from the same or from the opposite side relative to the metal atom. Groups such as $Fe(CO)_3$ and $Cr(CO)_3$ usually exert pronounced steric effects, and can thus affect the stereochemistry of reactions occurring elsewhere in the molecule.

A related feature is the ability of asymmetric complexes to induce chiral reactions, e.g., in catalytic hydrogenation (reviewed by Birch and Williamson, 1976). Asymmetric synthesis using homogeneous transition metal catalysts has recently been reviewed (Bogdanović, 1973).

A. Formation of Diene Complexes

The least sterically hindred complex is favored in dienes containing alkyl groups. For example, the tricarbonyliron complexes of α-phellandrene are obtained with the $Fe(CO)_3$ group on the same and opposite faces relative to the isopropyl group in the ratio of about 3:7, respectively (A.J. Birch, B. J. Chauncy and I. D. Jenkins, unpublished work.) Initial complexation [of $Fe(CO)_4$ for example] to a group such as COOMe may result in the metal atom being directed to the same side of the molecule [Eq. (184) (A. J. Birch and B. J. Chauncy, unpublished work)].

Ph COOMe → Ph COOMe, $Fe(CO)_3$ (184)

(XXVII)

The structure of **XXVII** has been confirmed by X-ray crystallography (B. Anderson, personal communication).

A similar effect of COOMe has been reported [Eq. (185) (Whitesides *et al.*, 1974)]. This kind of directive influence may be a rather general one with carbonyl groups and possibly even with olefins.

MeOOC, MeOOC → $Fe(CO)_4$, MeOOC, MeOOC (185)

B. Acid-Catalyzed Isomerizations

With diene tricarbonyliron complexes, such reactions involve addition to the same face as the iron. Some examples of the stereospecific introduction of "allylic" deuterium are given in Section IV,B. This can be brought about stepwise in some instances [Eq. (186) (A. J. Birch, B. J. Chauncy, and D. J. Thompson, unpublished work). Such results require a number of reversible stages, so that the processes both of addition and elimination must be totally

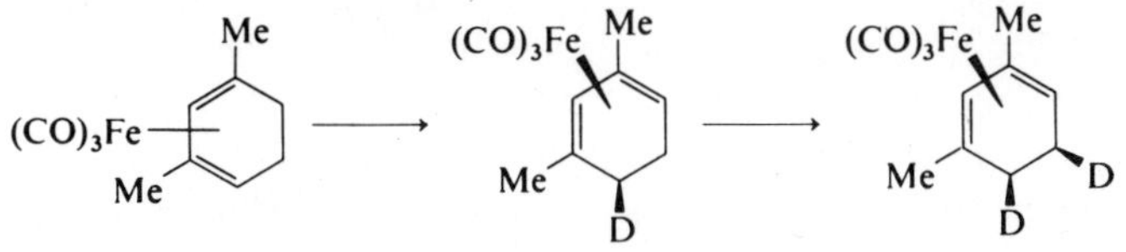

(186)

stereospecific (see Scheme 11). Through resolved complexes it should be possible to obtain optically active products.

C. Electrophilic Attack from the Rear

Tricarbonyl(cyclohexadiene)iron complexes are stereospecifically attacked by trityl cation on the rear face. If there is no hydrogen available on that face, hydride abstraction does not occur [Eqs. (187) and (188) A. J. Birch, B. J. Chauncy, and D. J. Thompson, unpublished work.)]

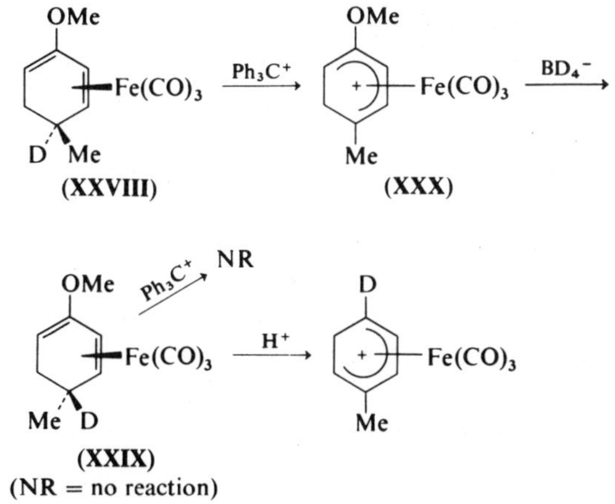

(XXVIII) (XXX)

(188)

(XXIX)
(NR = no reaction)

D. Stereospecific Attack on Cations

When the reaction is irreversible, the addition of nucleophiles to tricarbonyliron-stabilized cations such as **VII** normally takes place on the side of the molecule opposite the tricarbonyliron group [for example, Eqs. (34)–(42) and (114)–(120)].

Nucleophilic attack by borohydride, however, can result in attack on both sides. Treatment of cation **XXX** with $NaBD_4$ gives a mixture of complexes **XXVIII** and **XXIX**.

When addition is reversible, the product ratio appears to be thermodynamically determined, and in the example shown (Scheme 17) (A. J. Birch,

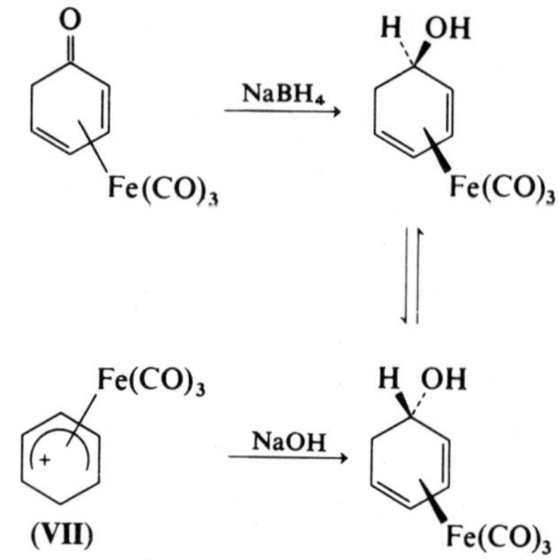

O
NaBH$_4$
H OH
$Fe(CO)_3$
$Fe(CO)_3$
$Fe(CO)_3$
NaOH
H OH
(VII)
$Fe(CO)_3$

Scheme 17

and B. J. Chauncy, unpublished work), the crystalline hydroxy isomers can be equilibrated to the same mixture.

Similar results have been obtained with methoxide addition to **VII**. (Hine *et al.*, 1975). The product formed initially (kinetic control) is that arising from addition to the side opposite the tricarbonyliron group. Under the conditions of the reaction (NaOMe in MeOH) or in the presence of an acid catalyst, this product is slowly isomerized to a mixture of the products arising from addition to both sides.

With acyclic cations, once again the kinetic product is formed when the reaction is irreversible, while reversible processes result in a more thermodynamically stable product [Eqs. (189) (A. J. Birch and A. J. Pearson, unpublished work) and (190) (Maglio and Palumbo, 1974)].

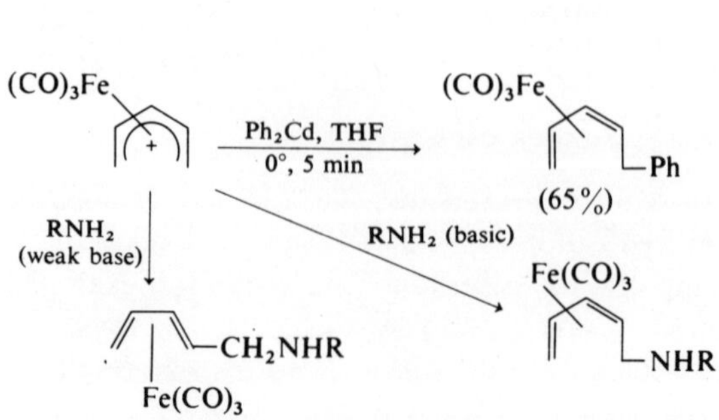

$(CO)_3Fe$
Ph_2Cd, THF
0°, 5 min
$(CO)_3Fe$
Ph
(65%)
(189)
RNH_2
(weak base)
RNH_2 (basic)
$Fe(CO)_3$
CH_2NHR
$Fe(CO)_3$
NHR
(190)

E. Stereospecific Attack on Complexed Double Bonds

In contrast to the trans addition observed for irreversible nucleophilic attack on tricarbonyliron dienylium cations and olefins complexed to Fp^+ (Section III,A), irreversible nucleophilic attack (e.g., by carbon nucleophiles) on olefins coordinated to Pt(II) and Pd(II) appears to result in cis addition, presumably via initial attack at the coordinatively unsaturated Pt(II) or Pd(II), followed by intramolecular transfer of the coordinated nucleophile to the double bond (Section III,A).

Reversible nucleophilic attack (e.g., by nitrogen and oxygen nucleophiles) usually results in trans addition, presumably reflecting thermodynamic control. The mechanism in this latter case could involve the "ping-pong" effect suggested in Section IV,A.

F. Steric Control in Adjacent Groups

The use of groups such as $Fe(CO)_3$ or $Cr(CO)_3$ as blocking substituents to direct attack on a particular side of a molecule has been mentioned earlier [cf. Eqs. (171) and (174)]. Since removal of the metal results in the loss of an element of asymmetry, the result is not of interest to the organic chemist unless a second chiral center is present or unless the initial complex is optically active. Some further examples of directed attack are shown in Eqs. (191) (Kuhn and Lillya, 1972) and (192) (see also Besançon *et al.*, 1973; Jaouen, 1973; Caro and Jaouen, 1974; Simonneaux *et al.*, 1975; Meyer and Dabard, 1972).

Ph, $Fe(CO)_3$, O —[$NaBH_4$, MeOH; RT, 3 hr]→ Ph, $Fe(CO)_3$, HO, H (191)

(85%, endo product only)

$(CO)_3Cr$, O, H, Me —[MeMgX]→ $(CO)_3Cr$, OH, H, Me, Me —[sunlight, air]→ OH, H, Me, Me (192)

(100%)

G. Optically Active Complexes

An early, illustrative example of the use of chiral organometallic complexes in organic chemistry was the resolution of *trans*-cyclooctene by fractional crystallization of the diastereoisomeric *trans*-dichloro[(+)- or (−)-α-methylbenzylamine]platinum(II) complexes of this olefin followed by removal of the metal with aqueous potassium cyanide (Cope *et al.*, 1963).

A similar technique has been used to resolve dicyclopentadiene [Eq. (193) (Paiaro *et al.*, 1966; Panunzi *et al.*, 1967)].

(S)-PhCH(Me)NH$_2$ → PhC*HNH$_2$(Me)—Pt(Cl)$_2$(dicyclopentadiene)

separated diastereoisomers —(1) HCl, (2) NaCN→ (*R*) or (*S*) (193)

Other examples of the use of chiral organometallic reagents in organic synthesis are: asymmetric hydroformylation [Eq. (194) Consiglio *et al.*, 1973)] and the synthesis of D-(+)-phenylalanine [Eq. (195) Chenard *et al.*, 1972], chiral 3-vinylcyclooctene [Eq. (196) (Bogdanović *et al.*, 1972)], (*R*)- or (*S*)-citronellol from isoprene [Eq. (197) (Hidai *et al.*, 1975)], and *N*,*N'*-diethyl-*sec*-butylamine [Eq. (198) (Panunzi *et al.*, 1970)].

+ Ph$_3$P—Rh(H)(CO)(PPh$_2$)$_2$ —CO, H$_2$, 1 atm, 20°→ CHO (194)

(−)-bisphosphine-Rh complex (catalyst) (27% optically pure)

PhCH(Me)N=CHCOOEt [Fe(CO)$_4$] —PhCH$_2$Br, EtOH, 35°, 2 hr→ PhCH(Me)N(BrFe(CO)$_4$)—CH(CH$_2$Ph)COOEt

L-(+)-imine

(1) H$_2$/Pd
(2) HO$^-$

PhCH$_2$CH(NH$_2$)COOH (195)

(53%) (77% optical yield)

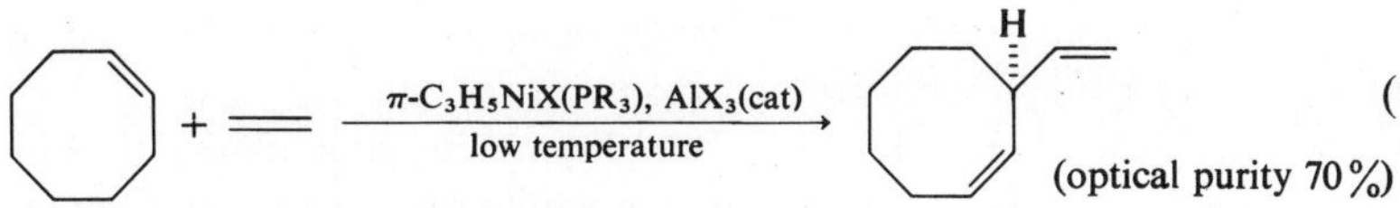

π-$C_3H_5NiX(PR_3)$, AlX_3(cat)
low temperature
(optical purity 70%)
(196)

(PR_3 = (−)-dimenthylmethylphosphine)

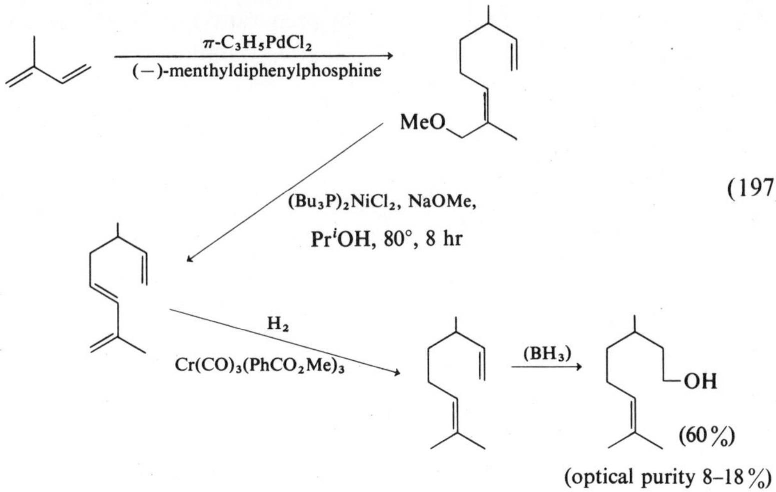

π-$C_3H_5PdCl_2$
(−)-menthyldiphenylphosphine
MeO
$(Bu_3P)_2NiCl_2$, NaOMe,
Pr^iOH, 80°, 8 hr
(197) 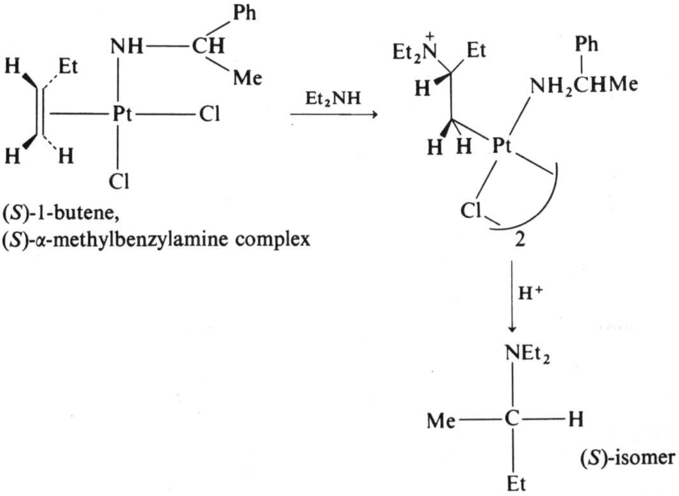
H_2
$Cr(CO)_3(PhCO_2Me)_3$
(BH_3)
OH
(60%)
(optical purity 8–18%)

(S)-1-butene,
(S)-α-methylbenzylamine complex
Et_2NH
H^+
(S)-isomer
(198)

Standard methods of resolution of complexes can be used when appropriate groups are present. For example, tetracarbonyliron complexes of acrylic and fumaric acids can be resolved by crystallization of the diastereoisomeric brucine salts (Paiaro *et al.*, 1965; Paiaro and Palumbo, 1967). Tricarbonyl(*trans-trans*-2,4-hexadienoic acid)iron is similarly resolved via the diastereoisomeric

(*S*)-α-methylbenzylamine salts, and the enantiomeric complexes obtained by treating the salts with hydrochloric acid (Musco *et al.*, 1971).

In the vast majority of reactions involving asymmetric complexes, the chirality has resided either in the molecule as a whole or in chiral ligands attached to the metal atom. There are a few examples, however, of complexes containing an asymmetric metal atom. The preparation of these complexes is illustrated by Eqs. (199) (Brunner, 1971) and (200) (Simmonneaux *et al.*, 1975).

$Fe^+(OC)(CO)(PPh_3)$ PF_6^- + NaO— ⟶ CO—Fe—CO—O— (PPh$_3$) (199)

(diastereoisomers resolved)

$(CO)_3Cr$ (Me, CO_2Me) ⟶ Cr(OC)(CS)(CO) —P(OPh)$_3$⟶ Cr(OC)(CS)(P(OPh)$_3$)

P(OPh)$_3$ ⟶ $(PhO)_3P$ Cr (CS)(CO) (Me, CO_2Me) ⟶ (*S*)-isomer $(PhO)_3P$ Cr (CS)(CO) (Me, Me) (200)

The synthesis of optically pure compounds still remains a challenging problem to the organic chemist and the examples given here are only the beginning of wide developments. It is once again worth remarking that the most useful organic processes will be either those involving a cheap metal or catalytic ones.

In the introduction to this chapter it was noted that many transition metals are essential for biological processes. Since biological processes often involve chiral reactions, the potential use, to the synthetic organic chemist, of asymmetric catalysts or reagents modeled on enzyme systems is apparent.

REFERENCES

Adkins, H., and Krsek, G. (1949). *J. Am. Chem. Soc.* **71**, 3051.
Agar, J., Kaplan, F., and Roberts, B. W. (1974). *J. Org. Chem.* **39**, 3451.
Agnes, G., Chiusoli, G. P., and Cometti, G. (1968). *Chem. Commun.* p. 1515.
Aguilo, A. (1967). *Adv. Organomet. Chem.* **5**, 321.
Åkermark, B., Bäckvall, J. E., Siirala-Hansen, K., Sjöberg, K., and Zetterberg, K. (1974). *Tetrahedron Lett.* p. 1363.
Alderson, T., Jenner, E. L., and Lindsey, R. V. (1965). *J. Am. Chem. Soc.* **87**, 5638.
Allenmark, S. (1974). *Tetrahedron Lett.* p. 371.
Alper, H. (1975). *J. Organomet. Chem.* **96**, 95.
Alper, H., and Edward, J. T. (1968). *J. Organomet. Chem.* **14**, 411.
Alper, H., and Huang, C.-C. (1973). *J. Organomet. Chem.* **50**, 213.
Alper, H., and Keung, E. C.-H. (1972). *J. Am. Chem. Soc.* **94**, 2144.
Alper, H., LePort, P. C., and Wolfe, S. (1969). *J. Am. Chem. Soc.* **91**, 7553.
Alper, H., Dinkes, L. S., and Lennon, P. J. (1973). *J. Organomet. Chem.* **57**, C12.
Anderson, M., Clague, A. D. H., Blaauw, L. P., and Couperus, P. A. (1973). *J. Organomet. Chem.* **56**, 307.
Arnet, J. E., and Pettit, R. (1961). *J. Am. Chem. Soc.* **83**, 2954.
Atkins, K. E., Walker, W. E., and Manyik, R. M. (1970). *Tetrahedron Lett.* p. 3821.
Attridge, C. J., Baker, S. J., and Parkins, A. W. (1970). *Organomet. Chem. Synth.* **1**, 183.
Aumann, R., and Averbeck, H. (1975). *J. Organomet. Chem.* **85**, C4.
Baird, W. C. (1966). *J. Org. Chem.* **31**, 2411.
Banks, R. L. (1972). *Fortschr. Chem. Forsch.* **25**, 39.
Banthorpe, D. V., Fitton, H., and Lewis, J. (1973). *J. Chem. Soc., Perkin Trans. 1* p. 2051.
Barborak, M., and Pettit, R. (1967). *J. Am. Chem. Soc.* **89**, 3080.
Bennett, M. A., Johnson, R. N., and Tomkins, I. B. (1973). *J. Organomet. Chem.* **54**, C48.
Bertelo, C. A., and A. Schwartz, J. (1975). *J. Am. Chem. Soc.* **97**, 228.
Besançon, J., Tirouflet, J., Card, A., and Dusausoy, Y. (1973). *J. Organomet. Chem.* **59**, 267.
Binger, P. (1973). *Synthesis* p. 427.
Binger, P., Schroth, G., and McMeeking, J. (1974). *Angew. Chem., Int. Ed. Engl.* **13**, 465.
Bingham, D., Webster, D. E., and Wells, P. B. (1974). *J. Chem. Soc., Dalton Trans.* p. 1514.
Birch, A. J., and Chamberlain, K. B. (1973). *Org. Synth.* **53**, 1859.
Birch, A. J., and Fitton, H. (1969). *Aust. J. Chem.* **22**, 971.
Birch, A. J., and Haas, M. A. (1971). *J. Chem. Soc. C*, p. 2465.
Birch, A. J., and Jenkins, I. D. (1975). *Tetrahedron Lett.* p. 119.
Birch, A. J., and Pearson, A. J. (1975). *Tetrahedron Lett.* p. 2379.
Birch, A. J. and Subba Rao, G. S. R. (1968). *Tetrahedron Lett.* p. 3797.
Birch, A. J., and Williamson, D. H. (1973). *J. Chem. Soc., Perkin Trans. 1* p. 1892.
Birch, A. J., and Williamson, D. H. (1976). *Org. React.* (in press).
Birch, A. J., Cross, P. E., and Fitton, H. (1965). *Chem. Commun.* p. 366.
Birch, A. J., Cross, P. E., Conner, D. T., and Subba Rao, G. S. R. (1966). *J. Chem. Soc.* p. 54.
Birch, A. J., Cross, P. E., Lewis, J., White, D. A., and Wild, S. B. (1968). *J. Chem. Soc. A* p. 332.
Birch, A. J., Chamberlain, K. B., Haas, M. A., and Thompson, D. J. (1973a). *J. Chem. Soc., Perkin Trans. 1* p. 1882.

Birch, A. J., Chamberlain, K. B., and Thompson, D. J. (1973b). *J. Chem. Soc., Perkin Trans. 1* p. 1900.

Birch, A. J., Jenkins, I. D., and Liepa, A. J. (1975). *Tetrahedron Lett.* p. 1723.

Bird, C. W. (1972). *Chem. Ind. (London)* p. 520.

Bittler, K., Kutepow, N. V., Neubauer, D., and Reis, H. (1968). *Angew. Chem., Int. Ed. Engl.* **7**, 329.

Bogdanović, B. (1973). *Angew. Chem., Int. Ed. Engl.* **12**, 954.

Bogdanović, P., Heimbach, P., Kröner, M., Wilke, G., Hoffmann, E. G., and Brandt, J. (1969). *Justus Liebigs Ann. Chem.* **727**, 143.

Bogdanović, B., Henc, B., Meister, B., Pauling, H., and Wilke, G. (1972). *Angew. Chem., Int. Ed., Engl.* **11**, 1023.

Bott, K. (1973). *Angew. Chem., Int. Ed. Engl.* **12**, 851.

Braye E. H., and Hübel, W. (1966). *Inorg. Synth.* **8**, 178.

Brenner, W., Heimbach, P., Hey, H. J., Müller, E. W., and Wilke, G. (1969). *Justus Liebigs Ann. Chem.* **727**, 161.

Brookes, A., Howard, J., Knox, S. A. R., and Woodward, P. (1973). *Chem. Commun.* p. 589.

Brown, C. K., and Wilkinson, G. (1970). *J. Chem. Soc. A* p. 2753.

Brown, E. S. (1974). *In* "Aspects of Homogeneous Catalysis" (R. Ugo, ed.), Vol. II, p. 57. Reidel Publ., Dordrecht, Netherlands.

Bruce, M. I. (1968). *Adv. Organomet. Chem.* **6**, 273.

Brunner, H. (1971). *Angew. Chem., Int. Ed. Engl.* **10**, 249.

Buchholz, H., Heimbach, P., Hey, H., Selbeck, H., and Wiese, W. (1972). *Coord. Chem. Rev.* **8**, 129.

Bunnett, J. F., and Hermann, H. (1971). *J. Org. Chem.* **36**, 4081.

Busetto, L., Pallazzi, A., Ros, R., and Belluco, U. (1970). *J. Organomet. Chem.* **25**, 207.

Cais, M., and Maoz, N. (1966). *J. Organomet. Chem.* **5**, 370.

Calderon, N. (1972). *Acc. Chem. Res.* **5**, 127.

Calderon, N., Ofstead, E. A., Ward, J. P., Judy, W. A., and Scott, K. W. (1968). *J. Am. Chem. Soc.* **90**, 4133.

Card, R. J., and Trahanovsky, W. S. (1973). *Tetrahedron Lett.* p. 3823.

Cardin, D. J., Cetinkaya, B., Doyle, M. J., and Lappert, M. F. (1973). *Chem. Soc. Rev.* **2**, 99.

Caro, B., and Jaouen, G. (1974). *Tetrahedron Lett.* p. 3539.

Casey, C. P., and Burkhardt, T. J. (1974). *J. Am. Chem. Soc.* **96**, 7808.

Chatt, J., and Duncanson, L. A. (1953). *J. Chem. Soc.* p. 2939.

Chenard, J. Y., Commereuc, D., and Chauvin, Y. (1972). *Chem. Commun.* p. 750.

Chini, P., Martinengo, S., and Garlaschelli, G. (1972). *Chem. Commun.* p. 709.

Chung, S. K., and Scott, A. I. (1975). *Tetrahedron Lett.* p. 49.

Consiglio, G., Botteghi, C., Salomon, C., and Pino, P. (1973). *Angew. Chem., Int. Ed. Engl.* **12**, 669.

Cope, A. C., Ganellin, C. R., Johnson, H. W., Van Auken, T. V., and Winkler, H. J. S. (1963). *J. Am. Chem. Soc.* **85**, 3276.

Corey, E. J., and Moinet, G. (1973). *J. Am. Chem. Soc.* **95**, 7185.

Corey, E. J., and Suggs, J. W. (1973). *J. Org. Chem.* **38**, 3224.

Cotton, F. A., and Deganello, G. (1972). *J. Organomet. Chem.* **38**, 147.

Cowles, R. J. H., Johnson, B. F. G., Lewis, J., and Parkins, A. W. (1972). *J. Chem. Soc., Dalton Trans.* p. 1768.

Cramer, R., and Lindsey, R. V. (1966). *J. Am. Chem. Soc.* **88**, 3534.

Cruickshank, B., and Davies, N. R. (1966). *Aust. J. Chem.* **19**, 815.

Dauben, H. J., and Bertelli, D. J. (1961). *J. Am. Chem. Soc.* **83**, 497.
Davies, N. R. (1964). *Aust. J. Chem.* **17**, 212.
Davies, N. R. (1967). *Rev. Pure Appl. Chem.* **17**, 83.
Davis, R. E., and Pettit, R. (1970). *J. Am. Chem. Soc.* **92**, 716.
Davis, R. E., Dodds, T. E., Hseu, T. H., Wagnon, J. C., Devon, T., Tancrede, J., McKennis, J. S., and Pettit, R. (1974). *J. Am. Chem. Soc.* **96**, 7562.
Day, A. C., and Powell, J. T. (1968). *Chem. Commun.* p. 1241.
Deeming, A. J., Ullah, S. S., Domingos, A. J. P., Johnson, B. F. G., and Lewis, J. (1974). *J. Chem. Soc., Dalton Trans.* p. 2093.
Dewar, M. J. S. (1951). *Bull. Soc. Chim. Fr.* **18**, C79.
Dixon, J. A., and Fishman, D. H. (1963). *J. Am. Chem. Soc.* **85**, 1356.
Dowd, P. (1972). *Acc. Chem. Res.* **5**, 242.
Dudley, C. W., Read, G., and Walker, P. J. C. (1974). *J. Chem. Soc., Dalton Trans.* p. 1926.
Eaborn, C., and Bott, R. W. (1968). *In* "Organometallic Compounds of Group IV Elements (A. G. MacDiarmid, ed.), Vol. 1, p. 105. Dekker, New York.
Edwards, R., Howell, J. A. S., Johnson, B. F. G., and Lewis, J. (1974). *J. Chem. Soc., Dalton Trans.* p. 2105.
Emerson, G. F., Ehlich, K., Giering, W. P., and Lauterbur, P. C. (1966). *J. Am. Chem. Soc.* **88**, 3172.
Evans, J., Howe, D. V., Johnson, B. F. G., and Lewis, J. (1973). *J. Organomet. Chem.* **61**, C48.
Falbe, J. (1971). *In* "Newer Methods of Preparative Organic Chemistry" (W. Foerst, ed.). Vol. 6, p. 193. Academic Press, New York.
Falbe, J., and Korte, F., (1965). *Chem. Ber.* **98**, 1928.
Fischer, E. O., and Fischer, R. D. (1960). *Angew. Chem.* **72**, 919.
Fischer, E. O. and Moser, E. (1970). *Inorg. Synth.* **12**, 35.
Fischer, F., Jonas, K., Misbach, P., Stabba, R., and Wilke, G. (1973). *Angew. Chem., Int. Ed. Engl.* **12**, 943.
Fitzpatrick, J. D., Watts, L., Emerson, G. F., and Pettit, R. (1965). *J. Am. Chem. Soc.* **87**, 3254.
Gibson, D. H., and Vonnahme, R. L. (1974). *J. Organomet. Chem.* **70**, C33.
Gibson, D. H., Vonnahme, R. L., and McKiernan, J. E. (1971). *Chem. Commun.* p. 720.
Giering, W. P., and Rosenblum, M. (1971). *Chem. Commun.* p. 441.
Giering, W. P., Rosenblum, M., and Tancrede, J. (1972). *J. Am. Chem. Soc.* **94**, 7170.
Gill, G. B., Gourlay, N., Johnson, A. W., and Mahendron, M. (1969). *Chem. Commun.* p. 631.
Graf, R. E., and Lillya, C. P. (1972). *J. Am. Chem. Soc.* **94**, 8282.
Graf, R. E., and Lillya, C. P. (1973). *Chem. Commun.* p. 271.
Greaves, E. O., Knox, G. R., and Pauson, P. L. (1969). *Chem. Commun.* p. 1124.
Greaves, E. O., Knox, G. R., Pauson, P. L., Toma, S., Sim, G. A., and Woodhouse, D. I. (1974). *Chem. Commun.* p. 257.
Green, M., Heathcock, S., and Wood, D. C. (1973). *J. Chem. Soc., Dalton Trans.* p. 1564.
Green, M. L. H., and Nagy, P. L. I. (1963). *J. Chem. Soc.* p. 189.
Grimme, W. (1972). *J. Am. Chem. Soc.* **94**, 2525.
Grubbs, R. H., Pancoast, T. A., and Grey, R. A. (1974). *Tetrahedron Lett.* p. 2425.
Guiochon, G., and Pommier, C. (1973). "Gas Chromatography in Inorganics and Organometallics." Chichester-Wiley.
Haines, R. J., and Leigh, G. J. (1975). *Chem. Soc. Rev.* **4**, 155.
Haque, F., Miller, J., Pauson, P. L., and Tripathi, J. B. P. (1971). *J. Chem. Soc., C* p. 743.
Hart, D. W., and Schwartz, J. (1974). *J. Am. Chem. Soc.* **96**, 8115.

Hartley, F. R. (1969). *Chem. Rev.* **69**, 799.
Hartley, F. R. (1972). *Angew Chem., Int. Ed. Engl.* **11**, 596.
Hartley, F. R. (1973). "The Chemistry of Platinum and Palladium." Appl. Sci. Publ., London.
Hashmi, M. A., Munro, J. D., Pauson, P. L., and Williamson, J. M. (1967). *J. Chem. Soc. A* p. 240.
Hauser, C. R., and Lindsay, J. K. (1957). *J. Org. Chem.* **22**, 482.
Heck, R. F. (1968). *J. Am. Chem. Soc.* **90**, 5518–5535.
Heck, R. F. (1969). *J. Am. Chem. Soc.* **91**, 6707.
Heck, R. F. (1972). *Organomet. Chem. Synth* **1**, 455.
Heil, V., Johnson, B. F. G., Lewis, J., and Thompson, D. J. (1974). *Chem. Commun.* p. 270.
Heimbach, P. (1973). *Angew. Chem., Int. Ed. Engl.* **12**, 975.
Heimbach, P., and Wilke, G. (1969). *Justus Liebigs Ann. Chem.* **727**, 183.
Helling, J. K., and Braitsch, D. M. (1970). *J. Am. Chem. Soc.* **92**, 7207.
Helling, J. F., and Cash, G. G. (1974). *J. Organomet. Chem.* **73**, C10.
Henrici-Olivé, G., and Olivé, S. (1972). *J. Organomet. Chem.* **35**, 381.
Henry, P. M. (1972). *J. Am. Chem. Soc.* **94**, 5200.
Henry, P. M. (1973). *Acc. Chem. Res.* **6**, 16.
Henry, P. M. (1974). *J. Org. Chem.* **39**, 3871.
Herberhold, M. (1974). "Metal π-Complexes," Vol. II. Elsevier, Amsterdam.
Herberich, G. E., and Müller, H. (1971). *Chem. Ber.* **104**, 2781.
Hidai, M., Ishiwatari, H., Yagi, H., Tanaka, E., Onozawa, K., and Uchida, Y. (1975). *Chem. Commun.* p. 170.
Hine, K. E., Johnson, B. F. G., and Lewis, J. (1975). *Chem. Commun.* p. 81.
Holmes, J. D., and Pettit, R. (1963). *J. Am. Chem. Soc.* **85**, 2531.
Howell, J. A. S., Johnson, B. F. G., Josty, P. L., and Lewis, J. (1972). *J. Organomet. Chem.* **39**, 329.
Hubert, A. J., and Reimlinger, H. (1970). *Synthesis* p. 405.
Hubert, A. J., Georis, A., Warin, R., and Teyssié, P. (1972). *J. Chem. Soc. Perkin Trans. II* p. 366.
Hubert, A. J., Moniotte, P., Goebbels, G., Warin, R., and Teyssié, P. (1973). *J. Chem. Soc. Perkin Trans. II.* p. 1954.
Hughes, W. B. (1972). *Organomet. Chem. Synth.* **1**, 341.
Hunt, D. F., Farrant, G. C., and Rodeheaver, G. T. (1972). *J. Organomet. Chem.* **38**, 349.
Iqbal, A. F. M. (1971). *Helv. Chim. Acta* **54**, 1440.
Ireland, R. E., Brown, G. G., Stainford, R. H., and McKenzie, T. C. (1974). *J. Org. Chem.* **39**, 51.
Jaouen, G. (1973). *Tetrahedron Lett.* p. 1753.
Johnson, B. F. G., Lewis, J., Parkins, A. W., and Randall, G. L. P. (1969). *Chem. Commun.* p. 595.
Johnson, B. F. G., Lewis, J., and Randall, G. L. P. (1971). *J. Chem. Soc. A* p. 422.
Johnson, B. F. G., Lewis, J., McArdle, P., and Randall, G. L. P. (1972). *J. Chem. Soc., Dalton Trans.* pp. 456 and 2076.
Johnson, B. F. G., Lewis, J., and Thompson, D. J. (1974). *Tetrahedron Lett.* p. 3789.
Jolly, P. W., Stone, F. G. A., and MacKenzie, K. (1965). *J. Chem. Soc.* p. 6416.
Jolly, W. L. (1968). *Inorg. Synth.* **11**, 120.
Jones, D., Pratt, L., and Wilkinson, G. (1962). *J. Chem. Soc.* p. 4458.
Kane-Maguire, L. A. P., and Mansfield, C. A. (1973). *Chem. Commun.* p. 540.
Kemmitt, R. D. W. (1972). *MTP Int. Rev. Sci. Inorg. Chem., Ser. 1*, 1972 **6**, 239.
Kerber, R. C., and Ehntholt, D. J. (1973). *J. Am. Chem. Soc.* **95**, 2927.

Ketley, A. D., ed. (1967). "The Stereochemistry of Macromolecules," Vol. I. Dekker, New York.
Khand, I. U., Pauson, P. L., and Watts, W. E. (1968). *J. Chem. Soc. C* pp. 2257 and 2261.
Khand, I. U., Pauson, P. L., and Watts W. E. (1969). *J. Chem. Soc. C* p. 116.
Kiji, J., Masui, K., and Furukawa, J. (1970a). *Chem. Commun.* p. 1310.
Kiji, J., Masui, K., and Furukawa, J. (1970b). *Tetrahedron Lett.* p. 2561.
King, R. B., and Stone, F. G. A. (1961). *Inorg. Synth.* **7**, 193.
Kitching, W., Rappoport, Z., Winstein, S., and Young, W. G. (1966). *J. Am. Chem. Soc.* **88**, 2054.
Klopman, G., and Calderazzo, F. (1967). *Inorg. Chem.* **6**, 977.
Knoche, H. (1970). German Patent 2,024,835.
Knox, G. R., Leppard, D. G., Pauson, P. L., and Watts, W. E. (1972). *J. Organomet. Chem.* **34**, 347.
Knox, S. A. R., and Stone, F. G. A. (1974). *Acc. Chem. Res.* **7**, 321.
Kohll, C. F., and Van Helden, R. (1968). *Rec. Trav. Chim. Pays-Bas* **87**, 481.
Koser, G. F., and St. Cyr, D. R. (1974). *Tetrahedron Lett.* p. 3015.
Krausz, P., Garinier, F., and Dubois, J. E. (1975). *J. Am. Chem. Soc.* **97**, 437.
Kricka, L. J., and Ledwith, A. (1974). *Synthesis* p. 539.
Kuhn, D. E., and Lillya, C. P. (1972). *J. Am. Chem. Soc.* **94**, 1682.
Kurkov, V. P., Pasky, J. Z., and Lavigne, J. B. (1968). *J. Am. Chem. Soc.* **90**, 4743.
Kutney, J. P., Greenhouse, R., and Ridaura, V. E. (1974). *J. Am. Chem. Soc.* **96**, 7364.
Labunskaya, V. I., Shebaldova, A. D., and Khidekel, M. L. (1974). *Russ. Chem. Rev. (Engl. Transl.)* **43**, 1.
Landesberg, J. M., and Sieczkowski, J. (1968). *J. Am. Chem. Soc.* **90**, 1655.
Landesberg, J. M., and Sieczkowski, J. (1969). *J. Am. Chem. Soc.* **91**, 2120.
Lewis, J., and Johnson, B. F. G. (1968). *Acc. Chem. Res.* **1**, 245.
Lloyd, W. G., and Luberoff, B. J. (1969). *J. Org. Chem.* **34**, 3949.
McArdle, P., and Sherlock, H. (1973). *J. Organomet. Chem.* **52**, C29.
McFarlane, W., and Wilkinson, C. (1966). *Inorg. Syntheses* **8**, 181.
Maddox, M. L., Stafford, S. L., and Kaesz, H. D. (1965). *Adv. Organomet. Chem.* **3**, 1.
Maglio, G., and Palumbo, R. (1974). *J. Organomet. Chem.* **76**, 367.
Mahler, J. E., and Pettit, R. (1963). *J. Am. Chem. Soc.*, **85**, 3955.
Maitlis, P. M. (1966). *Adv. Organomet. Chem.* **4**, 95.
Maitlis, P. M. (1971). "The Organic Chemistry of Palladium," Vol. 2. Academic Press, New York.
Mansfield, C. A., Al-Kathumi, K. M., and Kane-Maguire, L. A. P. (1974). *J. Organomet. Chem.* **71**, C11.
Mantzaris, J., and Weissberger, E. (1974). *J. Am. Chem. Soc.* **96**, 1873, and 1880.
Manuel, T. A. (1962). *J. Org. Chem.* **27**, 3941.
Markó, L. (1974). *In* "Aspects of Homogeneous Catalysis (R. Ugo, ed.), Vol. II. Reidel Publ., Dordrecht, Netherlands.
Markó, L., and Bakos, J. (1974). *J. Organomet. Chem.* **81**, 411.
Mathews, C. N., Magee, T. A., and Wotiz, J. H. (1959). *J. Am. Chem. Soc.* **81**, 2273.
Meinwald, J., and Mioduski, J. (1974). *Tetrahedron Lett.* p. 3839.
Meyer, A., and Dabard, R. (1972). *J. Organomet. Chem.* **36**, C38.
Mole, T., and Jeffrey, E. A. (1972). *In* "Organoaluminium Compounds," p. 405. Elsevier, Amsterdam.
Mori, Y., and Tsuji, J. (1972). *Tetrahedron* **28**, 29.
Moser, G. A., and Rausch, M. D. (1974). *Synth. React. Inorg. Met.-Org. Chem.* **4**, 37.
Munro, J. D., and Pauson, P. L. (1961). *J. Chem. Soc.* pp. 3475, 3479, and 3484.

Musco, A., Palumbo, R., and Paiaro, G. (1971). *Inorg. Chim. Acta* **5**, 157.
Natta, G. (1955). *J. Polym. Sci.* **16**, 143.
Natta, G. (1956). *Angew. Chem.* **68**, 393.
Natta, G. (1957). *Chem. Ind.* (*London*) p. 1520.
Nesmeyanov, A. N., Kursanov, D. N., Setkina, V. N., Kislyakova, N. V. and Kochetkova, N. S. (1961). *Tetrahedron Lett.* p. 41.
Nesmeyanov, A. N., Rybin, L. V., Gubenko, N. T., Rybinskaya, M. I., and Petrovskii, P. V. (1974). *J. Organomet. Chem.* **71**, 271.
Nicholas, K. M., and Rosan, A. M. (1975). *J. Organomet. Chem.* **84**, 351.
Nicholls, B., and Whiting, M. C. (1959). *J. Chem. Soc.* p. 551.
Noyori, R., Ishigami, T., Hayashi, N., and Takaya, H. (1973). *J. Am. Chem. Soc.* **95**, 1674.
Nyholm, R. S., Tobe, M. L., and Phillip, A. T. (1969). *Proc. Int. Conf. Coord. Chem.*, 12*th* 1969.
Orchin, M., and Rupilius, W. (1972). *Catal. Rev.* **6**, 85.
Paiaro, G., and Palumbo, R. (1967). *Gazz. Chim. Ital.* **97**, 267.
Paiaro, G., Palumbo, R., Musco, A., and Panunzi, A. (1965). *Tetrahedron Lett.* p. 1067.
Paiaro, G., Panunzi, A., and De Renzi, A. (1966). *Tetrahedron Lett.* p. 3905.
Paiaro, G., De Renzi, A., and Palumbo, R. (1967). *Chem. Commun.* p. 1150.
Panunzi, A., De Renzi, A., and Paiaro, G. (1967). *Inorg. Chim. Acta* **1**, 475.
Panunzi, A., De Renzi, A., and Paiaro, G. (1970). *J. Am. Chem. Soc.* **92**, 3488.
Paquette, L. A. (1974). *Trans. N.Y. Acad. Sci.* [2] **36**, 357.
Paquette, L. A., and Wise, L. D. (1967). *J. Amer. Chem. Soc.* **89**, 6659.
Paulik, F. E. (1972). *Catal. Rev.* **6**, 49.
Pauson, P. L., Smith, G. H., and Valentine, J. H. (1967). *J. Chem. Soc. C* p. 1061.
Pelter, A., Gould, K. J., and Kane-Maguire, L. A. P. (1974). *Chem. Commun.* p. 1029.
Perevalova, E. G., and Nikitina, T. V. (1972). *Organomet. React.* **4**, 163–419.
Pettit, R., and Henery, J. (1970). *Org. Synth.* **50**, 21.
Pettit, R., Emerson, G., and Mahler, J. (1963). *J. Chem. Educ.* **40**, 175.
Pino, P., and Botteghi, C. (1973). *Org. Synth.* **53**, 162.
Pittman, C. U., and Smith, L. R. (1975). *J. Am. Chem. Soc.* **97**, 341.
Pittman, C. U., and Hanes, R. M. (1974). *Ann. N.Y. Acad. Sci.* **239**, 76.
Platbrood, G., and Wilputte-Steinert, L. (1974a). *J. Organomet. Chem.* **70**, 407.
Platbrood, G., and Wilputte Steinert, L. (1974b). *Tetrahedron Lett.* p. 2507.
Rausch, M. D. (1974). *J. Org. Chem.* **39**, 1787.
Reeves, P., Henery, J., and Pettit, R. (1969). *J. Am. Chem. Soc.* **91**, 5888.
Rinehart, R. E., and Lasky, J. S. (1964). *J. Amer. Chem. Soc.* **86**, 2516.
Roberts, B. W., Wissner, A., and Rimmerman, R. A. (1969). *J. Am. Chem. Soc.* **91**, 6208.
Rosan, A., Rosenblum, M., and Tancrede, J. (1973). *J. Am. Chem. Soc.* **95**, 3062.
Rosenblum, M. (1974). *Acc. Chem. Res.* **7**, 122 (and references therein).
Rosenblum, M., and Woodward, R. B. (1958). *J. Am. Chem. Soc.* **80**, 5443.
Rosenblum, M., Banerjee, A. K., Danieli, N., Fish, R. W., and Schlatter, V. (1963a). *J. Am. Chem. Soc.* **85**, 316.
Rosenblum, M., Santer, J. O., and Howells, W. J. (1963b). *J. Am. Chem. Soc.* **85**, 1450.
Rosenthal, A. (1968). *Adv. Carbohydr. Chem.* **23**, 59.
Rosenthal, A., and Koch, H. J. (1965). *Can. J. Chem.* **43**, 1375.
Rylander, P. N. (1973). "Organic Syntheses with Noble Metal Catalysts." Academic Press, New York.
Salzer, A., and Werner, H. (1975). *J. Organomet. Chem.* **87**, 101.
Sanders, A., and Giering, W. P. (1975). *J. Am. Chem. Soc.* **97**, 919.
Sanders, A., Magatti, C. V., and Giering, W. P. (1974). *J. Am. Chem. Soc.* **96**, 1610.

Sasson, Y., and Rempel, G. L. (1974). *Tetrahedron Lett.* p. 4133.
Schmidt, E. K. G. (1973). *Angew. Chem., Int. Ed. Engl.* **12**, 777 (and references therein).
Schmidt, J., Hafner, W., Jira, R., Sieber, R., Sedlmeir, J., and Sabel, A. (1962). *Angew. Chem., Int. Ed. Engl.* **1**, 80.
Schrauzer, G. N., and Glockner, P. (1064). *Chem. Ber.* **97**, 2451.
Schrauzer, G. N., Ho, R. K. Y., and Schlesinger, G. (1970). *Tetrahedron Lett.* p. 543.
Schroeder, M. A., and Wrighton, M. S. (1974). *J. Organomet. Chem.* **74**, C29.
Semmelhack, M. F., and Hall, H. T. (1974). *J. Am. Chem. Soc.* **96**, 7091 and 7092.
Semmelhack, M. F., Hall, H. T., Yoshifuji, M., and Clark, G. (1975). *J. Am. Chem. Soc.* **97**, 1247.
Shvo, Y., and Hazum, E. (1974). *Chem. Commun.* p. 336.
Simonneaux, G., Meyer, A., and Jaouen, G. (1975). *Chem. Commun.* p. 69.
Singer, H. (1974). *Synthesis* p. 189.
Stern, E. W., and Spector, M. L. (1961). *Proc. Chem. Soc., London* p. 370.
Stille, J. K., and Fox, D. B. (1970). *J. Am. Chem. Soc.* **92**, 1274.
Stille, J. K., and James, D. E. (1975). *J. Am. Chem. Soc.* **97**, 674.
Stille, J. K., and Morgan, R. A. (1966). *J. Am. Chem. Soc.* **88**, 5135.
Strohmeier, W., and Weigelt, L. (1975). *J. Organomet. Chem.* **86**, C17.
Takahashi, H., and Tsuji, J. (1968). *J. Am. Chem. Soc.* **90**, 2389.
Tolman, C. A. (1972). *Chem. Soc. Rev.* **1**, 337.
Trahanovsky, W. S. and Baumann, C. R. (1974). *J. Org. Chem.* **39**, 1924.
Trifan, D. S., and Nicholas, L. (1957). *J. Am. Chem. Soc.*, **79**, 2746.
Tsuji, J. (1972). *Top. Curr. Chem.* **28**, 41.
Tsuji, S. (1969). *Acc. Chem. Res.* **2**, 144.
Tsutsui, M. (1971). "Characterization of Organometallic Compounds," Parts I and II. Wiley (Interscience), New York.
Van Meurs, F., Metselaar, F. W., Post, A. J. A., Van Rossum, J. A. A. M., Van Wijk, A. M., and Van Bekkum, H. (1975). *J. Organomet. Chem.* **84**, C22.
Vedejs, E., and Weeks, P. D. (1974). *Tetrahedron Lett.* p. 3207.
Venanzi, L. M. (1964). *Angew. Chem., Int. Ed. Engl.* **3**, 453.
Walker, P. J. C., and Mawby, R. J. (1973a). *J. Chem. Soc., Dalton Trans.* p. 622.
Walker, P. J. C., and Mawby, R. J. (1973b). *Inorg. Chim. Acta* **7**, 621.
Walker, P. J. C., and Mawby, R. J. (1973c). *J. Organomet. Chem.* **55**, C39.
Ward, J. S., and Pettit, R. (1970). *Chem. Commun.* p. 1419.
Watts, L., and Pettit, R. (1966). *Adv. Chem. Ser.* ("*Werner Centennial*") **62**, 549.
Watts, L., Fitzpatrick, J. D., and Pettit, R. (1965). *J. Am. Chem. Soc.* **87**, 3253.
Watts, L., Fitzpatrick, J. D., and Pettit, R. (1966). *J. Am. Chem. Soc.* **88**, 623.
Wender, I., Metlin, S., Ergun, S., Sternberg, H. W., and Greenfield, H. (1956). *J. Am. Chem. Soc.* **78**, 5401.
Whitesides, T. H., and Arhart, R. W. (1971). *J. Am. Chem. Soc.* **93**, 5296.
Whitesides, T. H., and Nielan, J. P. (1975). *J. Am. Chem. Soc.* **97**, 907.
Whitesides, T. H., and Shelley, J. (1975). *Abstr. Nat. Meet. 169th Am. Chem. Soc. Orgn.* p. 86.
Whitesides, T. H., Arhart, R. W., and Slaven, R. W. (1973). *J. Am. Chem. Soc.* **95**, 5792.
Whitesides, T. H., Slaven, R. W., and Calabrese, J. C. (1974). *Inorg. Chem.* **13**, 1895.
Whyman, R. (1974). *J. Organomet. Chem.* **81**, 97.
Wilke, G. (1963). *Angew. Chem., Int. Ed. Engl.* **2**, 105.
Winstein, S., Kaesz, H. D., Kreiter, C. G., and Freidrich, E. C. (1965). *J. Am. Chem. Soc.* **87**, 3267.
Wrighton, M., Hammond, G. S., and Gray, H. B. (1974). *J. Organomet. Chem.* **70**, 283.

Wrighton, M. S., and Schroeder, M. A. (1974). *J. Am. Chem. Soc.* **96**, 6235.
Ziegler, K., Holzkamp, E., Breil, H., and Martin, H. (1955). *Angew. Chem.* **67**, 541.
Zuech, E. A., Hughes, W. B., Kubicek, D. H., and Kittleman, E. T. (1970). *J. Am. Chem. Soc.* **92**, 528.

2

COUPLING REACTIONS VIA TRANSITION METAL COMPLEXES

R. NOYORI

The formation of carbon–carbon bonds is one of the most fundamental problems in organic chemistry, and the proper choice of this kind of reaction is the key in effecting elegant organic syntheses. Organotransition metal chemistry has provided novel and unique methods for accomplishing this task. Within the last decade, numerous selective reaction processes have been realized with transition metal reagents or catalysts, and several books treating this subject have already been published (Bird, 1967; Wender and Pino, 1968; Falbe, 1970; Candlin *et al.*, 1968; Maitlis, 1971; Rylander, 1973; James, 1973; Heck, 1974; Tsuji, 1975). This chapter surveys topics on carbon–carbon bond forming reactions via transition metal complexes with emphasis on the utility of σ-bonded organometallics and π-allyls. Certain reactions, however, may not involve species having a carbon–metal bond as the reactive intermediate. The mechanistic considerations, including the structures of the reactive species, are in many cases still based on empirical facts.

I. COUPLING REACTIONS OF σ- AND π-BONDED ORGANIC LIGANDS

A. Coupling between σ-Bonded Carbon Moieties

The title reactions have been frequently postulated as a key step in various transition metal promoted processes (cf. Section II). However, transition metal complexes having a carbon–metal σ bond are usually both thermodynamically and kinetically unstable, and hence detailed studies of their properties have only been made in a few cases.

1. Thermal Reactions

Organocopper compounds are thus far the most extensively examined σ-bonded complexes because of their ready availability and high synthetic utility. Among various procedures, Gilman's method is the most conventional for the synthesis of organocopper compounds. The reaction of copper(I) salts and organolithium or Grignard reagents in ethereal solvents generally leads to the formation of organocopper species, but the nature of the products is highly dependent on the method of preparation and the stoichiometry. These organocopper compounds are considered to be polymeric, and are

$$\mathrm{RLi + CuX \longrightarrow RCu + LiX}$$

$$\mathrm{RMgX + CuX \longrightarrow RCu + MgX_2}$$

R = alkyl, allyl, vinyl, or aryl

insoluble in ether. However, addition of a further equivalent of alkyllithium can solubilize the complex. For example, reaction of yellow, ether-insoluble methylcopper with methyllithium gives a clear colorless solution of lithium dimethylcuprate, $LiCu(CH_3)_2$ (Gilman *et al.*, 1952). Thus dialkylcuprates can be readily prepared from copper(I) salts and two equivalents of organolithium reagents. Because of their thermal instability, they are usually generated and used *in situ*. It is doubtful that the products derived from copper(I) salts and

$$RCu + R'Li \longrightarrow LiCuRR'$$

Grignard reagents have a similar structure (Whitesides *et al.*, 1967, 1969b). Little is known about the exact structure of organocoppers and organocuprates (van Koten and Noltes, 1972; van Koton *et al.*, 1975a,b). A low-temperature nmr study has suggested that lithium dimethylcuprate is a tetrahedral metal cluster with face-centered bridging methyl groups (House *et al.*, 1966). Organocuprates and organocoppers coordinated with suitable ligands such as phosphines, phosphites, amines, and sulfides have higher solubility and stability as compared to organocopper compounds of stoichiometric composition RCu. Their stability is also improved by the presence of metal halides, MX; the compounds may be represented as M(RCuX). Reactivity of the copper reagents is profoundly affected by solvents and by added ligands.

Thermal decomposition of alkylcopper compounds gives rise to dimeric coupling products, along with products arising from hydrogen abstraction and disproportionation. The reaction course, as well as the mechanism, is markedly influenced by the structure of the alkyl groups and metallic species present in the reaction system. The dimerization of alkyl groups having no α-hydrogen atoms probably proceeds by a radical mechanism. When a mixture

$$RCu \xrightarrow{\Delta} R{-}R + R{-}H + \cdots$$

of two different alkylcopper compounds is decomposed, cross-coupled dimers are obtained. Decomposition of the organocopper (**I**) gives various organic products; rearrangement of the alkyl group strongly suggests a radical mechanism (Whitesides *et al.*, 1972).

$$C_6H_5{-}C(CH_3)_2{-}CH_2Cu{\cdot}P(n\text{-}C_4H_9)_3 \xrightarrow[-Cu(0)\ \ -P(n\text{-}C_4H_9)_3]{\Delta} C_6H_5{-}C(CH_3)_2{-}CH_2{\cdot} \longrightarrow$$

(**I**)

$$\cdot C(CH_3)_2{-}CH_2{-}C_6H_5 \longrightarrow \text{products}$$

Dimerization of aryl- and vinylcopper compounds is selective and hence synthetically useful (Reich, 1923; Gilman and Kirby, 1929; Hashimoto and Nakano, 1966; Nilsson and Wennerström, 1969; Seitz and Madl, 1972). A non-free radical mechanism has been advanced for the reaction of pentafluorophenylcopper and *o*-trifluoromethylphenylcopper tetramers (Cairncross *et al.*, 1971) and *m*-trifluoromethylphenylcopper octamer (Cairncross and Sheppard, 1971). These reactions proceed on the metal cluster, and the Cu(I)–Cu(0) cluster compound is, in fact, isolable.

$$R_8Cu_8 \longrightarrow R_2 + R_6Cu_8 \longrightarrow R_2 + R_4Cu_8$$
$$R = m\text{-}CF_3C_6H_4$$

Certain primary and secondary alkyl(tri-*n*-butylphosphine)silver(I) complexes can be prepared by reaction between the corresponding organolithium or organomagnesium reagents and halo(tri-*n*-butylphosphine)-silver(I) at $-78°$. The thermal decomposition of these substances occurs rapidly at $-50°$ to $0°$. The *n*-butyl derivative decomposes by a process in which carbon–carbon bond formation, generating octane, is concerted with carbon–silver bond breaking. The thermal decomposition of the *sec*-butyl complex takes place by a different path, possibly involving silver(I) hydride elimination (Whitesides *et al.*, 1974).

$$n\text{-}C_4H_9AgP(n\text{-}C_4H_9)_3 \xrightarrow[\text{ether, } <5\text{ min}]{20°} \underset{(93\%)}{n\text{-}C_8H_{18}} + \underset{(5\%)}{n\text{-}C_4H_{10}} + \underset{(2\%)}{1\text{-}C_4H_8} + \underset{(94\%)}{P(n\text{-}C_4H_9)_3} + \underset{(92\%)}{Ag(0)}$$

Alkyl(triphenylphosphine)gold(I) complexes are stable and exist in solution as monomeric species (Whitesides *et al.*, 1971, 1974). Reductive coupling of *n*-alkylgold(I) gives the corresponding dimeric alkanes in high yields. A detailed study with the methylgold complex has demonstrated operation of a

$$2C_2H_5AuL \longrightarrow n\text{-}C_4H_{10} + 2Au(0) + 2L$$

molecular mechanism involving a rate-limiting ligand elimination process. Binuclear complexes such as $CH_3AuAuCH_3(L)$ or $Au(CH_3)_2AuL$ are suggested as the likely intermediates (Tamaki and Kochi, 1973).

$$CH_3AuL \longrightarrow CH_3Au + L$$
$$CH_3Au + CH_3AuL \longrightarrow C_2H_6 + 2Au(0) + L$$

In a series of trialkyl(triphenylphosphine)gold(III) complexes, a pair of *cis*-alkyl groups are eliminated through a first-order process to give ligand coupling products (Tamaki *et al.*, 1973). The absence of alkenes and products derived from attack on solvent rules out a radical mechanism. Analogous

$$C_2H_5Au(CH_3)_2P(C_6H_5)_3 \longrightarrow C_3H_8 + CH_3AuP(C_6H_5)_3$$

reductive eliminations have been observed in platinum(IV) (Brown *et al.*, 1973) and nickel(II) complexes (Yamamoto *et al.*, 1971). The latter decomposition is facilitated by the addition of olefinic ligands.

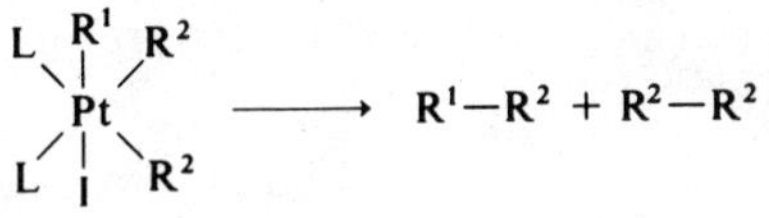

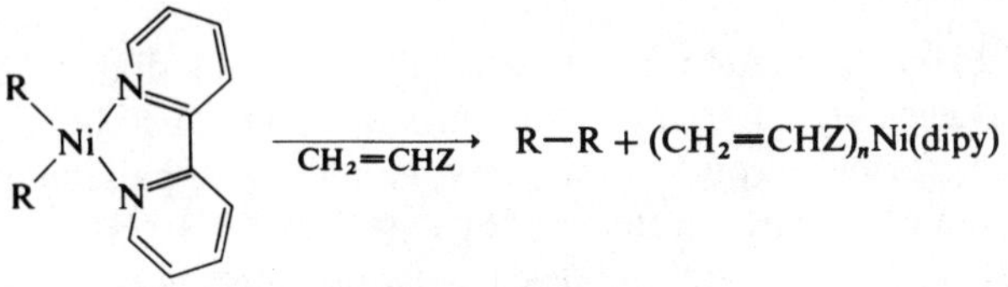

$$L_2Pt(R^1)(R^2)_2I \longrightarrow R^1{-}R^2 + R^2{-}R^2$$

$$R_2Ni(dipy) \xrightarrow{CH_2{=}CHZ} R{-}R + (CH_2{=}CHZ)_nNi(dipy)$$

Z = electron-withdrawing groups

$n = 1$ or 2

Decomposition of *cis*- and *trans*-1-propenylcoppers and the phosphine complexes gives hexadienes with retention of stereochemistry about the double bond (Whitesides and Casey, 1966; Whitesides *et al.*, 1971). 2-Butenylcopper behaves similarly. A free-radical mechanism is unlikely to be operative

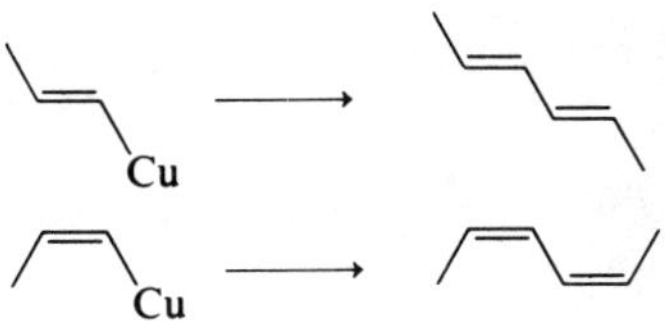

Cu

Cu

in view of the high stereospecificity of the reaction. Rather, a mechanism involving the four-center transition state **(II)** or intermediate **(III)** seems more probable. The vinylcopper intermediate **(IV)** formed by a metal exchange reaction of the vinylaluminum compound gives the dimeric 1,3-diene **(V)** with high stereospecificity (Zweifel and Miller, 1970).

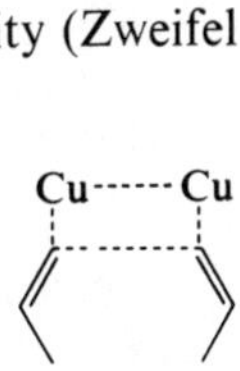

Cu------Cu

(II)

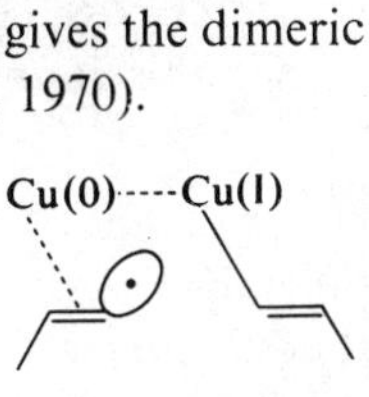

Cu(0)-----Cu(I)

(III)

$$\underset{\text{H}}{\overset{\text{R}}{>}}\text{C}=\text{C}\underset{\text{AlR}'_2}{\overset{\text{R}}{<}} \xrightarrow{\text{CuCl}} \left[\underset{\text{H}}{\overset{\text{R}}{>}}\text{C}=\text{C}\underset{\text{Cu}}{\overset{\text{R}}{<}}\right] \longrightarrow \left(\underset{\text{H}}{\overset{\text{R}}{>}}\text{C}=\text{C}\overset{\text{R}}{<}\right)_2$$

(IV) (V)

Aromatic hydrocarbons dimerize in the presence of palladium(II) salts, at $>100°$, with loss of hydrogens (van Helden and Verberg, 1965; Iataaki and Yoshimoto, 1973). When appropriate oxidants are present in the reaction

$$2\,C_6H_6 + PdCl_2 \longrightarrow C_6H_5{-}C_6H_5 + 2HCl + Pd$$

system, the biaryl formation becomes catalytic in palladium. The reaction probably proceeds via an arylpalladium intermediate. Unstable organopalladium species, formed from arylmercury halides and palladium chloride, decompose readily at room temperature to give biaryls (Heck, 1968b; Henry, 1968). The "dimerization" mechanism remains uncertain. It could be a bimolecular palladium reaction where the aryl group from one palladium complex is exchanged with an aryl anion from another, so that a diarylpalladium species is formed which decomposes to the biaryl and palladium. Alternatively, the coupling could arise from the reaction of the arylpalladium and the aryl mercurial. The second pathway may be more probable. Bis(triphenylarsine)arylpalladium halides are known to decompose at 100°–160° to produce biaryls.

$$2ArPdL_2X \xrightarrow{-PdL_2X_2} Ar{-}Ar + Pd + 2L$$

$$ArPdL_2X + ArHgX \xrightarrow{-HgX_2} Ar{-}Ar + Pd + 2L$$

$X = Cl$, $S = $ solvent

Thermodynamic data on the strengths of transition metal–carbon bonds are scarce, but the dissociation energies are considered to be relatively high. Recently theoretical treatment of the ligand coupling reactions has been given by Braterman and Cross (1973). Molecular orbital considerations suggest the ligand coupling reactions, at least in certain cases, are concerted.

2. Oxidative Reactions

σ-Bonded organometallics are usually sensitive to oxidants. Oxidative coupling of terminal acetylenes via copper acetylide intermediates is synthetically useful and is well known as the Glaser reaction; review articles have been published by Eglinton and McCrae (1963) and Sladkov and Ukhin

(1968). Thermal and oxidative dimerizations of aryl and vinyl groups via copper reagents have been extensively studied by Kauffmann and his associates (Kauffmann and Sahm, 1967; Kauffmann, 1974).

Whitesides and co-workers (1967) have reported that oxidative treatment of alkyl- (primary and secondary), vinyl-, and arylcuprates in tetrahydrofuran or dimethoxyethane affords the dimeric hydrocarbons in high yields. Evidence against the intervention of long-lived free radicals would suggest an

$$\mathrm{LiCuR_2} \xrightarrow{[\mathrm{O}]} \mathrm{R{-}R}$$

alternative mechanism involving dialkylcopper(II) intermediates. With cuprates of type LiCuRR′, a mixture of three products, R–R, R–R′, and

$$2[\mathrm{R_2Cu(I)}]^- \xrightarrow{-4e} 2\mathrm{R_2Cu(II)} \longrightarrow \mathrm{R{-}R} + 2\mathrm{RCu(I)}$$

R′–R′, is formed (Sheppard, 1970). Therefore, reaction of certain organocoppers with excess cuprates, followed by oxidation, can lead to selective cross-coupling (Corey and Katzenellenbogen, 1969; Whitesides *et al.*, 1969b). Reverse addition of organolithium compounds or Grignard reagents

$$\mathrm{RR'C{=}C(COOCH_3)Cu} \xrightarrow[2.\ \mathrm{O_2}]{1.\ \mathrm{LiCu(CH_3)_2}} \mathrm{RR'C{=}C(COOCH_3)CH_3}$$

to excess copper(II) salts leads to the dimeric products (Kauffmann *et al.*, 1972a,b). This reaction using aromatic dilithium or magnesium compounds provides a useful way of preparing polyphenylene derivatives (Rapson *et al.*, 1943; Wittig and Klar, 1967; Braünling *et al.*, 1967; Staab and Binning, 1967a,b).

$$\mathrm{R{-}C_6H_4{-}M} + \mathrm{CuX_2} \longrightarrow \mathrm{R{-}C_6H_4{-}C_6H_4{-}R}$$

M = Li or Mg

Alkyl- and aryllithium reagents add readily to metal carbonyls to give thermally unstable anionic acyl or aroyl complexes (Section V,D,2). The aroylnickel complex **(VI)**, when heated at 50°–60° or treated with bromine, gives the dimeric α-diketone in good yield (Ryang *et al.*, 1966). Reactivity of

$$\mathrm{RLi} + \mathrm{M(CO)}_n \longrightarrow \mathrm{Li[RCOM(CO)}_{n-1}]$$

$$\underset{\textbf{(VI)}}{\mathrm{Li[ArCONi(CO)_3]}} \xrightarrow{\Delta \text{ or } \mathrm{Br_2}} \mathrm{ArCOCOAr}$$

the acetylenic metal carbonylate **(VII)** is markedly dependent on the central metal. Treatment of the nickel complex with iodine–methanol in tetrahydrofuran gives mainly the diyne **(VIII)**, whereas the iron complex affords the acetylenic ester **(IX)** (Rhee *et al.*, 1968, 1969).

$$C_6H_5C{\equiv}CLi + M(CO)_n \longrightarrow Li[C_6H_5C{\equiv}CM(CO)_m]\ \textbf{(VII)}$$

$$\textbf{(VII)} \xrightarrow[M = Ni]{I_2\text{–}CH_3OH} C_6H_5C{\equiv}C{-}C{\equiv}CC_6H_5\ \textbf{(VIII)}$$

$$\textbf{(VII)} \xrightarrow{M = Fe} Li[C_6H_5C{\equiv}CCOM(CO)_{m-1}] \xrightarrow{I_2\text{–}CH_3OH} C_6H_5C{\equiv}CCOOCH_3\ \textbf{(IX)}$$

B. Reactions of Bis(π-allyl)nickel(0) Complexes

1. Ligand Dimerization

Wilke and Bogdanović (1961) first prepared bis(π-allyl)nickel(0) **(X)** from allylmagnesium bromide and nickel(II) bromide. The 16-electron complex is yellow and quite volatile. Its nmr spectrum suggests the planar delocalized structure for the allylic ligand (Wilke *et al.*, 1966; Bönnemann *et al.*, 1967). X-Ray analysis of the complexes established the "sandwich" arrangement (Wilke and Bogdanović, 1961; Dietrich and Uttech, 1963, 1965). A monograph written by Jolly and Wilke (1974, 1975), as well as a review by Fischer *et al.* (1973), describe characteristic features of the nickel complexes. The synthetic utility of these complexes has been reviewed by Heimbach *et al.* (1970).

$$2\ CH_2{=}CHCH_2MgBr + NiBr_2 \longrightarrow (\pi\text{-}C_3H_5)_2Ni\ \textbf{(X)} \xrightarrow{2L} [L_2Ni(CH_2CH{=}CH_2)_2]\ \textbf{(XI)} \xrightarrow[-NiL_4]{2L} CH_2{=}CHCH_2CH_2CH{=}CH_2$$

Under the influence of suitable ligands such as carbon monoxide, the allyl groups of **X** couple to form 1,5-hexadiene. Probably the bis(σ-allyl) species **(XI)** is involved as a reactive intermediate (Walter and Wilke, 1966). Reaction of bis(π-crotyl)nickel(0) with an atmospheric pressure of carbon monoxide at $-40°$ gives *trans,trans*-2,6-octadiene (98%), where the coupling is taking place selectively at the nonalkylated termini (Wilke *et al.*, 1966).

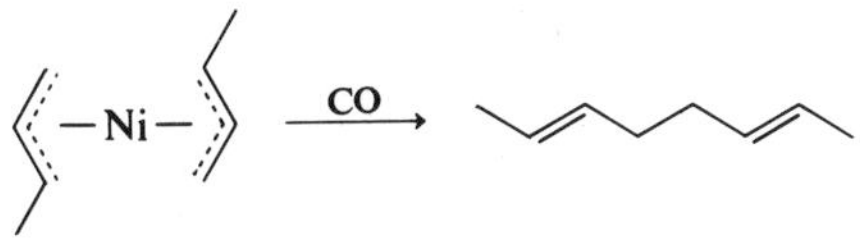

The nickel(0) complexes are effective in cyclooligomerization of butadienes. The complexes **XII** and **XIII** have been demonstrated to be intermediates in the catalytic oligomerization of butadiene; the ligand displacement reaction of **XII** gives a mixture of 1,2-divinylcyclobutane, 4-vinylcyclohexene, and 1,5-cyclooctadiene (Brenner *et al.*, 1969; Heimbach and Wilke, 1969), whereas complex **XIII** affords 1,5,9-cyclododecatriene (Wilke, 1963; Bogdanović *et al.*, 1969). Similarly, a number of bis(π-cyclo-

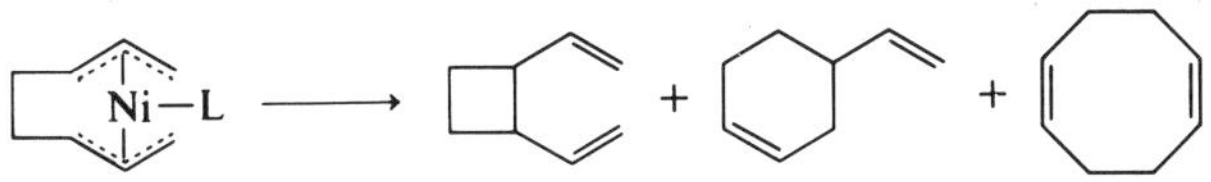

(XII)
L = phosphite ligand

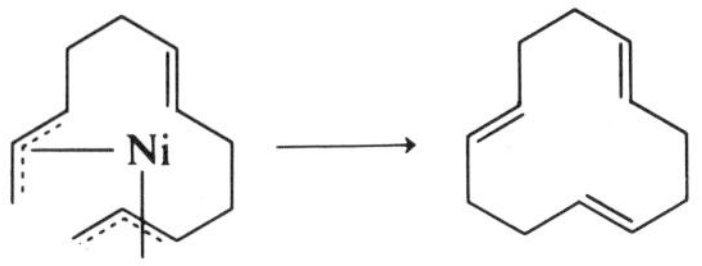

(XIII)

alkenyl)nickel(0) derivatives have been synthesized and allowed to react with carbon monoxide (Heimbach *et al.*, 1970). The zerovalent complex **XIV** has been suggested as an intermediate in the cyclotetramerization of butadiene (Miyake *et al.*, 1971a,b).

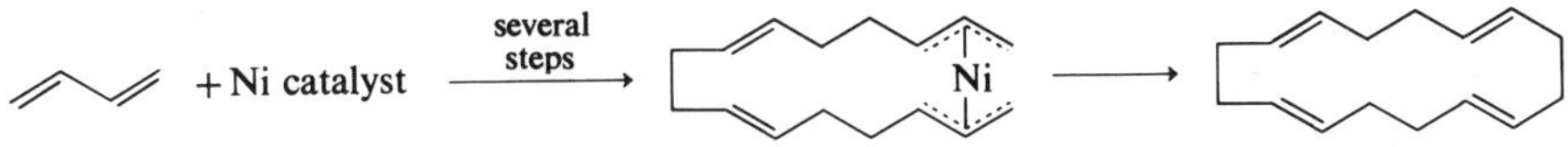

+ Ni catalyst —several steps→

(XIV)

Treatment of the nickel complex **(XV)** (L = tricyclohexylphosphine) with carbon monoxide at $-30°$ gives rise to *d,l*-limonene **(XVI)** in over 90% yield (Barnett *et al.*, 1972). Recently, synthesis of grandisol **(XVII)**, a pheromone

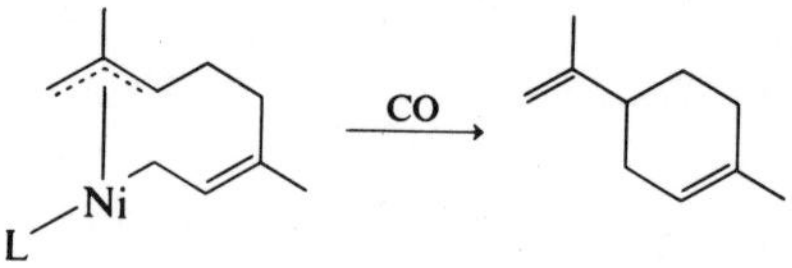

CO

(XV) (XVI)

of male boll weevil, has been achieved starting with isoprene (Billups *et al.*, 1973).

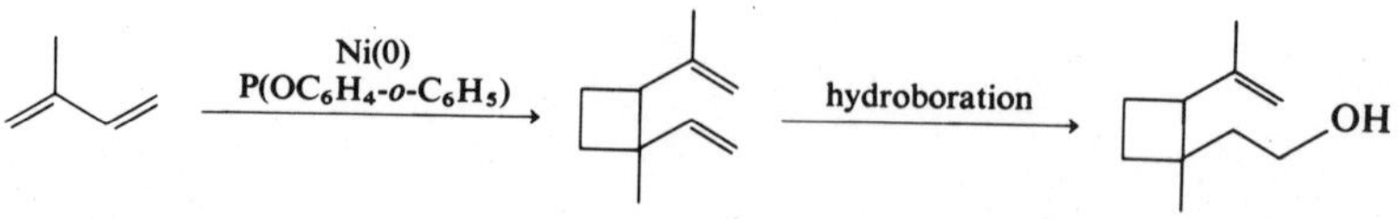

Ni(0)
$P(OC_6H_4\text{-}o\text{-}C_6H_5)$

hydroboration

OH

(12%)

(XVII)

2. Coupling Reaction via Carbonyl Insertion

Certain bis(π-allyl)nickel(0) complexes can be converted to symmetrical ketones by interaction with carbon monoxide (Corey *et al.*, 1968b). When complex **XVIII** is treated with carbon monoxide at $-40°$, the 11-membered

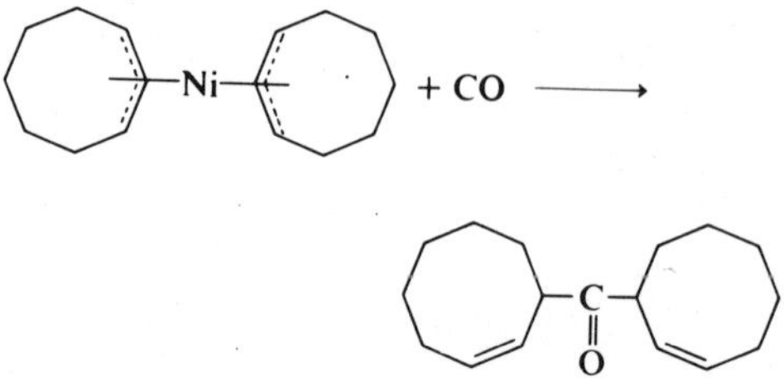

+ CO

unsaturated ketone **(XX)** is formed through the acyl π-allyl intermediate **(XIX)**. The reaction with an isocyanide gives a 13-membered carbocycle as well (Breil and Wilke, 1970). Reaction with allene followed by carbonylation

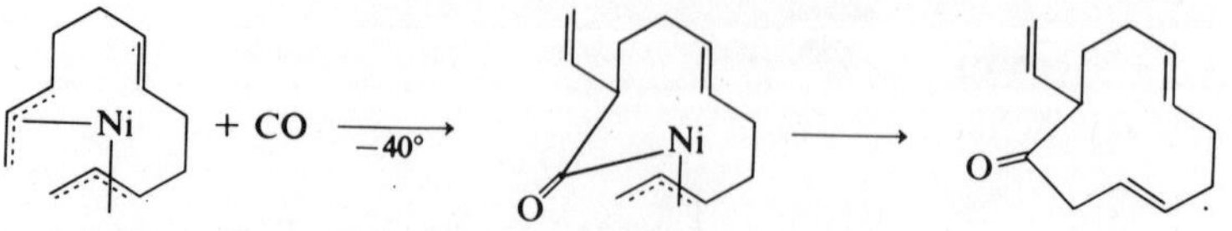

+ CO $\xrightarrow{-40°}$

(XVIII) (XIX) (XX)

(XVIII) + RNC $\xrightarrow{-Ni(RNC)_4}$ RN + RN → (XX)

gives a complex mixture which contains a small amount of the unsaturated ketone **(XXII)**, a precursor of *d,l*-muscone **(XXIII)** (Baker *et al.*, 1972c). Probably a new bis(π-allyl) species **(XXI)** is involved in the reaction. The yield of the 15-membered ring product is improved by using an isocyanide at $-20°$ in place of carbon monoxide (Baker *et al.*, 1974a).

(XVIII) + $CH_2{=}C{=}CH_2$ $\xrightarrow{-20° \text{ to } -10°}$ Ni

(XXI)

CO | 0°–10°

H_2

O O

(XXIII) **(XXII)**

C. Transition Metal-Catalyzed Reactions of Olefins, Dienes, and Acetylenes

Cyclooligomerizations and linear oligomerizations of olefins or dienes with transition metal catalysts are now one of the most important carbon–carbon bond-forming reactions. Many of these types of transformations are considered, and demonstrated in certain cases, to involve reactions of organometallic intermediates such as σ-alkyls, σ- or π-allyls, and metal hydrides. The ligand coupling reactions produce the organic products and at the same time regenerate the active transition metal catalysts. Progress in the field of homogeneous catalysis has made catalytic asymmetric synthesis possible. Remarkably, codimerization of norbornene and ethylene with a nickel catalyst containing a chiral phosphine ligand can result in optical yields of 80.6% (for a review of this subject see Bogdanović, 1973).

$$+ CH_2{=}CH_2 \xrightarrow{NiPR_3^*}$$

H

Since excellent reviews have been presented by Maitlis (1971), Heimbach (1973), Bönnemann (1973), and Heck (1974), we will deal with only recent topics in this area.

1. [2 + 2] Cycloadditions Involving a Bis(σ-alkyl) Complex Intermediate

When two alkyl groups are coordinated to the same metal atom with a cis relationship, the complex is expected to readily release the alkyl coupling product. Although a variety of mechanisms may be conceived for the metal-catalyzed [2 + 2] cycloaddition of olefins (reviews: Mango and Schachtschneider, 1971; Kricka and Ledwith, 1974), definite evidence for the intervention of a metallocyclic intermediate has been provided. On reacting [Ir(1,5-cyclooctadiene)Cl]$_2$ with excess norbornadiene in acetone, an insoluble complex is

$$\| + \| \xrightarrow{M} \quad M \quad \xrightarrow{-M} \square$$

produced with the empirical formula Ir(norbornadiene)$_3$Cl. The structure has been established by converting the product to the more soluble acetylacetonate (acac) derivative **(XXIV)**. On reaction with excess triphenylphosphine the norbornadiene dimer **(XXV)** is produced (Fraser *et al.*, 1973).

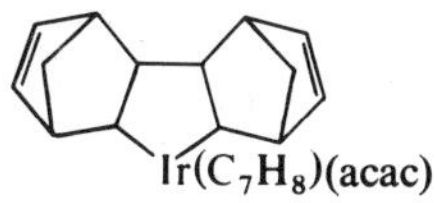

$Ir(C_7H_8)(acac)$

(XXIV) 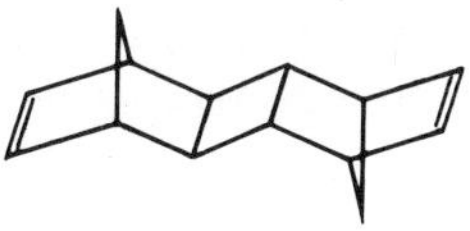 **(XXV)**

Irradiation of (methyl acrylate)iron tetracarbonyl in the presence of excess methyl acrylate at 20° forms a ferracyclopentane product. Thermal reaction of the metallocyclopentane with carbon monoxide or triphenylphosphine affords a cyclopentanone derivative (Grevels *et al.*, 1974). Similar ferracyclopentane complexes may be involved as intermediates in the well-known cyclopentanone formation from iron carbonyls and strained olefins (for example, Grandjean *et al.*, 1974; Mantzaris and Weissberger, 1974).

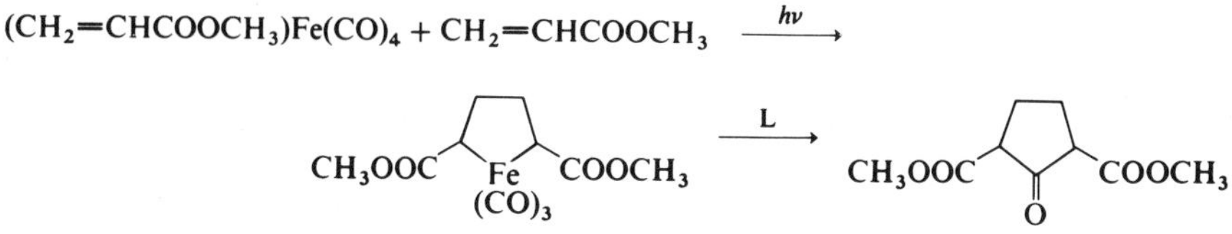

$$(CH_2{=}CHCOOCH_3)Fe(CO)_4 + CH_2{=}CHCOOCH_3 \xrightarrow{h\nu}$$

CH_3OOC Fe $(CO)_3$ $COOCH_3$ $\xrightarrow{L}$ CH_3OOC O $COOCH_3$

In addition, cyclopentanone is formed from ethylene and carbon monoxide via a titanium(IV) metallocycle (McDermott and Whitesides, 1974).

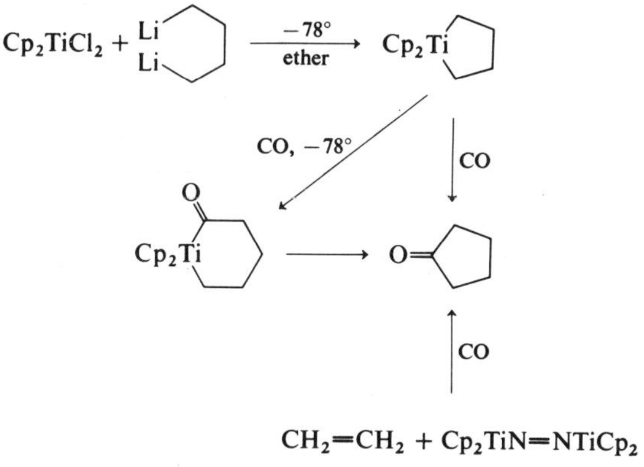

Cp_2TiCl_2 + Li Li $\xrightarrow[\text{ether}]{-78°}$ Cp_2Ti

CO, −78° CO

O Cp_2Ti ⟶ O=

CO

$CH_2{=}CH_2 + Cp_2TiN{=}NTiCp_2$

In the presence of transition metal complexes, certain strained hydrocarbon systems are activated under mild thermal conditions and undergo characteristic transformations. Methylenecyclopropane **(XXVI)** (Noyori *et al.*, 1970, 1972b), bicyclo[2.1.0]pentane **(XXVII)** (Noyori *et al.*, 1971c, 1974a), and quadricyclane **(XXVIII)** (Noyori *et al.*, 1975a) add to electron-deficient olefins with the aid of a nickel(0) catalyst such as bis(1,5-cyclooctadiene)-nickel(0) or bis(acrylonitrile)nickel(0). These cycloaddition reactions proceed with complete or a high degree of stereospecificity with respect to the olefinic substrate. The reactions involving cyclopropane cleavage have been postulated

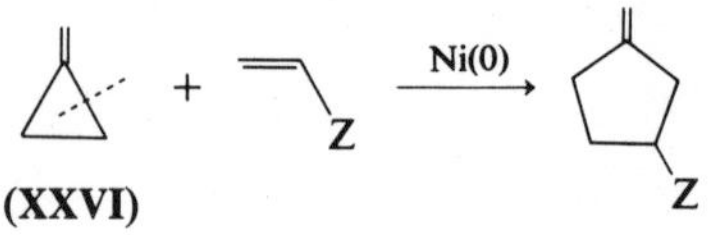

(XXVI)

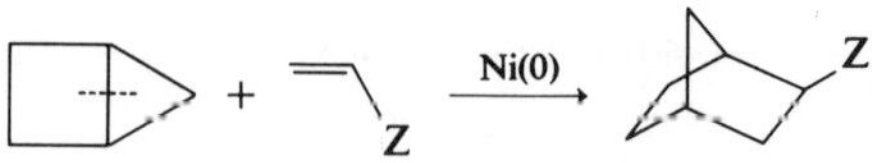

Ni(0)

Z

(XXVII)

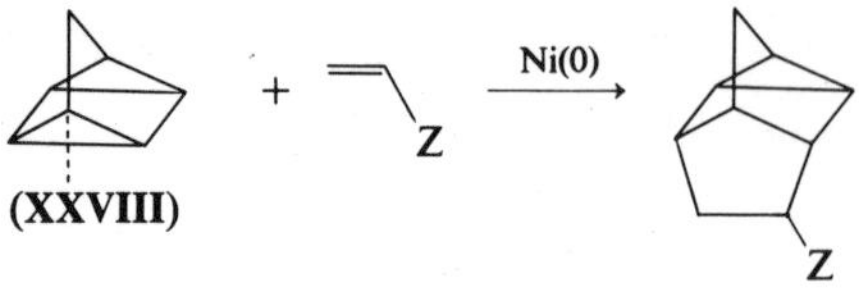

Ni(0)

Z

(XXVIII)

Ni(0)

Z

Z = CN, $COOCH_3$, etc.

to proceed via (1) oxidative addition of a strained carbon–carbon σ bond onto a nickel(0) atom giving the nickelacyclobutane; (2) insertion of the coordinated olefin in the carbon–metal σ bond to produce the nickelacyclohexane; and (3) reductive elimination of the organic moiety yielding the final five-membered ring product, as outlined in Scheme 1.

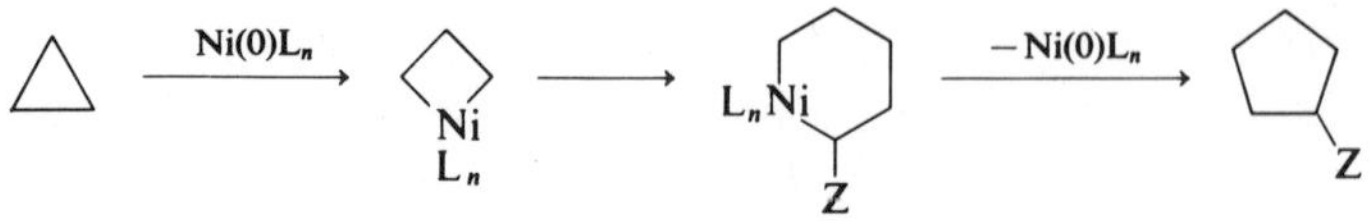

$Ni(0)L_n$ — Ni L_n — L_nNi Z — $-Ni(0)L_n$ — Z

L = CH_2=CHZ

Scheme 1

It would be worthwhile to compare the behavior of bicyclo[1.1.0]butane **(XXIX)** with that of the hydrocarbons described above. The smallest bicyclic hydrocarbon **(XXIX)** suffers two-bond cleavage by a nickel(0) catalyst to produce the allylcarbene–nickel(0) intermediate **(XXX)**, which is trapped with coordinated olefins, giving the allylcyclopropane derivatives **(XXXI)** (Noyori *et al.*, 1971b, 1974d; Noyori, 1973, 1975). The catalytic reaction of 3,3-

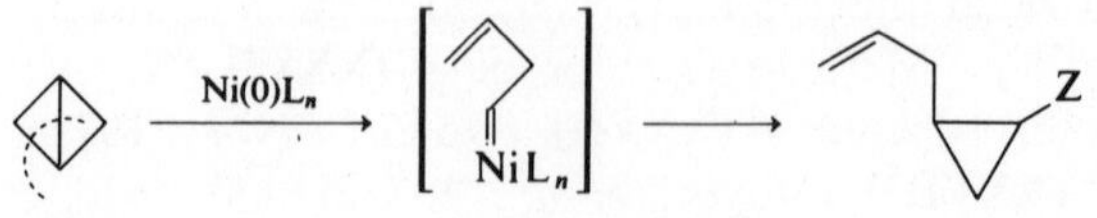

$Ni(0)L_n$ — NiL_n — Z

(XXIX) **(XXX)** **(XXXI)**

L = CH_2=CHCN, CH_2=$CHCOOCH_3$, etc.

dimethylcyclopropene **(XXXII)** with electron-deficient olefins affords vinylcyclopropane derivatives **(XXXIII)**. This coupling reaction may involve a vinylcarbene–nickel(0) intermediate (Binger and McMeeking, 1974).

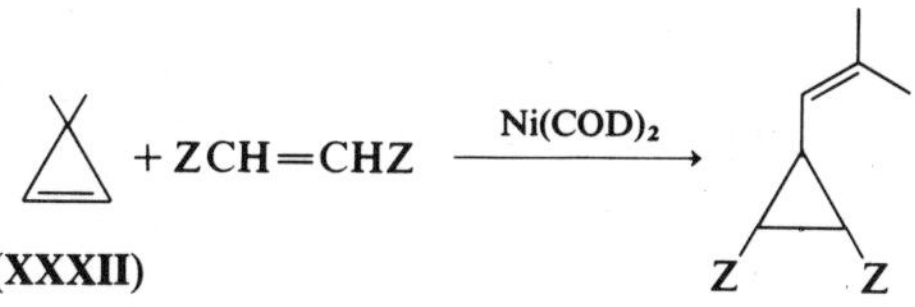

(XXXII) + ZCH=CHZ $\xrightarrow{Ni(COD)_2}$ **(XXXIII)**

Z = COOR

2. Cyclotrimerization of Acetylenes

The catalysis of the cyclotrimerization of acetylenes by transition metal complexes has been extensively studied (reviews: Bird, 1967; Hoogzand and Hübel, 1968; Maitlis, 1971, 1973; Heck, 1974). Various mechanistic studies have been done on this type of reaction (Blomquist and Maitlis, 1962; Meriwether *et al.*, 1962; Schrauzer, 1964; Collman *et al.*, 1968; Whitesides and Ehmann, 1969; Yamazaki and Hagihara, 1970; Müller and Beissner, 1973; Gardner *et al.*, 1973). The iridium-promoted trimerization of dimethyl acetylenedicarboxylate has been demonstrated to occur through an iridocycle as shown in Scheme 2 (Collman and Kang, 1967; Baddley and Tupper, 1974).

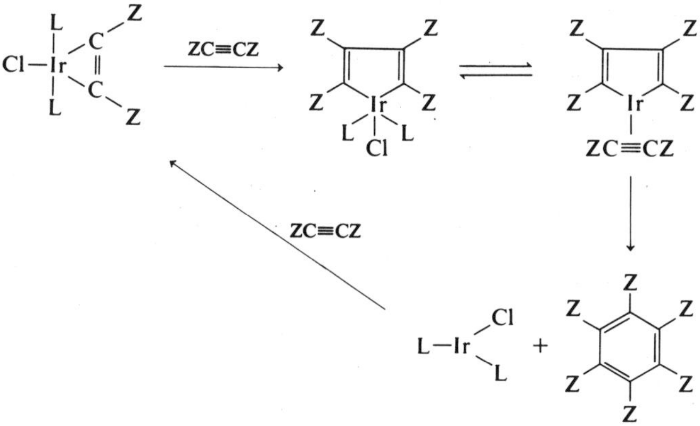

ZC≡CZ ZC≡CZ ZC≡CZ

$L = (C_6H_5)_2PCH_3$

$Z = COOCH_3$

Scheme 2

When the reaction of diphenylacetylene and acrylonitrile is carried out with π-$C_5H_5Co[(C_6H_5)_3P](C_6H_5C{\equiv}CC_6H_5)$, the linear cooligomerization products **(XXXIVa)** and **(XXXIVb)** are produced. The reaction of π-C_5H_5-$Co[(C_6H_5)_3P](C_6H_5C{\equiv}CCOOCH_3)$ with dimethyl maleate at room temperature gives the red crystalline complex **(XXXV)** in 30% yield. The isolable metallocycle **(XXXV)** reacts with acrylonitrile or diphenylacetylene yielding the diene complexes **(XXXVI)** and **(XXXVII)**, respectively (Wakatsuki *et al.*, 1974a).

(XXXIVa) **(XXXIVb)**

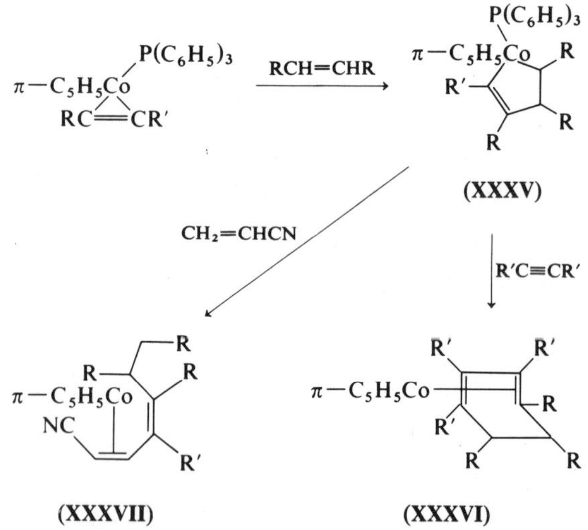

π-$C_5H_5Co(P(C_6H_5)_3)(RC{\equiv}CR')$ $\xrightarrow{RCH{=}CHR}$ **(XXXV)**

(XXXV) $\xrightarrow{CH_2{=}CHCN}$ **(XXXVII)**

(XXXV) $\xrightarrow{R'C{\equiv}CR'}$ **(XXXVI)**

$R = COOCH_3$
$R' = C_6H_5$

A variety of pyridine derivatives has been prepared by cooligomerization of acetylene and nitriles (Wakatsuki and Yamazaki, 1973a, 1976). 1,2-Dithiopyrones, *N*-methyl-2-thiopyridones, thiophenes, selenophenes, and pyrroles can be prepared from the cobaltacyclopentadiene and suitable unsaturated molecules containing a heteroatom or group (Wakatsuki and Yamazaki, 1973b; Wakatsuki *et al.*, 1974b).

Behavior of palladacyclopentadiene and platinacyclopentadiene systems has been extensively studied by Maitlis and his group (for example, Moseley and Maitlis, 1974; Roe *et al.*, 1975).

The diyne reactions of 1,4-, 1,5-, 1,6-, and 1,7-diynes via transition metal complexes are useful for the synthesis of new cyclic compounds (Müller, 1974; Wagner and Meier, 1974, 1975). π-$C_5H_5Co(CO)_2$ reacts catalytically with linear 1,*m*-diacetylenes **(XXXVIII)** to give trimers **(XXXIX)**, formation of which involves the interaction of six acetylene functions (Vollhardt and Bergman, 1974).

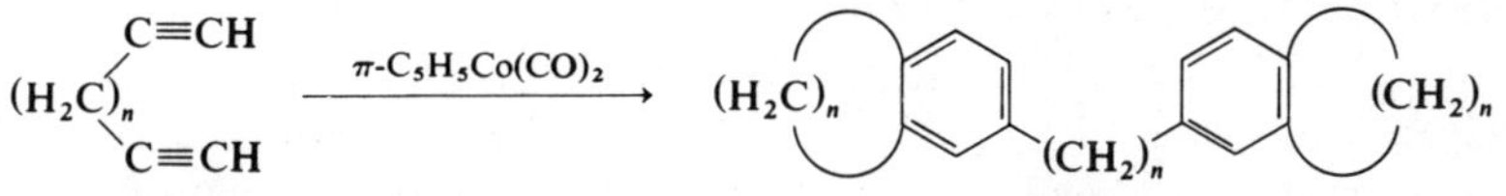

C≡CH
$(H_2C)_n$
C≡CH
$\xrightarrow{\pi\text{-}C_5H_5Co(CO)_2}$
$(H_2C)_n$ $(CH_2)_n$ $(CH_2)_n$

(XXXVIII) **(XXXIX)**

The ease with which these cyclic trimerizations take place is ascribed to the intermediacy of metallocycles which have two σ-bonded organic moieties in a cis orientation.

3. Ten-Membered Ring Formation from Dienes and Olefins or Acetylenes

The nickel(0)-catalyzed reaction of butadiene and alkenes or alkynes provides an excellent method for the preparation of 10-membered rings (reviews by Heimbach, 1973; Heimbach *et al.*, 1970). The reaction with alkynes has been believed to involve the bis(π-allyl)nickel(0) (**X**) and the

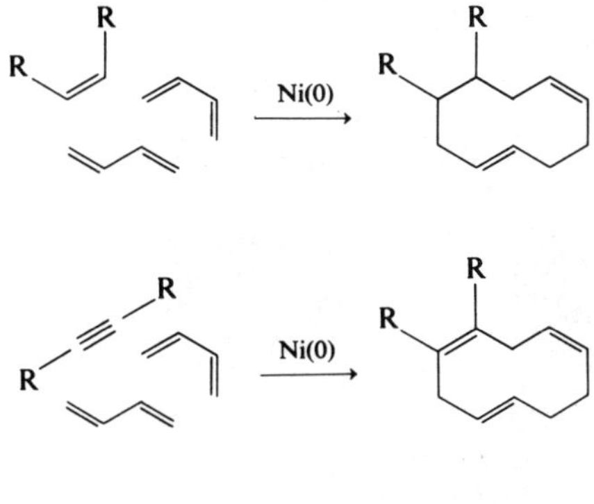

R R
Ni(0)
R R

R R
Ni(0)
R R

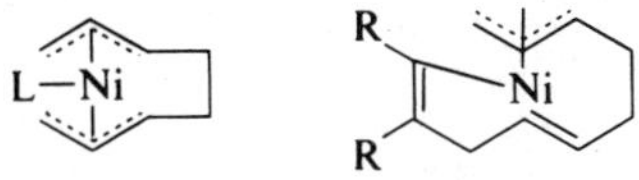

L—Ni

R R Ni

(X) **(XL)**

$$(C_6H_5)_3P{-}Ni\ (\mathbf{XLI}) + C_2H_5OOCC{\equiv}CCOOC_2H_5 \xrightarrow{-P(C_6H_5)_3} (\mathbf{XLII}) \xrightarrow[-Ni(CO)_4]{4CO} (\mathbf{XLIII})$$

(XLI) (XLII) (XLIII)

C_2H_5OOC, C_2H_5OOC, Ni, $COOC_2H_5$, C_2H_5OOC

σ-alkenyl π-allyl complexes **(XL)** derived therefrom. The yellow, monomeric complex **(XL)** was actually isolated from the reaction of **XLI** and diethyl acetylenecarboxylate at $-30°$ and characterized by nmr spectroscopy (Büssemeier *et al.*, 1974). The intermediate **(XLII)** absorbs 4 moles of carbon monoxide at $-20°$ to give the 10-membered triene **(XLIII)**.

4. Formal [4 + 2] Cycloaddition

Transition metal catalysts can exert an important influence on the regioselectivity of *formal* cycloaddition reactions. Cycloaddition of butadiene and **XLIV** in the absence of metal catalysts takes place at 135° giving a 20:1 mixture of **XLV** and **XLVI**. Catalysis by the Ni(acac)$_2$–triphenylphosphine–triethylaluminum system affords **XLVI**, **XLVII** as the major product, and only a trace of **XLV** (Garratt and Wyatt, 1974). A bis(π-allyl)nickel(0) complex **(XLVIII)** has been suggested as a reaction intermediate.

Butadiene + (XLIV) $\xrightarrow[60\%]{135°}$ (XLV) + (XLVI)

COOCH$_3$ COOCH$_3$ COOCH$_3$

(XLIV) (XLV) (XLVI)

(XLV):(XLVI) = 20:1

Butadiene + (XLIV) $\xrightarrow[90\%]{\text{Ni catalyst}}$ (XLV) + (XLVI) + (XLVII)
(trace)

COOCH$_3$ COOCH$_3$

(XLIV) (XLVII)

(XLVI):(XLVII) = 1:2

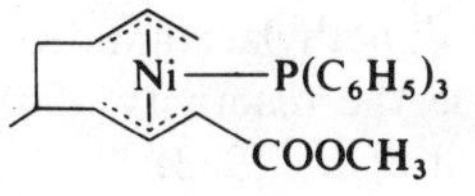

$$\text{Ni}—P(C_6H_5)_3 \quad COOCH_3$$

(XLVIII)

II. DISPLACEMENT REACTIONS WITH TRANSITION METAL COMPLEXES

The combination of two different organic groups is one of the most fundamental transformations in organic chemistry. An orthodox method may be offered by the coupling reaction between organometallic reagents and organic halides, but this process is not necessarily easy to attain with traditional synthetic techniques.

$$RM + R'X \longrightarrow R—R' + MX$$

A. Nucleophilic Displacements with Organocopper Reagents

Use of organocopper reagents is currently the most efficient method to solve this important problem. These reagents, particularly cuprate complexes, are in many cases more advantageous than organolithium or Grignard reagents. The synthetic utility of organocopper complexes has been reviewed by Bacon and Hill (1965), Coates (1968), Bähr and Burba (1970), Posner (1972, 1975), Normant (1972), and Jukes (1974).

1. Alkyl Halides and Related Compounds

Corey and Posner (1967, 1968) have reported that ether-soluble lithium dialkylcuprates react with a variety of organic halides to give the cross-coupled products in high yields. The method has broad applicability and some halides used include primary and secondary alkyl iodides, cyclopropyl bromide, allylic bromides, benzyl bromide, and phenyl and alkenyl iodides. The copper reagents allow the use of substrates containing functional groups

$$LiCu(CH_3)_2 + RX \longrightarrow CH_3—R$$

such as COOH or $CONR_2$. A 5 : 1 molar ratio of organocuprate to alkyl halide has been used in the coupling reactions. However, mixed cuprate(I) reagents of type (Het)RCuLi (Het = *tert*-C_4H_9O, C_6H_5O, *tert*-C_4H_9S, C_6H_5S, etc.) allow more economical use of the alkyl transfer agents (Posner and Whitten, 1973; Posner *et al.*, 1973). The method is particularly useful when the alkyl group is secondary or tertiary. Dipolar solvents, such as tetrahydrofuran (THF), dimethylacetamide, and hexamethylphosphoric triamide (HMPA), increase the reactivity of the cuprates (DePasquale and Tamborski, 1969; Normant *et al.*, 1973).

Whitesides *et al.* (1969a) have made a detailed study of the coupling reaction and found that (+)-(*R*)-2-bromobutane enters into the reaction with lithium diphenylcuprate with predominant inversion (84–92%) of stereochemistry. Alkyl tosylates can also be used as substrates (Whitesides *et al.*,

$$LiCu(C_6H_5)_2 + C_2H_5-\overset{H}{\underset{Br}{C}}-CH_3 \longrightarrow C_2H_5-\overset{C_6H_5}{\underset{H}{C}}-CH_3$$

1969a,b; Schaeffer and Zieger, 1969). The order of effectiveness of leaving groups in coupling reactions of lithium organocuprates is OTs > I ~ Br > Cl (Johnson and Dutra, 1973). The reaction of (+)-(*S*)-2-butyl tosylate or mesylate with lithium diphenylcuprate affords (−)-(*R*)-2-phenylbutane, indicating that the coupling reaction is proceeding with 100% inversion of configuration. A plausible mechanism based on these observations involves nucleophilic displacement of organocuprates on alkyl halides or tosylates to form oxidative addition intermediates and subsequent reductive elimination to give the coupled products. However, the mechanism of the substitution process seems to depend subtly on the substrates (Posner and Ting, 1974), and certain bromides and tosylates react with organocuprates to give rearranged products, suggesting the operation of different mechanisms.

$$R_2Cu^- \rightarrow R'\!-\!X \longrightarrow \left[\begin{matrix} R' \\ | \\ R-Cu-R \end{matrix}\right] \longrightarrow R-R' + RCu + X^-$$

$LiCu(CH_3)_2$ + 3-bromocyclooctene ⟶ (50%) + (20%) + (23%)

$LiCu(CH_3)_2$ + TsO– ⟶ +

Ts = Tosyl

Treatment of alkylmercuric halide (**XLIX**) with iodo(tri-*n*-butylphosphine)copper(I) and then with *tert*-butyllithium in tetrahydrofuran at −78°

leads to the reactive cuprate complex (**L**). This complex undergoes coupling with methyl iodide with ~95% stereospecificity, giving the hydrocarbon (**LI**), while oxidation with nitrobenzene causes ligand coupling to afford **LII** (cf. Section I,A,2). In certain instances, the overall conversion from alkylmercuric halide to product takes place with retention of configuration at the carbon originally bonded to mercury (Bergbreiter and Whitesides, 1974).

HgBr **(XLIX)** —[1. $ICuP(n\text{-}C_4H_9)_3$; 2. $3(CH_3)_3CLi$]→ $[CuLiHg][C(CH_3)_3]_3$ **(L)**

(L) —[CH_3I, ~70%]→ CH_3 **(LI)**

(L) —[~45% $C_6H_5NO_2$]→ $C(CH_3)_3$ **(LII)**

Fatty acid esters have been synthesized in good yield by reaction between cuprate complexes, formed from methylcopper(I) and primary or secondary Grignard reagents, and esters of primary iodoalkylcarboxylic acids. This procedure provides the most direct route presently available to a variety of representative classes of simple fatty acids (Bergbreiter and Whitesides, 1975).

$$CH_2{=}CH(CH_2)_8CH_2MgCl \xrightarrow[THF]{CH_3Cu(I)} \text{“}CH_2{=}CH(CH_2)_8CH_2Cu^-CH_3\cdot MgCl^+\text{”}$$

$$\xrightarrow{I(CH_2)_{10}COOC_2H_5,\ THF} CH_2{=}CH(CH_2)_{19}COOC_2H_5 \quad (79\%)$$

Certain geminal dihalides can be dialkylated by organocuprate reagents (Corey and Posner, 1967, 1968; Posner and Brunelle, 1972, 1973a), allowing net gem-alkylation of the carbonyl group. *d,l*-Glubulol **(LIV)** has been

$$R{-}\overset{O}{\overset{\|}{C}}{-}R \longrightarrow R{-}CX_2{-}R \xrightarrow{LiCuR'_2} R{-}CR'_2{-}R$$

prepared by the geminal dimethylation of the dibromocyclopropane **(LIII)** (Marshall and Ruth, 1974).

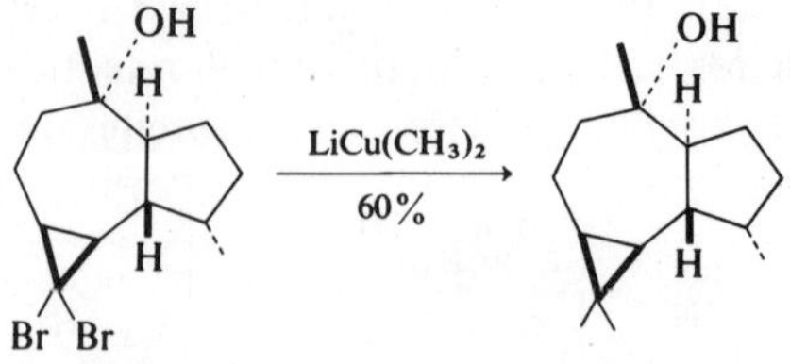

$LiCu(CH_3)_2$, 60%

(LIII) **(LIV)**

Anionic metal alkyls containing manganese, iron, or cobalt may be used in a way similar to cuprate complexes (Corey and Posner, 1970).

2. Allylic and Propargylic Substrates

Organocopper compounds of stoichiometric ratio, RCu, are not generally useful for coupling reactions with alkyl halides. However, allylic halides react with both organocuprates and organocopper compounds (Danehy *et al.*, 1936; Corey and Posner, 1967, 1968; Gump *et al.*, 1967; Corey and Jautelat, 1968; Vig *et al.*, 1968a; Castro *et al.*, 1969; Jukes *et al.*, 1970a; van Koten *et al.*, 1970, 1971; Normant and Bourgain, 1971; Smith *et al.*, 1971; Smith and Gilman, 1972; Coe and Milner, 1972; Kuwajima and Doi, 1972; Normant *et al.*, 1973).

Dialkyl- and divinylcuprates react with allylic bromides formally in an S_N2 manner (Corey *et al.*, 1967; Corey and Posner, 1968). The nonterminal, functionalized vinylcopper reagent **(LV)** reacts selectively with allylic bromides to afford the corresponding 1,4-dienes **(LVI)** in good yield (Marino and Floyd, 1974). The reaction offers a new route to α-methylene lactones

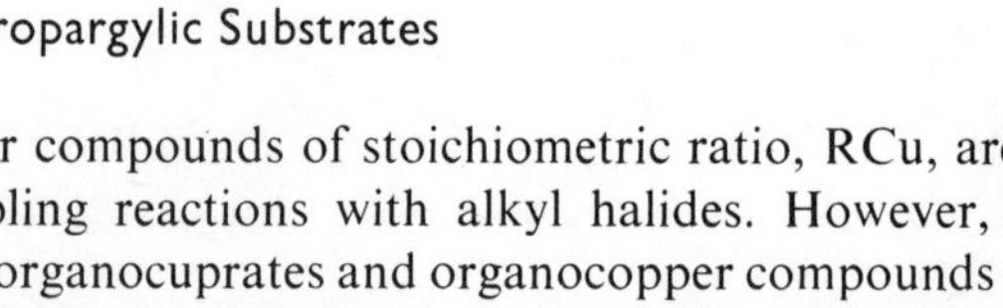

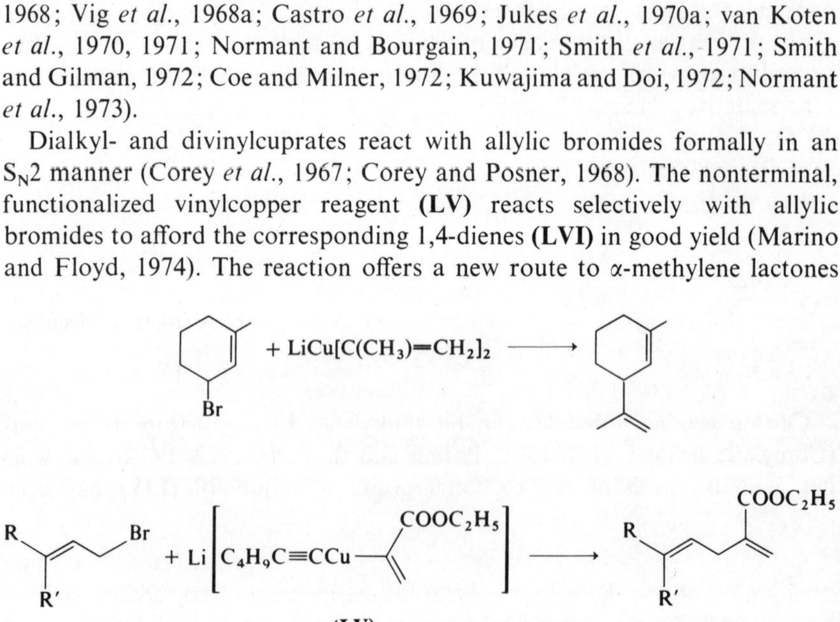

$+ LiCu[C(CH_3){=}CH_2]_2 \longrightarrow$

$+ Li[C_4H_9C{\equiv}CCu{-}C(=CH_2)COOC_2H_5] \longrightarrow$

(LV) **(LVI)**

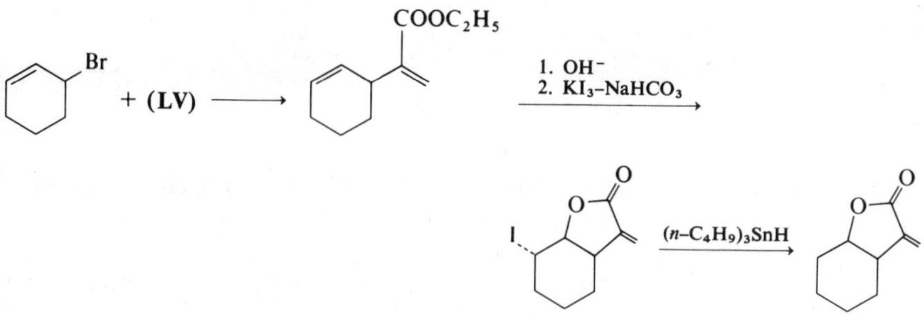

COOC$_2$H$_5$ Br + (LV) → 1. OH$^-$ 2. KI$_3$–NaHCO$_3$ → I O O (n–C$_4$H$_9$)$_3$SnH → O O

from allylic halides. Cyanomethylcopper is useful for the conversion of allylic halides to γ,δ-unsaturated nitriles. The reaction with *trans*-geranyl bromide **(LVII)** affords *trans*-homogeranyl cyanide **(LVIII)** in 92% yield (Corey and Kuwajima, 1972).

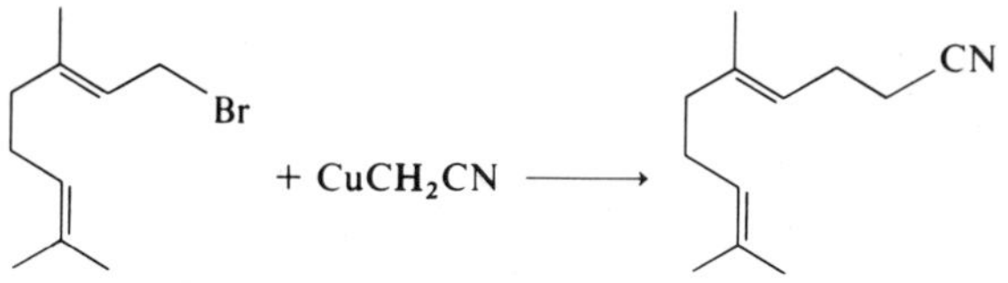

Br + CuCH$_2$CN → CN

(LVII) **(LVIII)**

In certain cases, the displacement reaction proceeds via an S$_N$2′ mechanism. Synthesis of *Cecropia* juvenile hormone **(LIX)** from *trans,trans*-farnesol has been achieved by this method (van Tamelen and McCormick, 1970).

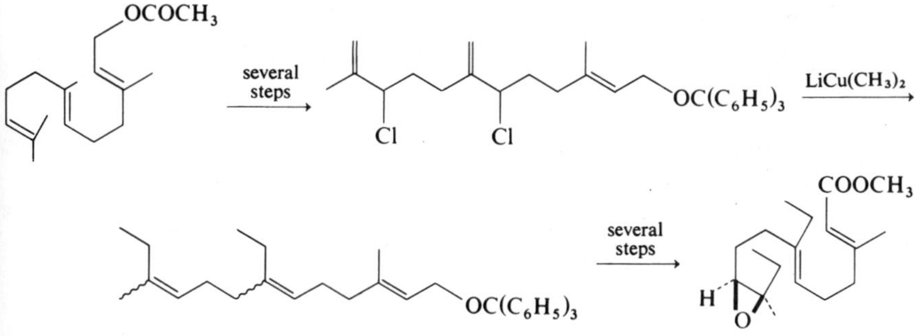

OCOCH$_3$ several steps → Cl Cl OC(C$_6$H$_5$)$_3$ LiCu(CH$_3$)$_2$ →

OC(C$_6$H$_5$)$_3$ several steps → COOCH$_3$ H O

(LIX)

Coupling of the sulfur-containing allylcopper reagent **(LX)** with allylic bromides occurs exclusively at the position gamma to the sulfur atom, again producing S$_N$2′ products (Oshima *et al.*, 1973).

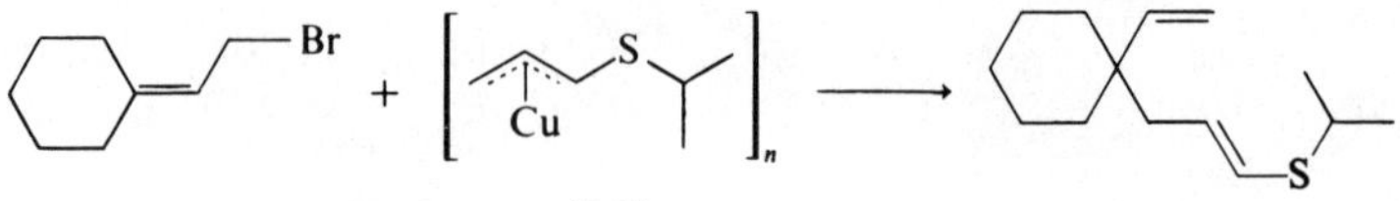

(LX)

Allylic acetates or related compounds can also be displaced by organocuprates. A synthetic approach to prostaglandins (PG's) via cross-coupling of a vinylic copper reagent with an allylic electrophile has been elaborated (Corey and Mann, 1973). Notably the reaction of **LXI** and **LXII** proceeds exclusively by an S_N2 pathway.

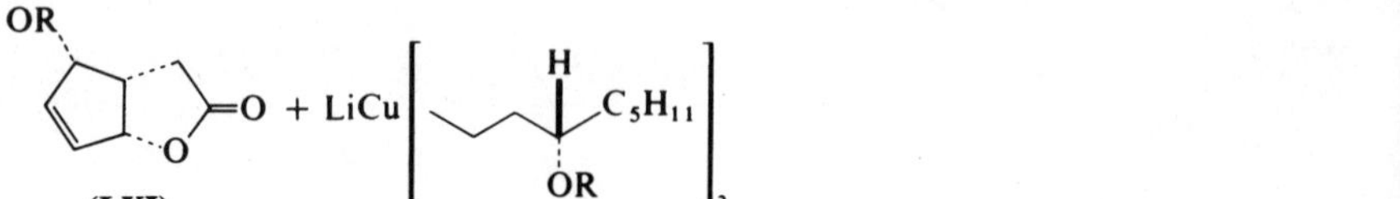

(LXI) (LXII)

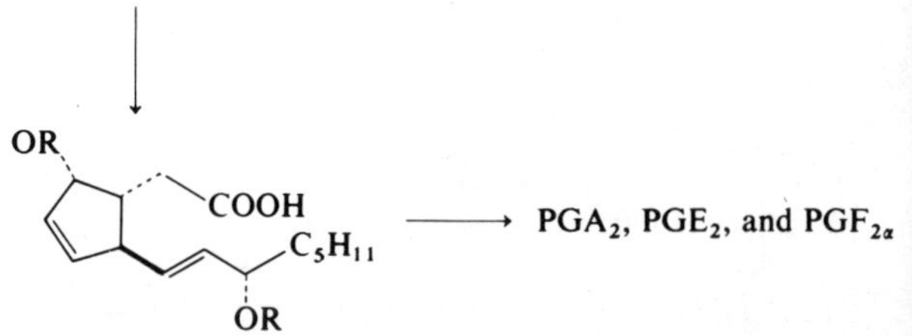

$\longrightarrow$ PGA_2, PGE_2, and $PGF_{2\alpha}$

By contrast, reaction of dialkylcuprates and allylic acetates **(LXIII)** gives the S_N2' displacement products **(LXIV)** in high yields (Anderson *et al.*, 1970, 1972). When $R^1 \leq R$ (entering group) and $R^3 = H$, trans trisubstituted olefins are formed stereoselectively. The coupling reaction may be used for the juvenile hormone synthesis.

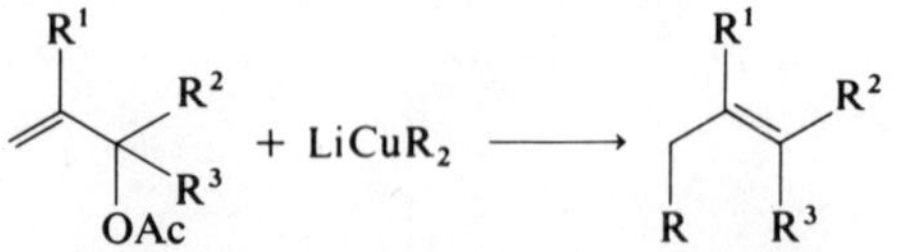

(LXIII) (LXIV)

$R^1, R^2, R^3 = H$ or alkyl

$R^2 > R^3$

An example of alkylation of an allylic ether has been reported as well (Katzenellenbogen and Corey, 1972).

Reaction of ethynylcopper reagents and propargyl halides gives 1,4-diynes and/or the allene isomers (Normant, 1972). The C_{12} acid **(LXVII)** can be

$$RC{\equiv}CCu + XCH_2C{\equiv}CR' \longrightarrow RC{\equiv}CCH_2C{\equiv}CR' + RC{\equiv}CCH{=}C{=}CHR'$$

prepared by the coupling of the propargylic bromide **(LXV)** and the acetylenic Grignard reagent **(LXVI)** in the presence of cuprous cyanide. The acid **(LXVII)** is then coupled with the C_8 fragment, 1-bromo-2-octyne, under the same conditions to form the tetraynic acid **(LXVIII)**, which is selectively hydrogenated to arachidonic acid **(LXIX)** (Fryer *et al.*, 1975).

Br + $BrMgC{\equiv}C$ COOMgX $\xrightarrow[THF]{CuCN}$

(LXV) **(LXVI)**

COOH

1. C_2H_5MgBr, THF
2. CuCN, $BrCH_2C{\equiv}CC_5H_{11}$-*n*

(LXVII)

COOH $\xrightarrow[\text{Lindlar catalyst}]{H_2}$ COOH

(LXVIII) **(LXIX)**

Diarylcuprates readily react with tertiary propargyl halides (Kalli *et al.*, 1972) and acetates (Rona and Crabbé, 1968, 1969; Descoins *et al.*, 1972) in an S_N2' manner to produce allene adducts. The latter compounds can also be obtained from haloallenes and cuprates.

$$R_2C(X){-}C{\equiv}CH + LiCuR'_2 \longrightarrow R_2C{=}C{=}C(H)R'$$

X = halogen or acetoxy group

3. Vinyl and Aryl Halides

Organocopper and cuprate compounds can couple with aryl halides in moderate to good yields (Stephens and Castro, 1963; Curtis and Taylor, 1966; Castro *et al.*, 1966, 1969; Rausch *et al.*, 1966, 1969; Malte and Castro, 1967; Corey and Posner, 1967, 1968; Atkinson *et al.*, 1967a,b,c, 1969; Vig *et al.*, 1969; Mładenovič and Castro, 1968; Whitesides *et al.*, 1969a;

Nilsson and Wahren, 1969; Nilsson and Wennerström, 1969, 1970; Nilsson and Ullenius, 1970, 1971; Nilsson *et al.*, 1970; Björklund *et al.*, 1970; Sheppard, 1970; Jukes *et al.*, 1970b; Cairncross *et al.*, 1970; Woo and Sondheimer, 1970; Owsley and Castro, 1972; Burdon *et al.*, 1972a,b).

Reaction of *cis*- or *trans*-β-bromostyrene and organocuprates proceeds with retention of configuration (Corey and Posner, 1967; Whitesides *et al.*, 1969a). Klein and Levene (1972) have accounted for the stereospecific substitution in terms of a mechanism with the transition state of type **LXX**; a nucleophile approaches perpendicularly to the plane of the molecule and the halide ion leaves perpendicularly to the plane from the opposite side. A

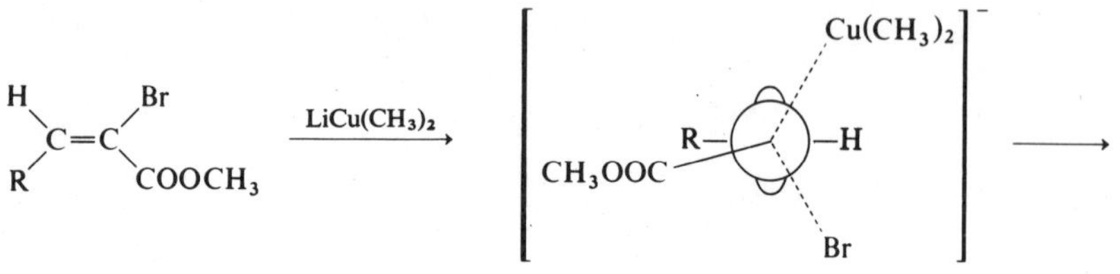

$$\text{(H)(R)C=C(Br)(COOCH}_3) \xrightarrow{\text{LiCu(CH}_3)_2} \left[\text{CH}_3\text{OOC, R, H; Cu(CH}_3)_2 \cdots \text{C} \cdots \text{Br} \right]^- \longrightarrow$$

(LXX)

$$\text{(H)(R)C=C(Cu(CH}_3)_2)(COOCH}_3) \longrightarrow \text{(H)(R)C=C(CH}_3)(COOCH}_3)$$

mechanism involving a four-center transition state **(LXXI)** has also been suggested for the reaction with aryl and vinyl halides (Burdon *et al.*, 1972b). Some reduction and elimination also occur in the reaction with isopropyl- or *tert*-butylcuprate (Worm and Brewster, 1970).

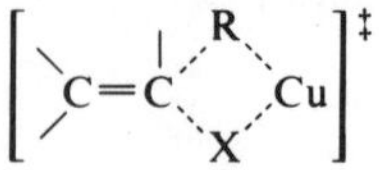

$$\left[\text{>C=C(R)(X)} \cdots \text{Cu} \right]^{\ddagger}$$

(LXXI)

Perfluoroalkylcopper compounds, derived from perfluoroalkyl halides and copper in dipolar aprotic solvents such as dimethyl sulfoxide and *N,N*-dimethylacetamide, are rather stable. These complexes couple effectively with aryl and vinyl halides (Burdon *et al.*, 1967, 1972a; McLoughlin and Thrower, 1969).

$$R_fI + Cu \longrightarrow R_fCu \xrightarrow{RX} R_f{-}R$$

R = aryl or vinyl

The Ullman coupling reaction is known to proceed via arylcopper intermediates. Pertinent reviews have appeared on this subject (Bacon and Hill, 1965; Fanta, 1964, 1974). Copper-induced coupling of vinyl halides has been found to occur with retention of configuration (Cohen and Poeth, 1972).

$$2ArX + Cu \longrightarrow Ar{-}Ar$$

Ethynylcopper reagents couple in dipolar solvents with various aromatic halides. The reaction can be used effectively for heterocyclic synthesis (Gump *et al.*, 1967; Castro *et al.*, 1966, 1969). The reaction with vinylic halides is also

I, QR + $CuC{\equiv}CR'$ $\longrightarrow$ $C{\equiv}CR'$, QR $\longrightarrow$ Q, R'

QR = OH, COOH, SH, NHR, etc.

known (Burdon *et al.*, 1967, 1972a; Jukes *et al.*, 1968; Soloski *et al.*, 1973).

Methylation of **LXXII** with lithium dimethylcuprate allows the synthesis of *trans,trans*-farnesol **(LXXIII)** from geranylacetone (Corey *et al.*, 1967).

O

several steps

$C{\equiv}CCH_2OH$

1. $LiAlH_4{-}CH_3O^-$
2. I_2

I

OH

(LXXII)

$LiCu(CH_3)_2$

OH

(LXXIII)

Similarly, sirenin **(LXXIV)** has been prepared by using this procedure as a key step (Corey *et al.*, 1969). Furthermore, *Cecropia* juvenile hormone **(LIX)**

OTHP
$LiCu(CH_3)_2$
OH
I
OTHP
several steps
OH
H
H
OH
OH
(LXXIV)

IHP = 2-tetrahydropyranyl

can be synthesized through the vinylic alkylation (Corey *et al.*, 1968c). Four possible stereoisomers of 7-methyl-3-propyl-2,6-decadien-l-ol, a tetrahomoterpene isolated from the codling moth, have been synthesized in a similar manner (Bowlus and Katzenellenbogen, 1973a). The reaction with dipropenylcuprate has been used for the synthesis of fulvoplumierin **(LXXV)** (Büchi and Carlson, 1968).

I
OH
$LiCu(C_2H_5)_2$
OH
several steps
OH
I
$LiCu(CH_3)_2$
OH
several steps
$COOCH_3$
O
(LIX)

Cl
O
O
$COOCH_3$
$LiCu(CH{=}CHCH_3)_2$
O
O
$COOCH_3$
(LXXV)

The coupling reaction of organocopper **(LXXVI)** and the iodoacetylene **(LXXVII)** is a key step in the total synthesis of the acetylenic sesquiterpene, freelingyne **(LXXVIII)** (Knight and Pattenden, 1974).

(LXXVI) (LXXVII)

(LXXVIII)

The oxygen atom of ketones can be replaced by an isopropylidene group in two steps. This method allows synthesis of 4(14),7(11)-selinadiene **(LXXX)** from a decalone derivative **(LXXIX)** (Posner *et al.*, 1975b).

$$R_2C{=}O \xrightarrow{(C_6H_5)_3P,\ CBr_4} R_2C{=}CBr_2 \xrightarrow{LiCu(CH_3)_2} R_2C{=}C(CH_3)_2$$

(LXXIX) —86%→ —$LiCu(CH_3)_2$, 93%→ (LXXX)

4. Acyl Halides and Related Substrates

Reaction of organocuprate reagents with acyl halides is extremely facile and provides a new technique for ketone synthesis (Vig *et al.*, 1968a; Jukes *et al.*, 1970a; Posner *et al.*, 1970, 1972, 1973; Jallabert *et al.*, 1970; Normant and Bourgain, 1970; Luong-Thi *et al.*, 1971). The reaction proceeds under conditions sufficiently mild that carbonyl, ester, or cyano functionalities are not affected.

$$RCOCl + LiCuR_2' \longrightarrow RCOR'$$

Organocopper compounds, RCu, though less reactive toward acid halides, couple to give the corresponding ketones as well. Reaction of alkynylcoppers with acyl halides produces acetylenic ketones in good yields (Normant and Bourgain, 1970).

$$RCOCl + R'C{\equiv}CCu \longrightarrow RCOC{\equiv}CR'$$

The coupling reaction with an organocuprate has been used in the synthesis of manicone **(LXXXI)** (Katzenellenbogen and Utawanit, 1974).

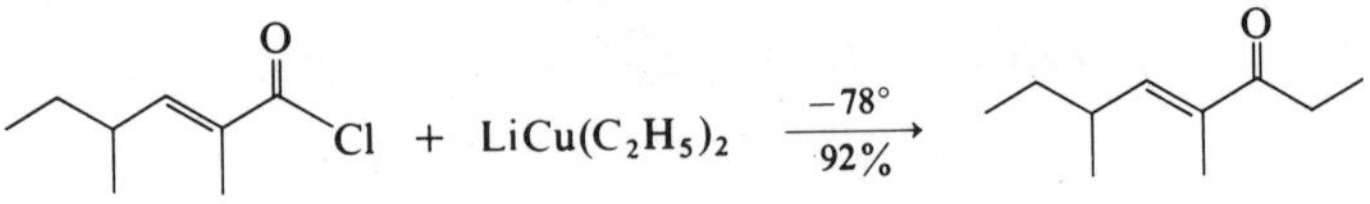

$$\text{RCOCl} + LiCu(C_2H_5)_2 \xrightarrow[92\%]{-78^\circ} \textbf{(LXXXI)}$$

The reaction with *S*-alkyl and *S*-aryl thioesters gives ketones in high yield with efficient utilization of the organocopper reagents (approximately stoichiometric amount required). The selectivity of this ketone synthesis is demonstrated by the conversion **(LXXXII)** → **(LXXXIII)** (Anderson *et al.*, 1974).

$$0.5LiCuR_2 + R'COSR'' \longrightarrow R'COR$$

$$RMgX \cdot CuI + R'COSR'' \longrightarrow R'COR$$

$$(RCu)_n \cdot LiI + R'COSR'' \longrightarrow R'COR$$

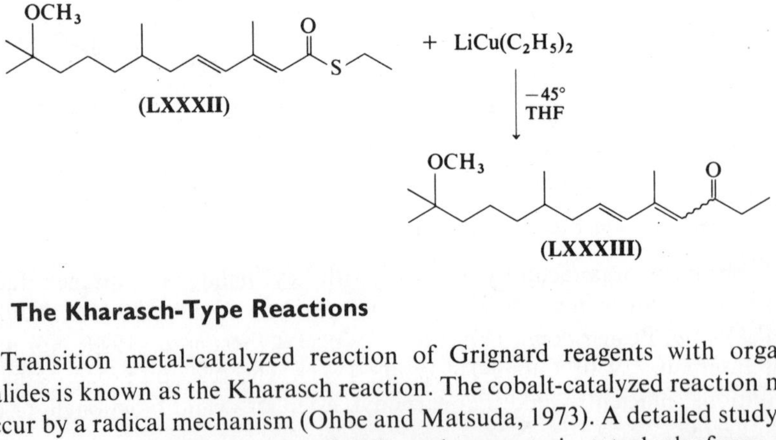

OCH3 **(LXXXII)** + $LiCu(C_2H_5)_2$ $\xrightarrow[\text{THF}]{-45^\circ}$ OCH3 **(LXXXIII)**

B. The Kharasch-Type Reactions

Transition metal-catalyzed reaction of Grignard reagents with organic halides is known as the Kharasch reaction. The cobalt-catalyzed reaction may occur by a radical mechanism (Ohbe and Matsuda, 1973). A detailed study on the copper-catalyzed reaction of Grignard reagents in tetrahydrofuran has revealed that the active species is an organocopper(I) compound (Tamura and Kochi, 1971a,b,c, 1972). The reaction kinetics are in accord with a mechanism involving S_N2-type displacement in the rate-determining step. The reaction proceeds effectively only with primary alkyl bromides and cannot be used for coupling with vinyl or aryl halides.

The dilithium tetrachlorocuprate (Li_2CuCl_4)-catalyzed alkylation of α,ω-dihalides with Grignard reagents allows simple halopolycarbon homologation of organic structures (Friedman and Shani, 1974). The reaction is useful for the preparation of intermediates of various pheromones and hormones. An improved carbon–carbon linking reaction has been attained by

$$RMgX + Br(CH_2)_nBr \xrightarrow{Cu(I)} R(CH_2)_nBr$$

$n = 3–6, 10$

R = primary, secondary, or tertiary alkyl, phenyl

controlled copper catalysis (Schlosser, 1974; Fouquet and Schlosser, 1974). Thus, reaction of primary alkyl tosylates and alkyl (primary, secondary, or tertiary) Grignard reagents proceeds smoothly in the presence of Li_2CuCl_4. The proposed mechanism is outlined in Scheme 3.

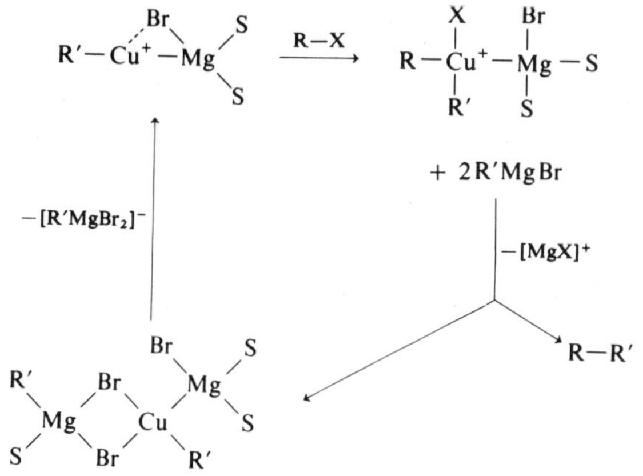

R′—Cu⁺(Br)—Mg(S)(S) —R—X→ R—Cu⁺(X)(R′)—Mg(Br)(S)—S

+ 2R′MgBr

−[MgX]⁺

R—R′

−[R′MgBr₂]⁻

R′(S)Mg(Br)(Br)Cu(R′)(Br)Mg(S)(S)

R = CH_3 or primary alkyl

R′ = alkyl, aryl, etc.

X = OTs, I

S = solvent

Scheme 3

Cuprous chloride coordinated with *N,N,N′,N′*-tetramethylethylenediamine is a good catalyst for the cross-coupling of aryl Grignard reagents and alkyl iodides (Onuma and Hashimoto, 1972).

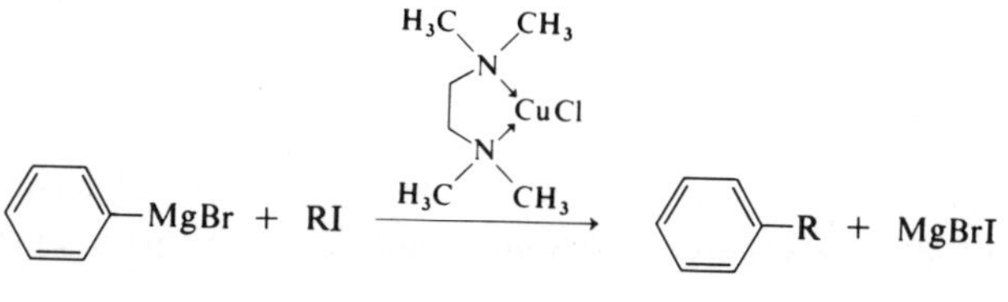

$$C_6H_5MgBr + RI \xrightarrow{(H_3C)_2N(CH_2)_2N(CH_3)_2 \cdot CuCl} C_6H_5R + MgBrI$$

Recently, a cuprous bromide-catalyzed coupling between β-keto ester anions and *o*-bromobenzoic acid has been reported (Bruggink and McKillop, 1974).

$$o\text{-BrC}_6\text{H}_4\text{COOH} + \text{H}_2\text{C(COR)(COOR)} \xrightarrow[80^\circ]{\text{NaH, CuBr}} o\text{-[CH(COR)(COOR)]C}_6\text{H}_4\text{COOH} \quad (>90\%)$$

Dubois and his associates have studied the copper-catalyzed reaction of Grignard reagents and acid halides (Dubois and Boussu, 1969, 1970, 1971; Dubois *et al.*, 1967a,b, 1969, 1971; MacPhee and Dubois, 1972). The reaction with a sterically hindered acid halide results in a symmetrical ketonic product, suggesting the operation of both ionic and free radical mechanisms.

$$\text{(t-Bu)}_2\text{CHCOCl} + C_2H_5MgBr \xrightarrow{\text{Cu/CuCl}} \text{(t-Bu)}_2\text{CHCOCH}_2\text{CH}_3 \ (41\%) + \text{(t-Bu)}_2\text{CHCOCH(t-Bu)}_2 \ (41\%)$$

The ease with which the coupling reaction takes place is profoundly influenced by the kind of organic moieties and the nature of the transition metal catalysts (Tamura and Kochi, 1971c; Kochi, 1974). Silver nitrate is an effective catalyst for the homocoupling reaction, whereas dilithium tetrachlorocuprate

$$\text{RMgX} + \text{RX} \xrightarrow{\text{Ag(I)}} \text{R—R} + \text{MgX}_2$$

$$\text{R}^1\text{MgX} + \text{R}^2\text{X} \xrightarrow{\text{Cu(I)}} \text{R}^1\text{—R}^2 + \text{MgX}_2$$

promotes the cross-coupling reaction of different organic groups. An organosilver species is the likely intermediate in the former catalysis. Facile, stereo-

$$\text{R'MgX} + \text{Ag(I)} \longrightarrow \text{R'Ag(I)} + \text{MgX}_2$$

$$\text{RAg(I)} + \text{R'Ag(I)} \longrightarrow [\text{R—R, R'—R, R'—R'}] + 2\text{Ag(0)}$$

$$\text{Ag(0)} + \text{RX} \longrightarrow \text{R}\cdot + \text{Ag(I)X}$$

$$\text{R}\cdot + \text{Ag(0)} \longrightarrow \text{RAg(I), etc.}$$

specific vinylation can be achieved with iron(III) chloride. This cross-coupling reaction can be employed as a synthetic route for alkenes, in which primary,

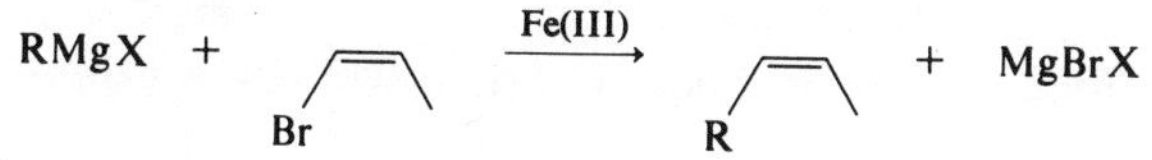

$$\text{RMgX} + \text{CH}_2{=}\text{CHBr} \xrightarrow{\text{Fe(III)}} \text{CH}_2{=}\text{CHR} + \text{MgBrX}$$

secondary, as well as tertiary alkyl groups are involved. The rearrangement of branched alkyl groups has not been observed. Among various iron(III) complexes tris(dibenzoylmethido)iron(III) is the most effective catalyst. Several mechanistic schemes are possible for this coupling reaction including oxidative addition of the alkenyl halide to a low-valent alkyliron species, followed by reductive elimination of the cross-coupled product (Scheme 4),

Initiation

$$\text{Fe(III)} + 2\text{RMgX} \longrightarrow \text{Fe(I)} + \text{R}_{\text{ox}}$$

Propagation

$$\text{Fe(I)} + \text{RMgX} \rightleftharpoons \text{RFe(I)}^- + \text{MgX}^+$$

$$\text{RFe(I)}^- + \text{R}'\text{Br} \longrightarrow \text{RR}'\text{Fe(III)} + \text{Br}^-$$

$$\text{RR}'\text{Fe(III)} \longrightarrow \text{R}{-}\text{R}' + \text{Fe(I)}, \text{ etc.}$$

Termination

$$n\text{Fe(I)} \longrightarrow [\text{Fe(I)}]_n$$

$$\text{Fe(I)} \xrightarrow{\text{ox}} \text{Fe(III)}$$

Scheme 4

and assistance by reduced iron in the concerted displacement of halide at the alkenyl center by the Grignard reagent (Scheme 5). The substitution process in Scheme 5, unlike that in Scheme 4, requires the reduced iron species to effect substitution by a coordination mechanism (Neumann and Kochi, 1975).

Propagation

$$\text{Fe(I)} + \text{R}'\text{Br} \rightleftharpoons \text{Fe(R}'\text{Br)}$$

$$\text{Fe(R}'\text{Br)} + \text{RMgX} \longrightarrow \text{R}{-}\text{R}' + \text{MgXBr} + \text{Fe(I)}, \text{ etc.}$$

Scheme 5

Nickel(II)–phosphine complexes catalyze selective carbon–carbon bond formation by cross-coupling of Grignard reagents with olefinic or aromatic halides (Corriu and Masse, 1972; Tamao *et al.*, 1972a,b, 1973; Kiso *et al.*, 1973). The yields are generally very high. Bidentate phosphine ligands exhibit this remarkable activity, while unidentate tertiary phosphines are much less effective. Scheme 6 outlines the overall catalytic cycle. When a *sec*-alkyl Grignard reagent is used, the coupling reaction is accompanied by secondary

$$L_2NiX_2 + 2RMgX \longrightarrow L_2NiR_2 + 2MgX_2$$
$$L_2NiR_2 + R'X \longrightarrow L_2NiR'X + R{-}R$$

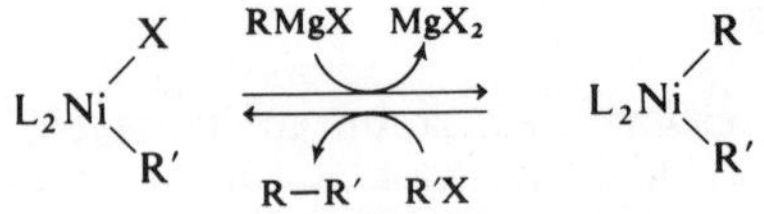

$$L_2Ni(X)(R') \underset{R'X \quad R{-}R'}{\overset{RMgX \quad MgX_2}{\rightleftarrows}} L_2Ni(R)(R')$$

Scheme 6

to primary alkyl group isomerization. The coupling reaction with monohalo olefins proceeds stereospecifically with retention of configuration, whereas the reaction involving 1,2-dihalo olefins occurs in a nonstereospecific manner. Acetylene would be formed as an intermediate in the latter case. When a racemic *sec*-alkyl Grignard reagent having magnesium attached to a chiral

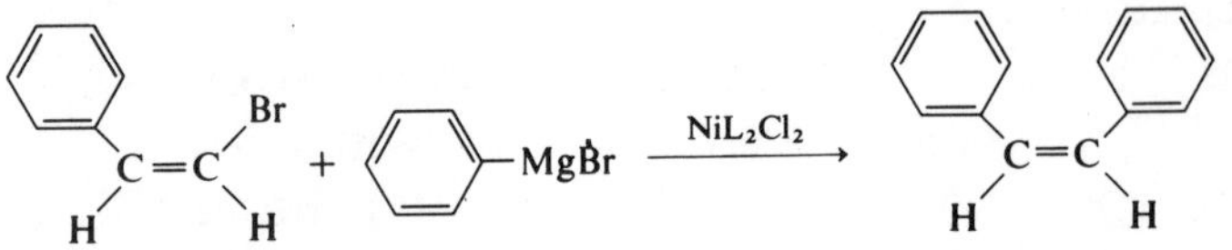

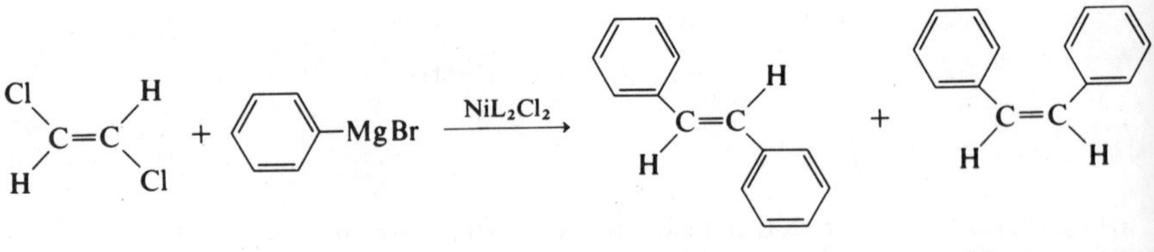

$$PhCH{=}CHBr + PhMgBr \xrightarrow{NiL_2Cl_2} PhCH{=}CHPh$$

$$ClCH{=}CHCl + PhMgBr \xrightarrow{NiL_2Cl_2} PhCH{=}CHPh\ (trans) + PhCH{=}CHPh\ (cis)$$

carbon is subjected to the reaction in the presence of a chiral nickel catalyst such as Ni[(−)-2,3-*O*-isopropylidene-2,3-dihydroxy-1,4-bis(diphenylphosphino)butane]Cl_2, an asymmetric cross-coupling can be achieved (Consiglio and Botteghi, 1973; Kiso *et al.*, 1974). Furthermore, a new asymmetric synthesis of molecular dissymmetric biaryls is possible by this method. The coupling reaction of **LXXXIV** and **LXXXV** with (*S*)-α-[(*R*)-1′,2-bis-(diphenylphosphino)ferrocenyl]ethyldimethylamine gives optically active **LXXXVI** in 4.6% enantiomeric excess (e.e.) (Tamao *et al.*, 1975).

$$R^1R^2CHMgX + R^3X \xrightarrow{Ni\ cat.^*} R^1R^2R^3CH^*$$

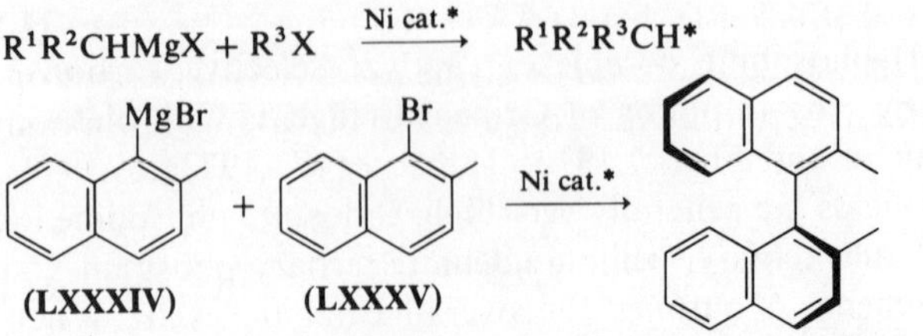

(LXXXIV) + **(LXXXV)** $\xrightarrow{Ni\ cat.^*}$ **(S)−(+)−(LXXXVI)**

C. Nucleophilic Displacement with π-Allylnickel(I) Halide Complexes

1. Allylic Halides

π-Allylnickel(I) halides are, in general, rather stable complexes, and have been widely used in organic synthesis as both discrete reagents and as species formed *in situ* from allyl halides and zerovalent nickel. Their synthetic utility has been surveyed by Semmelhack (1972) and Baker (1973). A general route to π-allyl complexes was first reported by Fischer and Bürger (1961). They employed allylic halides and nickel carbonyl as reactants and benzene as solvent. Later, Wilke *et al.* (1966) used bis(1,5-cyclooctadiene)nickel(0) in place of nickel carbonyl. The nickel(I) complex may also be derived from bis-(π-allyl)nickel(0) and hydrobromic acid.

$$2\ CH_2{=}CHCH_2X + Ni(0)L_4 \longrightarrow [(\pi\text{-}C_3H_5)Ni(\mu\text{-}X)]_2$$

The synthesis of 1,5-dienes via the reaction of allyl halides and nickel carbonyl is now an indispensable process in synthetic organic chemistry. To our knowledge, it originates from the observation by Webb and Borcherdt (1951) that the reaction of 1-chloro-2-butene or 3-chloro-1-butene with nickel carbonyl produced the dimeric 1,5-dienes in high yield. Later Corey and his

$$\begin{matrix} ClCH_2CH{=}CHCH_3 \\ \text{or} \\ CH_2{=}CHCHClCH_3 \end{matrix} + Ni(CO)_4 \xrightarrow[CH_3OH]{} \begin{matrix} CH_3CH{=}CHCH_2CH_2CH{=}CHCH_3 \\ + \\ CH_2{=}CHCH(CH_3)CH_2CH{=}CHCH_3 \end{matrix}$$

associates extended this reaction to a variety of systems and thereby accomplished elegant organic syntheses. Now, besides allylic halides, the acetates (Bauld, 1962) or tosylates may be used as a source of the three-carbon unit. Coordinating solvents are necessary for effecting reaction; the best of which are *N*,*N*-dimethylformamide (DMF) and *N*-methylpyrrolidone.

Corey *et al.* (1968a) demonstrated that these allylic coupling reactions take place via π-allylnickel(I) halides. The blood red complex **(LXXXVII)** is stable at 25° in tetrahydrofuran or glyme for several days, but on addition of allyl bromide it is quantitatively converted to biallyl in a few minutes. Interest-

$$[(\pi\text{-}C_3H_5)Ni(\mu\text{-}Br)]_2 + CH_2{=}CHCH_2Br \xrightarrow[100\%]{} CH_2{=}CHCH_2CH_2CH{=}CH_2$$

(LXXXVII)

ingly, reaction of **LXXXVII** and methallyl bromide gives rise to a mixture of C_6, C_7, and C_8 1,5-dienes. A similar mixture is produced by the reaction of the methallyl complex and allyl bromide. Free allyl bromide is formed when **LXXXVII** and methallyl bromide is mixed in *N*,*N*-dimethylformamide. A mechanism suggested by Corey is outlined in Scheme 7 (A^1 and A^2 = allylic

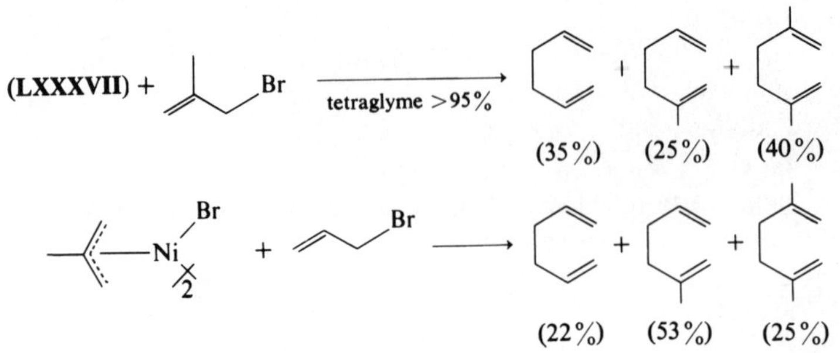

(LXXXVII) + Br —tetraglyme >95%→ + +

(35%) (25%) (40%)

Br Ni 2 + Br → + +

(22%) (53%) (25%)

moiety). π-Allylnickel(I) halides (dark red, nonvolatile) exist in equilibrium with the corresponding bis(π-allyl)nickel(0) complexes (yellow, sublimable)

$$2A^1Br + 2Ni(CO)_4 \rightleftharpoons 2\pi\text{-}A^1NiBr(CO) + 6CO$$

$$\updownarrow$$

$$[\pi\text{-}A^1NiBr]_2 + 8CO$$

$$\pi\text{-}A^1NiBr(CO) + A^2Br \rightleftharpoons A^1NiBr(CO)$$

$$\uparrow$$

$$A^2Br$$

(LXXXVIII)

$$\updownarrow$$

$$\pi\text{-}A^2NiBr(CO) + A^1Br \rightleftharpoons A^2NiBr(CO)$$

$$\uparrow$$

$$A^1Br$$

(LXXXIX)

(LXXXVIII) or **(LXXXIX)** —coordinating solvent→ $A_2 + NiBr_2 + CO$

(σ or π complex)

Scheme 7

and nickel(II) bromide, but the zerovalent species are of little importance in the allylic coupling reaction (Corey *et al.*, 1968b).

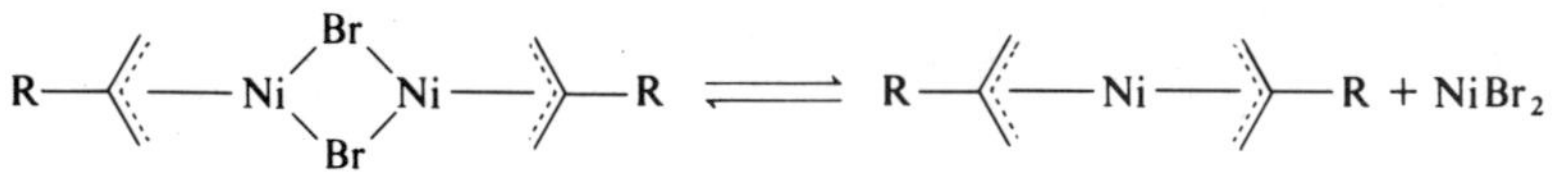

A number of terpenoid compounds have been synthesized by the reaction of π-allylnickel(I) halides with allyl bromides. Geranyl acetate **(XCI)** and farnesyl acetate **(XCII)** have been obtained from the nickel complex **(XC)** and appropriate allylic halides (Guerrieri and Chiusoli, 1969). The reaction of the complex **(XCIII)** and the bromides **(XCIV)** forms geranyl ether or ester

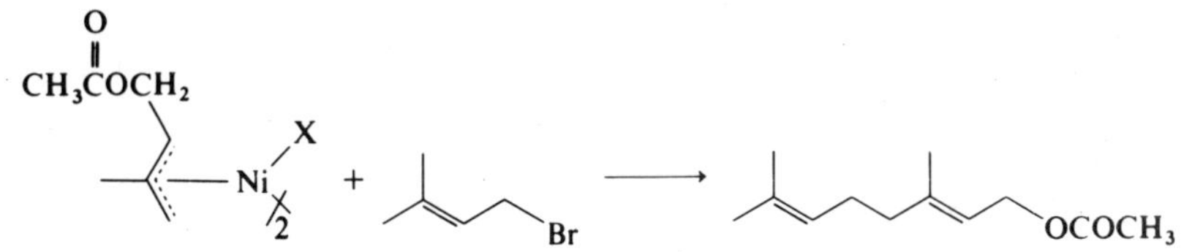

(XC) **(XCI)**

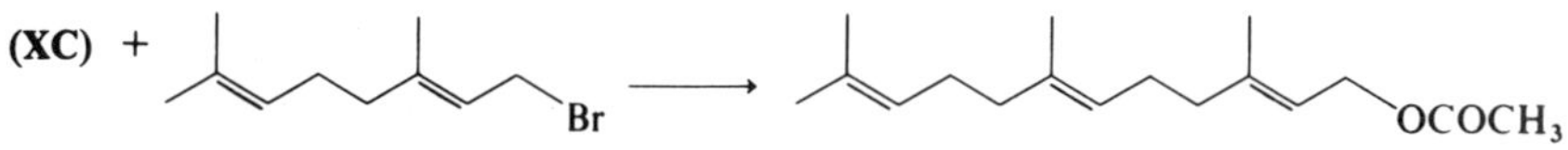

(XC) + **(XCII)**

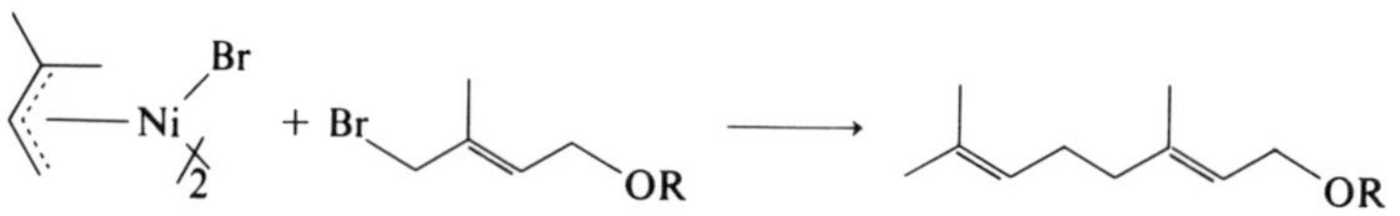

(XCIII) **(XCIVa)** R = C_2H_5 **(XCVa)** R = C_2H_5

(XCIVb) R = $COCH_3$ **(XCVb)** R = $COCH_3$

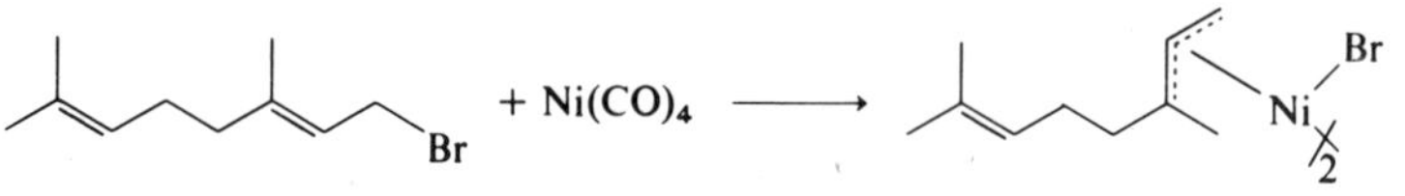

(XCVI)

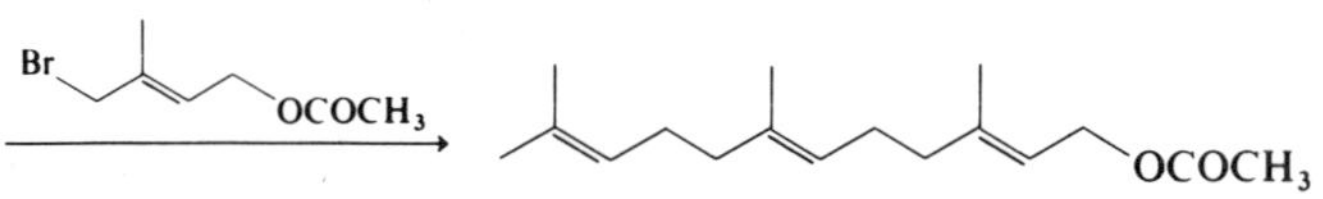

(XCII)

(XCV) (Sato *et al.*, 1972a). Farnesyl acetate **(XCII)** can be obtained via complex **XCVI**.

One of the problems to be solved is the lack of selectivity in the coupling reactions. In general, two allylic moieties, one from the π-allylnickel and the other from organic halide, combine randomly, leading to a mixture of homo- and cross-coupled products. Improvement of the selectivity in favor of the cross-coupled products has been observed by changing the leaving group

R—(π-allyl)—NiX + R'CH=CHCH$_2$X —cross-coupling→ R–CH=CH–CH$_2$–CH$_2$–CH=CH–R'

—exchange→ RCH=CHCH$_2$X + R'—(π-allyl)—NiX

from halide to dithiocarbamate (Semmelhack, 1967). This strongly coordinating group can inhibit the exchange reaction. For example, β-farnesenes **(XCIX)** and **(C)** were prepared from **XCVII** and **XCVIII** in 46% yield.

(XCVII) + **(XCVIII)** → **(XCIX)** + **(C)**

Further, the presence of an electron-withdrawing group on the π-allylic ligand was found to favor cross-coupling with allyl halides (Chiusoli, 1971).

Methyl farnesoate **(CIII)** has been synthesized stereoselectively by the coupling of the geranyl nickel compound **(CI)** (L = acetonitrile) with the allylic bromide **(CII)** (Chiusoli, 1971; Guerrieri *et al.*, 1974). The stereochemistry of the reaction of π-allyl complexes of type **CIV** with allyl chloride depends markedly on the nature of the solvents. Media of low polarity and low coordinating power favor the formation of the cis isomer **(CV)**, whereas in

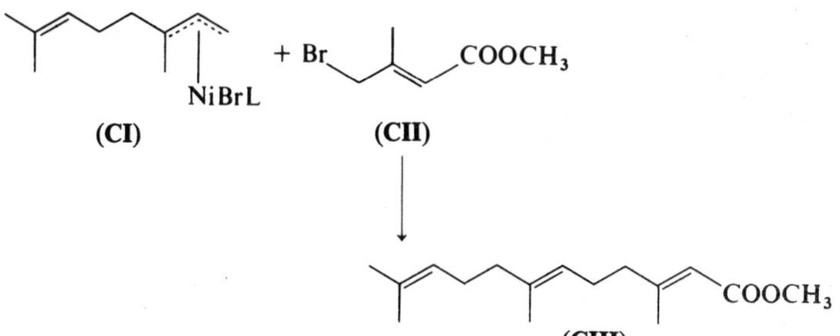

NiBrL + Br COOCH$_3$

(CI) **(CII)**

COOCH$_3$

(CIII)

more polar solvents and with aid of basic ligands, the trans adduct **(CVI)** is formed predominantly (Chiusoli, 1971).

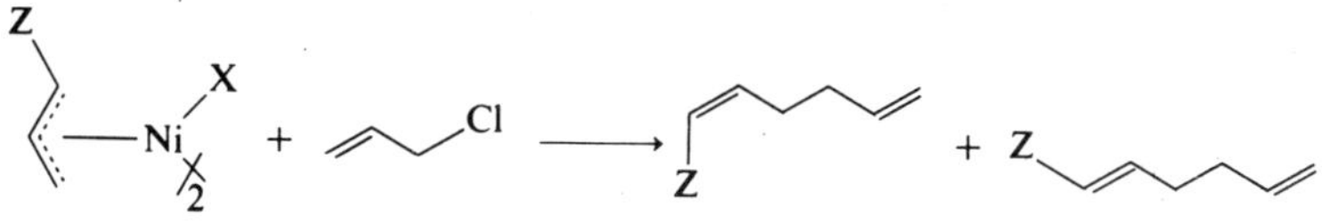

Z X Ni 2 + Cl ⟶ Z + Z

(CIV) **(CV)** **(CVI)**

Z = CN or COOCH$_3$

Application of this reaction to α,ω-bisallylic halides makes the synthesis of medium and large rings possible. Treatment of 1,1-bis(chloromethyl)-ethylene **(CVII)** with nickel carbonyl in tetrahydrofuran gives the dimer **CVIII** and trimer **CIX** in 11 and 54% yield, respectively (Corey and Semmelhack, 1966). These products are derived by intermolecular coupling, followed by intramolecular coupling of allylic chlorides. The nine-membered carbocycle **(CIX)** was produced by the coupling of **CVII** and the dichloride

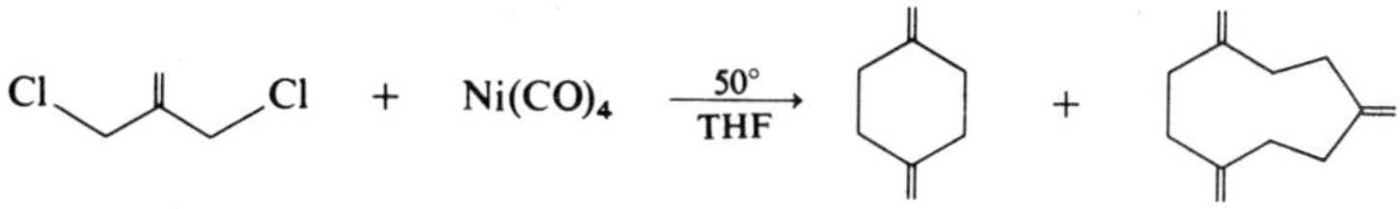

Cl Cl + Ni(CO)$_4$ $\xrightarrow[\text{THF}]{50°}$ +

(CVII) **(CVIII)** **(CIX)**

Cl Cl Cl Cl

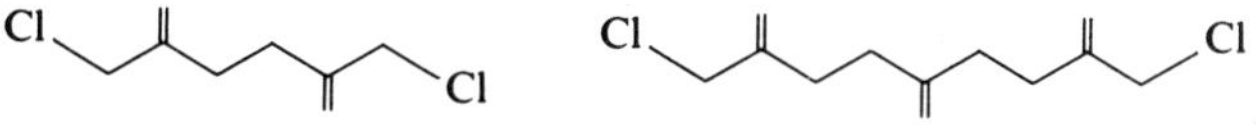

(CX) **(CXI)**

(CX) or by cyclization of the dichloride **(CXI)**. Various cyclic 1,5- dienes not readily accessible by other means have been obtained by this method (Corey and Wat, 1967).

Br Br (42%) (5%)

Br Br cis and trans

Br Br (59%)

Br Br (70–74%)

Br Br (76–84%)

This procedure has been applied to the synthesis of humulene **(CXII)**, the overall yield being 10% (Corey and Hamanaka, 1967). *d,l*-Elemol **(CXIII)**

Br Br + $Ni(CO)_4$ → $h\nu$ $(C_6H_5S)_2$ **(CXII)**

(Corey and Broger, 1969) and *d,l*-elemene **(CXIV)** have been prepared (Vig *et al.*, 1968b) by this method.

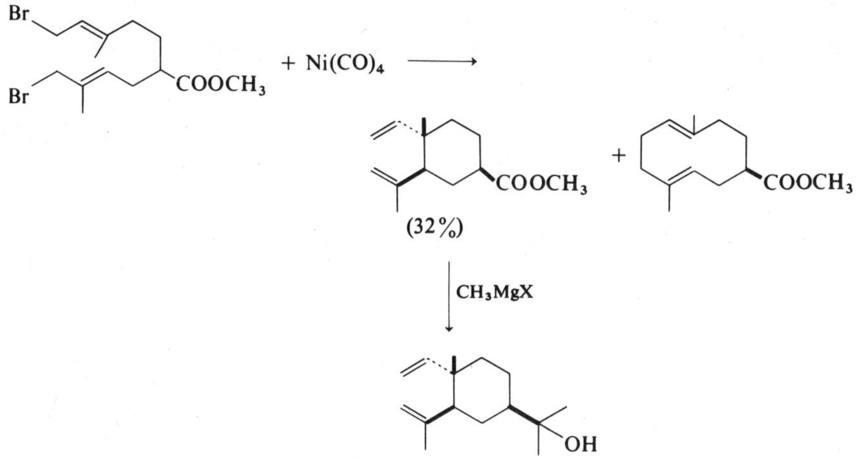

Br Br COOCH$_3$ + Ni(CO)$_4$ ⟶

COOCH$_3$ + COOCH$_3$

(32%)

CH$_3$MgX

OH

(CXIII)

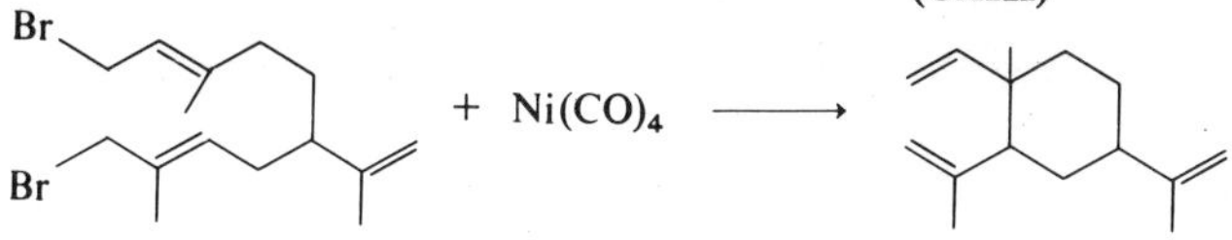

Br Br + Ni(CO)$_4$ ⟶

(CXIV)

The intramolecular coupling reaction has allowed the synthesis of cembrene **(CXV)**, a 14-membered diterpene (Dauben *et al.*, 1974).

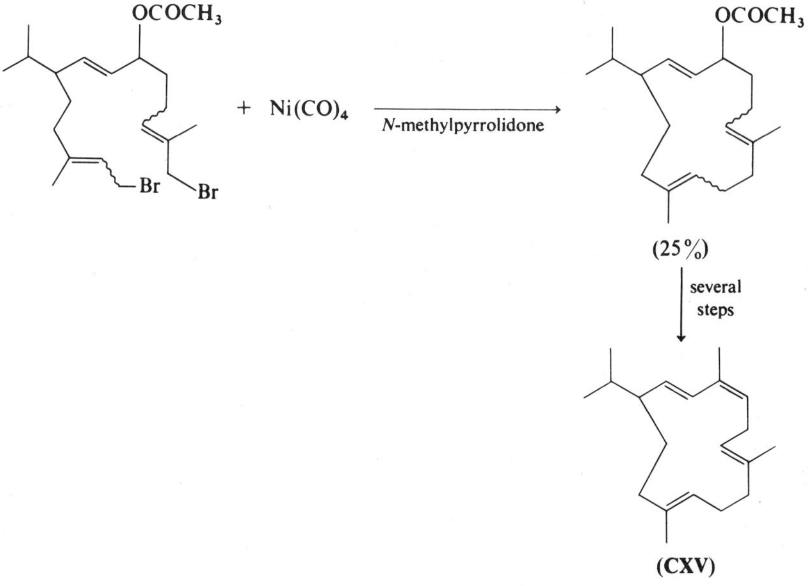

OCOCH$_3$ Br Br + Ni(CO)$_4$ —*N*-methylpyrrolidone⟶ OCOCH$_3$

(25%)

several steps

(CXV)

Since ester groups are stable under these cyclization conditions, this method may be used for the preparation of macrolides. Thus, the dibromide **(CXVI)** can be converted to **CXVII** in 76% yield (Corey and Kirst, 1972).

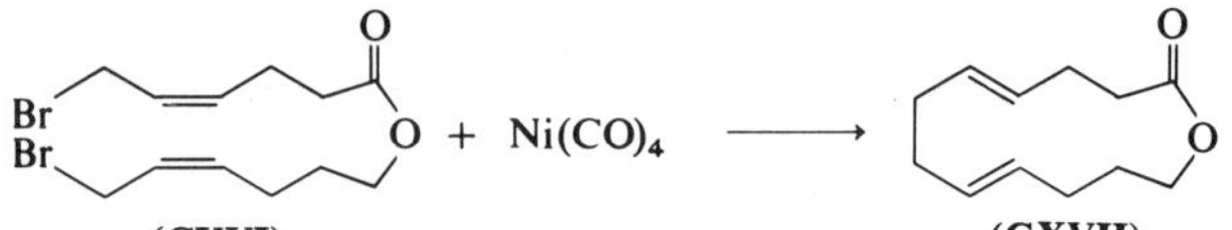

$$\text{(CXVI)} + Ni(CO)_4 \longrightarrow \text{(CXVII)}$$

2. Nonallylic Halides

π-Allyl ligands of the nickel(I) complexes also couple with ordinary non-allylic organic halides. Alkyl, alkenyl, and aryl halides are equally employable as substrates; the π-methallylnickel complex **(CXVIII)**, for example, couples in *N*,*N*-dimethylformamide (DMF) with methyl iodide (90% yield), methyl bromide (90%), cyclohexyl iodide (91%), *tert*-butyl iodide (25%), iodobenzene (98%), vinyl bromide (70%), benzyl bromide (91%), phenyl α-chloromethyl ester (50%), *p*-bromophenacyl bromide (75%), and chloroacetone (46%) (Corey and Semmelhack, 1967).

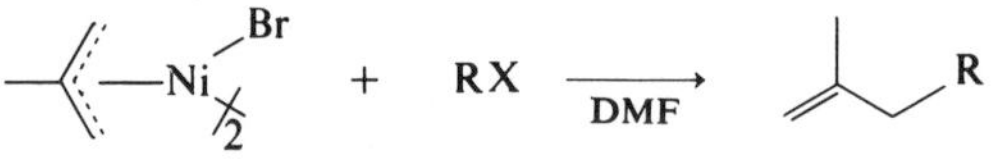

$$\text{(CXVIII)} + RX \xrightarrow{\text{DMF}}$$

A variety of mechanisms are conceivable for the coupling reaction. A mechanism (Scheme 8) originally proposed involved oxidative addition of the allylic halide to the central nickel atom (Corey and Semmelhack, 1967). Stereochemistry of the π-allylic ligands is not necessarily retained in the coupled products; the geranyl complex **(CXIX)** which has a trans geometry

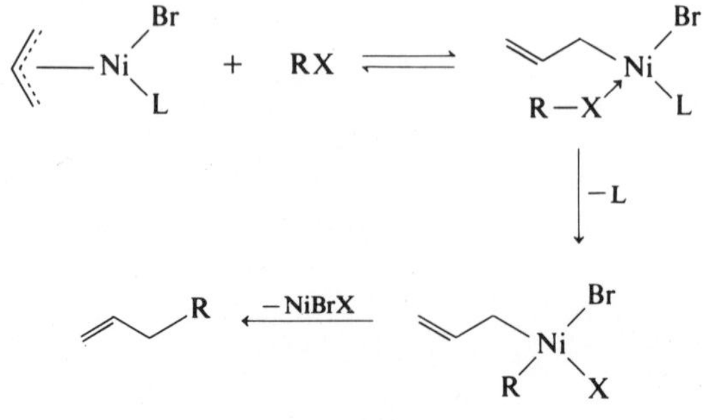

$$\pi\text{-allyl-Ni(Br)L} + RX \rightleftharpoons \text{allyl-Ni(Br)(L)} \leftarrow X{-}R \xrightarrow{-L} \text{allyl-Ni(Br)(R)(X)} \xrightarrow{-NiBrX} \text{allyl-R}$$

Scheme 8

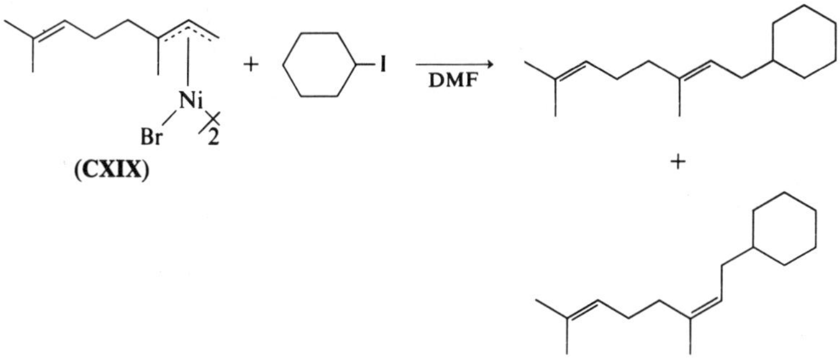
Ni
Br
2
DMF

(CXIX)

couples with cyclohexyl iodide to give a mixture of cis and trans isomers. This could be explained by assuming a π-allyl–σ-allyl equilibrium in the reaction intermediate. However, a recent study has presented evidence for a radical chain mechanism. First, the reaction of π-(2-methoxyallyl)nickel bromide with (*S*)-(+)-2-iodooctane produces a completely racemic product. Second, the coupling reaction is completely inhibited by the addition of less than 1 mole % *m*-dinitrobenzene, a potent radical anion scavenger. A new mechanism consistent with all these findings is outlined in Scheme 9 (Hegedus and Miller, 1975).

$$[(\text{allyl})\text{NiBr}] + \text{RX} \longrightarrow \text{RX}^{-}\cdot + [(\text{allyl})\text{NiBr}]^{+}\cdot$$
$$\text{RX}^{-}\cdot \longrightarrow \text{R}\cdot + \text{X}^{-}$$
$$\text{R}\cdot + [(\text{allyl})\text{NiBr}] \longrightarrow \text{R-allyl} + \text{NiBr}\cdot$$
$$\text{NiBr}\cdot + \text{RX} \longrightarrow \text{RX}^{-}\cdot + \text{NiBr}^{+}$$

Scheme 9

Cross-coupling of the complex **(XCIII)** and the iodide **(CXX)** leads to the formation of α-santalene **(CXXI)** in 88% yield (Corey and Semmelhack, 1967). β-Santalene **(CXXII)** has been also prepared by applying this process

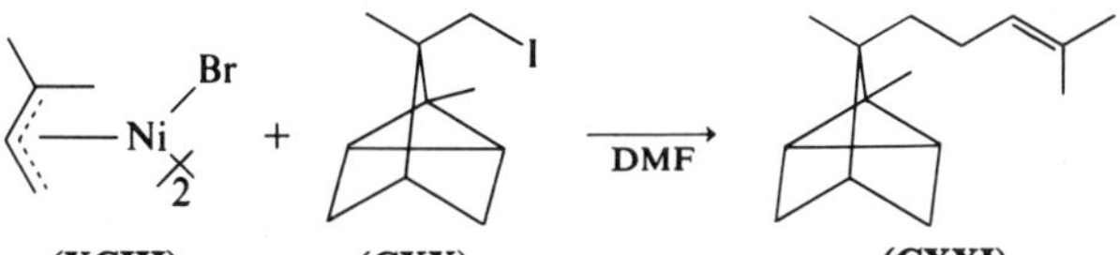
Br
Ni
2
I
DMF

(XCIII) **(CXX)** **(CXXI)**

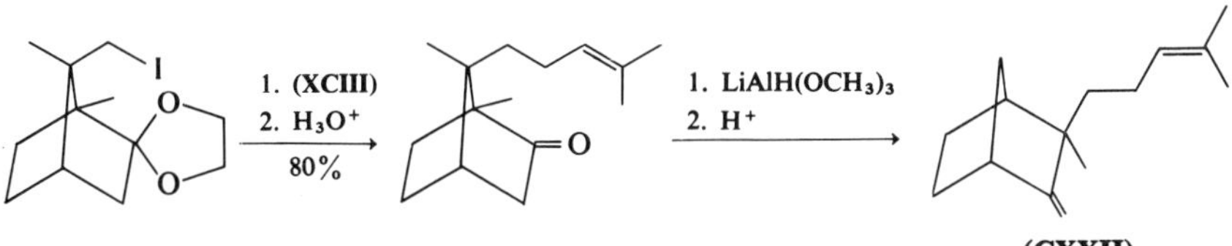
I
O
O
1. (XCIII)
2. H$_3$O$^+$
80%
O
1. LiAlH(OCH$_3$)$_3$
2. H$^+$

(CXXII)

(Hodgson *et al.*, 1973). When the π-allyl complex **(XCIII)** is allowed to react with the steroidal iodide **(CXXIII)**, the coupled product **(CXXIV)** is formed in 65% yield. The latter is readily convertible to desmosterol **(CXXV)** (Das-Gupta *et al.*, 1974).

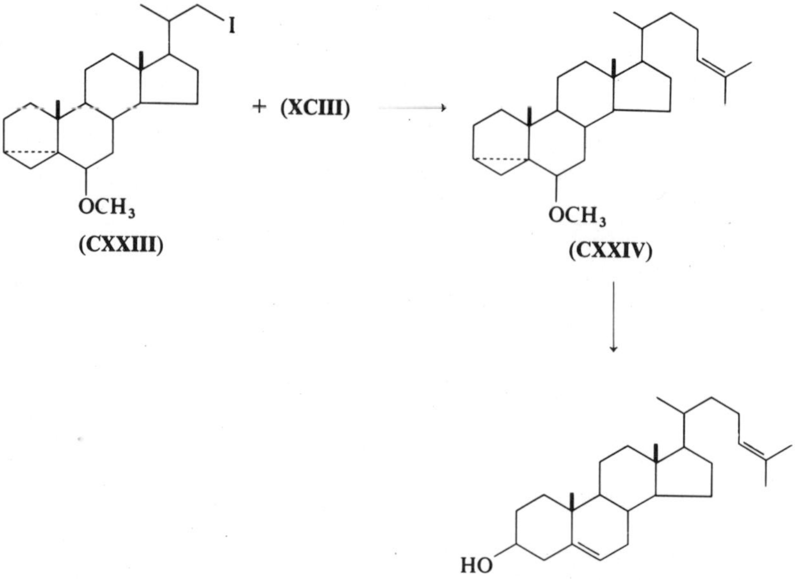

+ **(XCIII)**

OCH_3
(CXXIII)

OCH_3
(CXXIV)

HO
(CXXV)

An elegant method for the synthesis of coenzyme Q's, **(CXXVI)** and **(CXXVII)**, and vitamin K's **(CXXVIII)** has been developed by Sato *et al.* (1972b,c, 1973). Monomethyltocols and 2,2-dimethylchroman-6-ol models are also obtainable by this method (Inoue *et al.*, 1974).

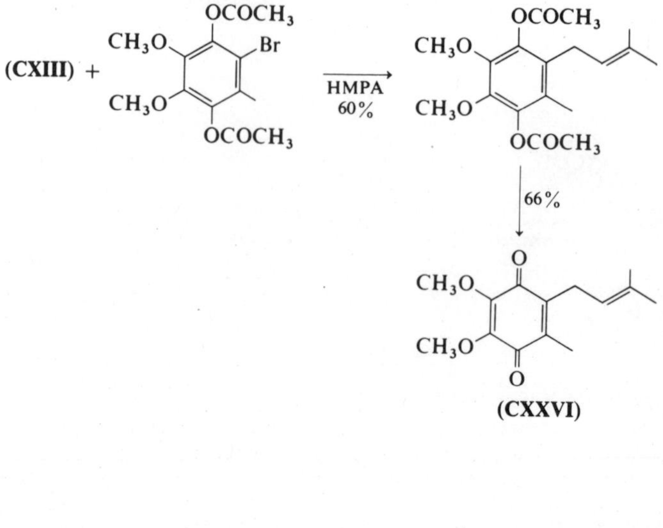

(CXIII) + $OCOCH_3$ CH_3O Br CH_3O $OCOCH_3$ → HMPA 60% $OCOCH_3$ CH_3O CH_3O $OCOCH_3$

66%

O CH_3O CH_3O O
(CXXVI)

$H(CH_2\overset{CH_3}{C}=CHCH_2)_9CH_2$ — Ni(Br)/2 + CH_3O, CH_3O, $OCOCH_3$, Br, $OCOCH_3$ $\xrightarrow{40\%}$

$OCOCH_3$, CH_3O, CH_3O, $(CH_2CH=\overset{CH_3}{C}CH_2)_{10}H$, $OCOCH_3$ ⟶ O, CH_3O, CH_3O, $(CH_2CH=\overset{CH_3}{C}CH_2)_{10}H$, O

(CXXVII)

R — Ni(Br)/2 + $OCOCH_3$, Br, $OCOCH_3$ ⟶

$OCOCH_3$, R, $OCOCH_3$ ⟶ O, R, O

(CXXVIII)

$R = H, C_{15}H_{25}$

The 2-methoxyallylnickel bromide **(CXXIX)** serves as a new reagent for the introduction of the acetonyl functional group (Hegedus and Stiverson, 1974). Halides containing both sp^2- and sp^3-hybridized carbons may be used. The reaction is facile and gives exclusively the cross-coupled products

CH_3O — Ni(Br)/2 + RX $\xrightarrow{DMF}$ OCH_3, R $\xrightarrow{H_3O^+}$ O, R

(CXXIX) **(CXXX)**

(CXXX). Reaction of **CXXIX** and the bromide **(CXXXI)** affords 6′-acetonylpapaverine **(CXXXII)** in 73% yield.

(CXXIX) + 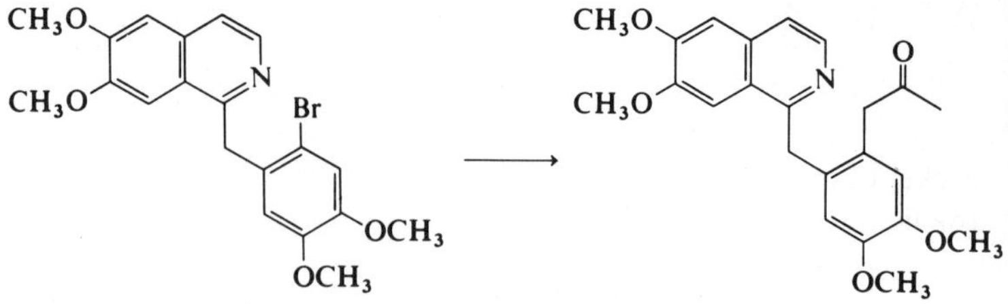 (CXXXI) ⟶ (CXXXII)

An allylnickel complex derived from the steroidal chloroketone **(CXXXIII)** and nickel carbonyl is methylated with methyl iodide to produce a mixture of **CXXXIV** and **CXXXV** (Harrison *et al.*, 1969).

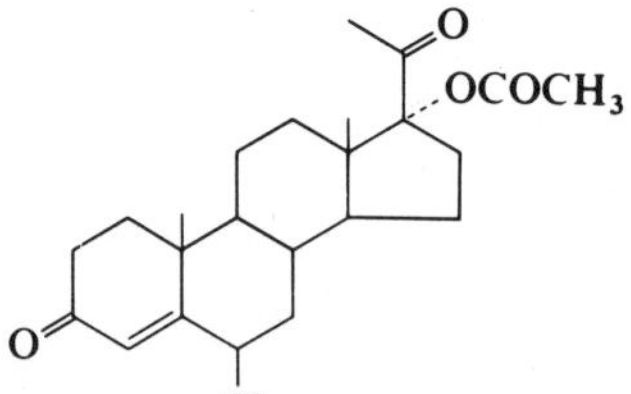

(CXXXIII) —[1. $Ni(CO)_4$; 2. CH_3I]→ (CXXXIV) 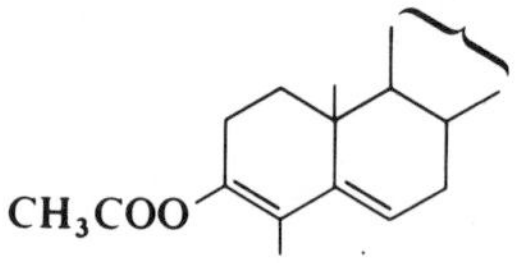 + (CXXXV)

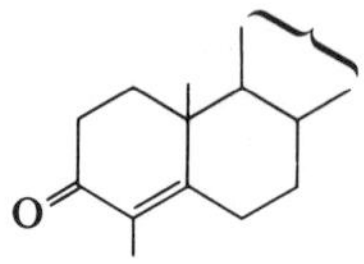

In the presence of bis(triphenylphosphine)nickel(II) chloride, allylic alcohols **(CXXXVI)** react with excess Grignard reagents to give a mixture of **CXXXVIII** and **CXXXIX** in high yield. Intervention of σ-alkyl-π-allylnickel intermediates of type **CXXXVII** has been suggested (Felkin and Swierczewski, 1972).

$R^1CH{=}CHCH(OH)R^2$ (CXXXVI) + RMgX —[$[P(C_6H_5)_3]_2NiCl_2$]→ [(CXXXVII)] → $R^1CH{=}CHCHRR^2$ (CXXXVIII) + $R^1CHRCH{=}CHR^2$ (CXXXIX)

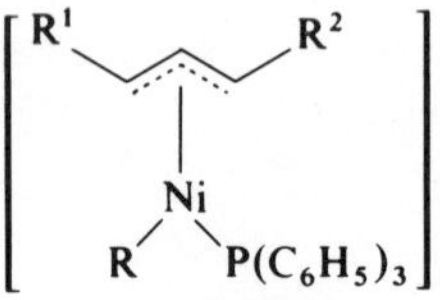

$R = CH_3, CH_2C_6H_5, C_6H_5$

D. Nickel(0)-Promoted Coupling of Organic Halides

Decomposition of organonickel compounds bearing two sp^2-hybridized carbon moieties is known to cause a ligand coupling reaction. The arylalkenylnickel compound **(CXL)**, on thermolysis or treatment with bromine, affords the ligand coupling product (Miller *et al.*, 1968) (see also Section I,A).

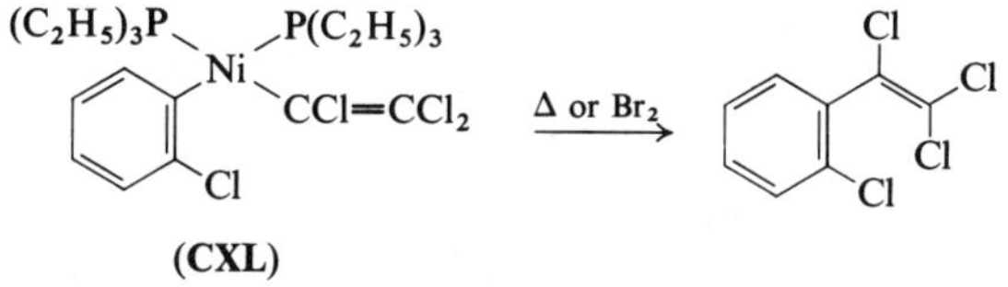

$(C_2H_5)_3P$, $P(C_2H_5)_3$, Ni, $CCl{=}CCl_2$, Cl $\xrightarrow{\Delta \text{ or } Br_2}$ Cl, Cl, Cl, Cl

(CXL)

Bis(1,5-cyclooctadiene)nickel(0) [$Ni(COD)_2$] reacts with a variety of aryl halides in dimethylformamide to produce biaryls in good yield. An arylnickel species has been suggested as an intermediate (Semmelhack *et al.*, 1971). By this method 1,*n*-bis(iodoaryl)alkanes are readily converted into the corresponding biphenyl derivatives. The total synthesis of alnusone dimethyl ether **(CXLI)** shows the usefulness of this procedure (Semmelhack and Ryono, 1975). Tris(triphenylphosphine)nickel is a superior coupling reagent in some instances (Kende *et al.*, 1975).

$$2ArX + Ni(COD)_2 \xrightarrow[DMF]{25^\circ\text{–}40^\circ} Ar{-}Ar$$

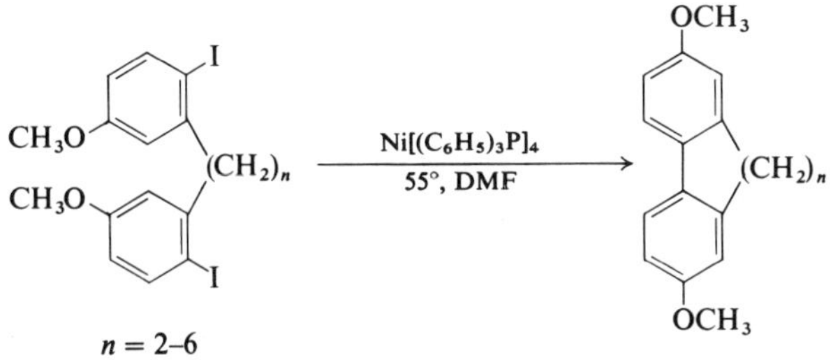

I, CH_3O, $(CH_2)_n$, CH_3O, I $\xrightarrow[55^\circ,\ DMF]{Ni[(C_6H_5)_3P]_4}$ OCH_3, $(CH_2)_n$, OCH_3

$n = 2$–6

(38–85%)

I, CH_3O, CH_3O, O, I $\longrightarrow$ CH_3O, CH_3O, O

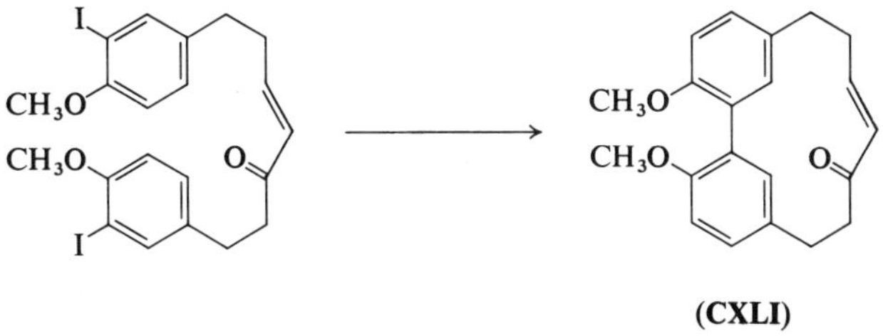

(CXLI)

(52%)

The efficiency of this reaction points to potential synthetic applications in inter- and intramolecular coupling of alkenyl halides (Semmelhack *et al.*, 1972). *trans*-β-Bromostyrene is converted selectively to *trans,trans*-1,4-diphenyl-1,3-butadiene. 2-Halo- and 3-haloacrylates react very rapidly

$$2\ \mathrm{Br{-}CH{=}CH{-}COOCH_3} + \mathrm{Ni(COD)_2} \xrightarrow[\text{ether}]{(C_6H_5)_3P} \mathrm{CH_3OOC{-}CH{=}CH{-}CH{=}CH{-}COOCH_3}$$

to give symmetrical unsaturated diesters in high yield with complete retention of configuration. In general, a donor ligand or coordinating solvent is necessary to promote smooth coupling reactions. The reaction of bromobenzene with tetrakis(triphenylphosphine)nickel(0) produces the phenylnickel(II) bromide **(CXLII)** in high yield, which couples with the enolate anion of acetophenone in dimethylformamide to give the phenylated ketone **(CXLIII)**

$$\mathrm{C_6H_5Br + NiL_4} \longrightarrow \underset{\textbf{(CXLII)}}{\mathrm{C_6H_5NiBrL_2}} \xrightarrow{{}^{-}\mathrm{CH_2COC_6H_5}} [\mathrm{C_6H_5COCH_2Ni(C_6H_5)L_2}] \longrightarrow \underset{\textbf{(CXLIII)}}{\mathrm{C_6H_5CH_2COC_6H_5}}$$

$$\mathrm{L = (C_6H_5)_3P}$$

in 65% yield. This procedure has been applied to the synthesis of cephalotaxinone **(CXLIV)** (Semmelhack *et al.*, 1973, 1975).

Ni(COD)₂

30%

(CXLIV)

It should be added that the nickel(0)–dipyridyl (dipy) complex is a powerful reagent for converting α,ω-dihaloalkanes to cycloalkanes (Takahashi *et al.*, 1974).

$$X(CH_2)_nX + Ni(COD)_2 \xrightarrow{\text{dipy}} \text{cyclo-}(CH_2)_n$$

$n = 3–6$

E. Coupling via Alkylrhodium Complexes

The reaction of the rhodium(I) complex **(CXLV)** and organolithium or organomagnesium compounds produces the alkylrhodiums **(CXLVI)**. Oxidative addition of an acid chloride to this intermediate, followed by reductive elimination, leads to unsymmetrical ketones in good yields. The alkylrhodium **(CXLVI)** does not react with aldehydes, esters, or nitriles. An additional benefit is that the starting rhodium complex **(CXLV)** is regenerated in reusable form in the last step (Hegedus *et al.*, 1973, 1975a). Reaction of

$$\underset{\textbf{(CXLV)}}{RhCl(CO)L_2} + RM \xrightarrow[\text{THF}, -78^\circ]{} \underset{\textbf{(CXLVI)}}{[RhR(CO)L_2]} + MCl$$

$$[RhR(CO)L_2] \xrightarrow[\text{THF}, -78^\circ]{R'COCl} [RhR(Cl)(R'CO)(CO)L_2]$$

$$[RhR(Cl)(R'CO)(CO)L_2] \longrightarrow \underset{\textbf{(CXLV)}}{RhCl(CO)L_2} + RCOR'$$

$L = (C_6H_5)_3P$

$M = $ Li or MgX

methyl Grignard reagent and the Wilkinson complex leads to the methylrhodium complex **(CXLVII)**, which couples readily with various sp^2-hybridized organic bromides and iodides (Semmelhack and Ryono, 1973).

$$RhClL_3 + CH_3MgX \longrightarrow \underset{\textbf{(CXLVII)}}{CH_3RhL_3}$$

$$CH_3RhL_3 \xrightarrow{C_6H_5I} [Rh(C_6H_5)(CH_3)(I)L_3]$$

$$[Rh(C_6H_5)(CH_3)(I)L_3] \longrightarrow C_6H_5CH_3 + RhIL_3$$

$L = (C_6H_5)_3P$

A new rhodium-based hydroacylation of olefins has been developed recently. Insertion of an olefin into the rhodium–hydride bond of complex **CXLVIII** gives the alkylrhodium complex **CXLIX**, which on interaction with acyl halides produces the ketonic products through the acylalkylrhodium(III) intermediate **(CL)** (Schwartz and Cannon, 1974).

$$\underset{\textbf{(CXLVIII)}}{HRh(CO)L_3} + CH_2{=}CH_2 \longrightarrow \underset{\textbf{(CXLIX)}}{CH_3CH_2Rh(CO)L_n}$$

$$\downarrow RCOCl$$

$$ClRh(CO)L_n + CH_3CH_2COR \longleftarrow \underset{\textbf{(CL)}}{(CH_3CH_2)(RC{=}O)Rh(CO)L_3}$$

$L = (C_6H_5)_3P$

The hydride complex **(CXLVIII)** can be used for monoalkylation of acetylenic derivatives as well. The reaction course consists of cis addition of the rhodium hydride across the carbon–carbon triple bond, oxidative addition of alkyl iodide, and reductive elimination of the organic moiety (Schwartz *et al.*, 1972).

$$\textbf{(CXLVIII)} + RC{\equiv}CR$$

$$\downarrow$$

$$(R)(H)C{=}C(R)Rh(CO)L_n \xrightarrow{CH_3I} (R)(H)C{=}C(R)Rh(I)(CH_3)(CO)L_n \longrightarrow (R)(H)C{=}C(R)CH_3 + IRh(CO)L_n$$

F. Coupling via Alkylzirconium Complexes

The hydrozirconium(IV) complex **(CLI)** is a new, versatile organometallic which is inexpensive, easy to prepare, and, in addition, requires only moderate care in handling (Hart and Schwartz, 1974; Bertelo and Schwartz, 1975). The complex **CLI** undergoes hydrozirconation rapidly with olefins, giving alkylzirconium(IV) species, in which the transition metal moiety is attached to the sterically least hindered and most accessible position of the olefin as a whole. The alkylzirconium **(CLII)** arising from **CLI** and 1- or 4-octene can be acetylated in high yield to give 2-decanone.

$$(\pi\text{-}C_5H_5)_2Zr(Cl)H \; \textbf{(CLI)} + \begin{matrix} CH_2{=}CH(CH_2)_5CH_3 \\ \text{or} \\ CH_3(CH_2)_2CH{=}CH(CH_2)_2CH_3 \end{matrix} \xrightarrow{\text{benzene}} (\pi\text{-}C_5H_5)_2Zr(Cl)CH_2(CH_2)_6CH_3 \; \textbf{(CLII)}$$

$$\textbf{(CLII)} \xrightarrow[80^\circ]{CH_3COCl} CH_3COC_8H_{17}\text{-}n + (\pi\text{-}C_5H_5)_2ZrCl_2$$

G. Reactions of Organic Halides and Transition Metal Carbonyl Complexes

Transition metal carbonyls are commercially available complexes that have great synthetic utility both as stoichiometric reagents and as catalysts in various carbon–carbon bond-forming reactions. Here we deal mainly with organic reactions using metal carbonyls as reducing agents of organic halides. Reviews covering this subject appeared several years ago (Ryang, 1970; Ryang and Tsutsumi, 1971).

1. Neutral Transition Metal Carbonyls

Metal carbonyl complexes readily displace halogen atoms of organic halides, producing reactive species which are useful in organic synthesis. Certain *gem*-dihalides are reduced with iron pentacarbonyl (Coffey, 1961) or dicobalt octacarbonyl (Seyferth and Millar, 1972) or molybdenum hexacarbonyl (Alper and Des Roches, 1976) to form dimeric olefin products. Ordinary monohalides are much less reactive, however.

$$RR'CX_2 + M(CO)_n \longrightarrow RR'C{=}CRR'$$

$X = Cl, Br$

$R, R' = C_6H_5, CN, COOR$, etc.

A remarkable solvent effect is observed in the reaction of α-bromo ketones with nickel carbonyl. In tetrahydrofuran, dimeric 1,4-diketones are produced, while in dimethylformamide furan derivatives are formed via epoxy ketone intermediates (Yoshisato and Tsutsumi, 1968b,c). Reaction of iron penta-

$$\mathrm{RCOCHBrR' + Ni(CO)_4 \longrightarrow \left[R{-}\underset{O}{C}{=}CHR'\right]Ni(CO)_nBr \xrightarrow{THF} RCOCHR'CHR'COR}$$

$$\downarrow \mathrm{DMF}$$

$$\mathrm{RCOCHR'CR{-}CHR'\ (epoxide\ O) \longrightarrow}$$

carbonyl with α-halo ketones also leads to 1,4-diketones and β-epoxy ketones (Alper and Keung, 1972).

Reaction of α,α′-dibromo ketones with iron carbonyls generates a reactive three-carbon unit. The first step of the reaction is the formation of the iron enolate **(CLIII)** via a two-electron reduction of the dibromo ketone or through an oxidative addition of the carbon–bromine bond onto iron(0). The second step is an S_N1 or Fe ion-assisted elimination of bromide ion to form the oxyallyl intermediate **(CLIV)**. Evidence for the production of

$\mathrm{Fe_2(CO)_9}$ $\quad -\mathrm{Br^-}$

OFeL$_n$ **(CLIII)** OFeL$_n$ **(CLIV)**

L = Br, CO, solvent, etc.

these ionic intermediates has been presented (Noyori *et al.*, 1972d). The validity of the enolate formation has been shown by the formation of α-*exo*-deuteriocamphor from α-bromocamphor in the deuterated solvent. The oxyallyl intermediate **(CLV)** generated from α,α′-dibromodibenzyl ketone

$\mathrm{Fe_2(CO)_9}$, DMF–$\mathrm{D_2O}$ → OFeL$_n$ → $\mathrm{D_2O}$

undergoes intramolecular electrophilic aromatic substitution giving the indanone derivative **(CLVI)**. Reaction of **CLVII** with diiron nonacarbonyl gives rise to the bicyclic enone **(CLIX)**. The rearranged product arises from [1*a*,4*s*] sigmatropy of the oxyallyl intermediate **(CLVIII)**. When the dibromo ketone **(CLX)** is subjected to the reaction with iron carbonyl, a neopentyl cation-type rearrangement affording a cyclobutanone is observed.

$Fe_2(CO)_9$

(CLV) (CLVI)

$Fe_2(CO)_9$

(CLVII) (CLVIII) (CLIX)

$Fe_2(CO)_9$

(CLX)

The coupling reaction of polybromo ketones and unsaturated substrates provides a new method for the construction of various cyclic organic structures. The oxyallyliron(II) species can be trapped efficiently by aromatic olefins to give 3-arylcyclopentanones (Noyori *et al.*, 1973d). β-*cis*-Deuteriostyrene, as olefinic substrate, undergoes stereospecific cycloaddition, though

$Fe_2(CO)_9$

the symmetry-restricted reaction is considered to occur in a stepwise fashion. The iron carbonyl-promoted reaction of secondary dibromo ketones and morpholinoenamines followed by simultaneous elimination of morpholine

$Fe_2(CO)_9$

provides a single-flask synthesis of cyclopentenones (Noyori *et al.*, 1972a). The reaction can be extended to the synthesis of azulenes and [4.*n*]spiroalkenones.

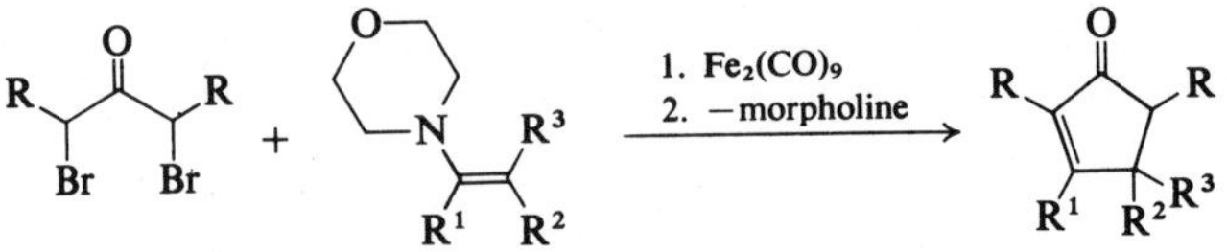

1. $Fe_2(CO)_9$
2. −morpholine

Cycloaddition of the oxyallyl intermediates with open-chain or cyclic 1,3-dienes leads to 4-cycloheptenones **(CLXI)** in fair to good yields (Noyori *et al.*, 1971a, 1973b). The seven-membered ring ketones thus obtained can be

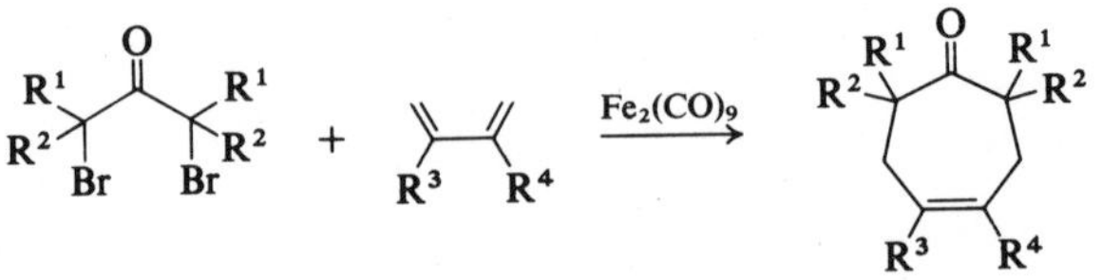

$Fe_2(CO)_9$

(CLXI)

easily converted to troponoid compounds such as tropones **(CLXII)**, γ-tropolones **(CLXIII)**, bridged tropones **(CLXIV)**, 4,5-homotropones **(CLXV)**, and hydroxyhomotropylium ion **(CLXVI)** (Noyori *et al.*, 1971a, 1973a,c). The limits of the new cyclocoupling reaction are defined clearly by the type of

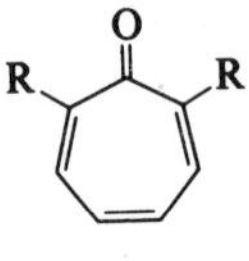

(CLXII) 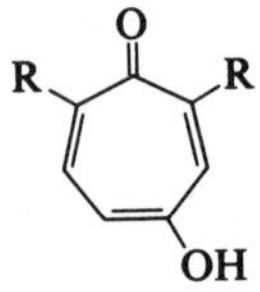 **(CLXIII)** 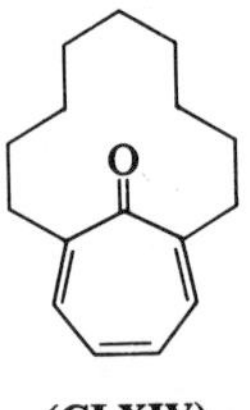 **(CLXIV)** 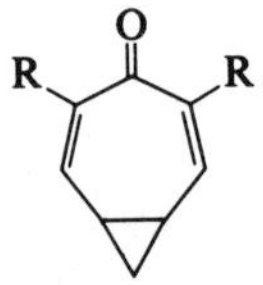 **(CLXV)** 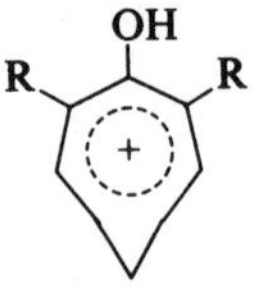 **(CLXVI)**

starting dibromides employed in the reaction. Although secondary and tertiary dibromo ketones react readily with unsaturated substrates to give the cyclocoupled products, the use of dibromides derived from acetone and other methyl ketones has been unsuccessful. Recently, however, a modified method surmounting this problem has appeared. The method consists of the generation of bromooxyallyls from polybromo ketones and the removal of bromine atom(s) from the cyclocoupled products. This modification has opened a new route to various naturally occurring substances. The iron carbonyl-promoted reaction of α,α,α′,α′-tetrabromoacetone and 3-isopropylfuran, followed by Zn–Cu couple reduction of the product, gives the bicyclic ketone **(CLXVII)**, which can be transformed readily to nezukone **(CLXVIII)** (Hayakawa *et al.*, 1975). In a similar fashion, the ketone **CLXIX**, a precursor of β-thujaplicin (hinokitiol) **(CLXX)**, can be obtained by this method. The coupling reaction between the tribromo ketone **(CLXXI)** and furan leads to the bicyclic ketone

(CLXXII), which serves as an intermediate in the synthesis of α-thujaplicin **(CLXXIII)** (Noyori *et al.*, 1975b).

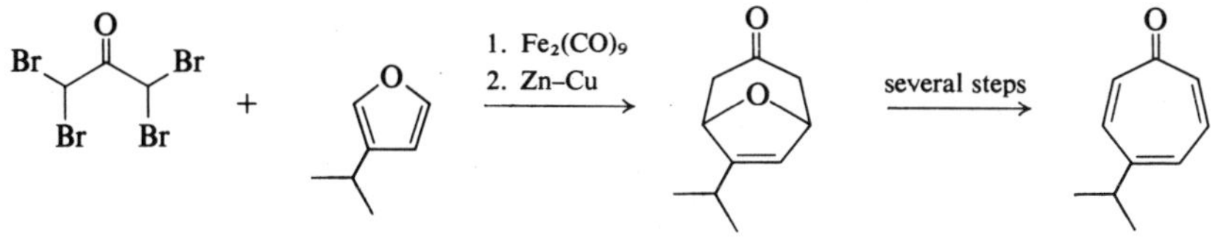

1. $Fe_2(CO)_9$
2. Zn–Cu
several steps

(CLXVII) **(CLXVIII)**

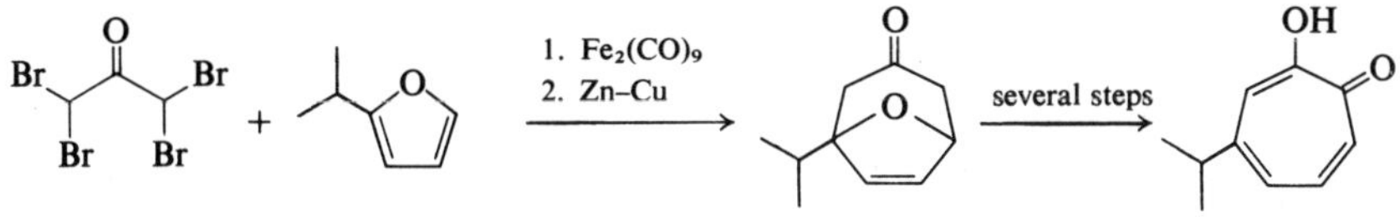

1. $Fe_2(CO)_9$
2. Zn–Cu
several steps

(CLXIX) **(CLXX)**

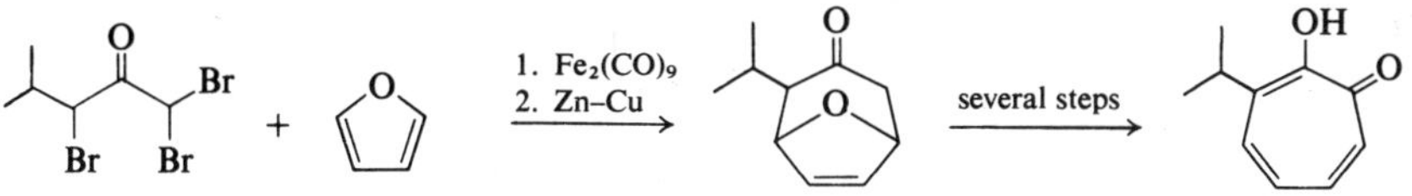

1. $Fe_2(CO)_9$
2. Zn–Cu
several steps

(CLXXI) **(CLXXII)** **(CLXXIII)**

Cyclopentadiene is also a good receptor of the oxyallyl species and the coupling reaction with the tribromide **(CLXXIV)** and subsequent Zn–Cu couple reduction gives the bicyclo[3.2.1]octene **(CLXXV)** in 83% yield. The latter is readily converted to carbocamphenilone **(CLXXVI)**, a terpenic α-diketone (Noyori *et al.*, 1975c).

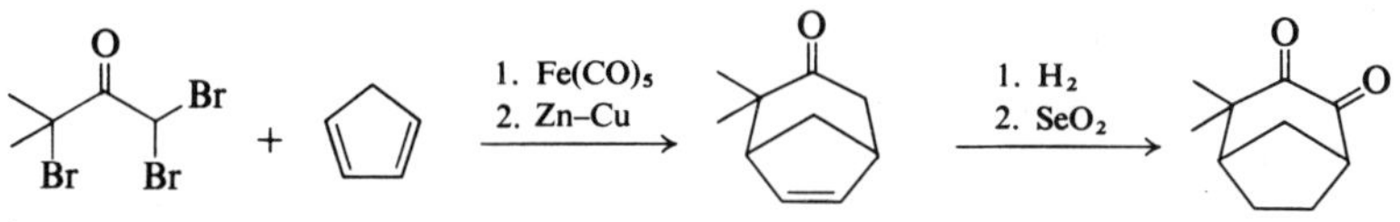

1. $Fe(CO)_5$
2. Zn–Cu
1. H_2
2. SeO_2

(CLXXIV) **(CLXXV)** **(CLXXVI)**

This cyclocoupling reaction has also been utilized in a new, general synthesis of tropane alkaloids (Noyori *et al.*, 1974b,c). The bicyclic ketone **(CLXXVII)** resulting from reaction of tetrabromoacetone, *N*-carbomethoxypyrrole, and iron carbonyl can be reduced with diisobutylaluminum hydride stereoselectively to the key intermediate **(CLXXVIII)**. This alcohol is convertible to all naturally occurring tropane alkaloids.

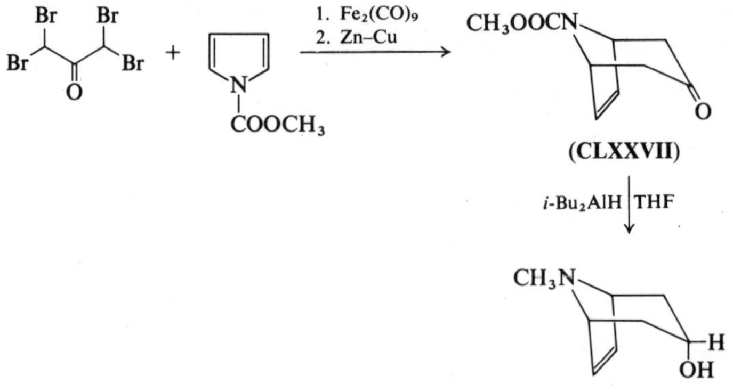

(CLXXVII)

i-Bu_2AlH | THF

(CLXXVIII)

The reduction of α,α'-dibromo ketones in the presence of *N,N*-dialkylated carboxamides generally leads to the formation of 3(2*H*)-furanones of type **CLXXIX**. *N,N*-Dimethylformamide, *N,N*-dimethylacetamide, *N,N*-dimethylbenzamide, and *N*-methylpyrrolidone can be used as carboxamide. A muscarine derivative **(CLXXXI)** has been prepared from **CLXXX** in several steps (Noyori *et al.*, 1973e).

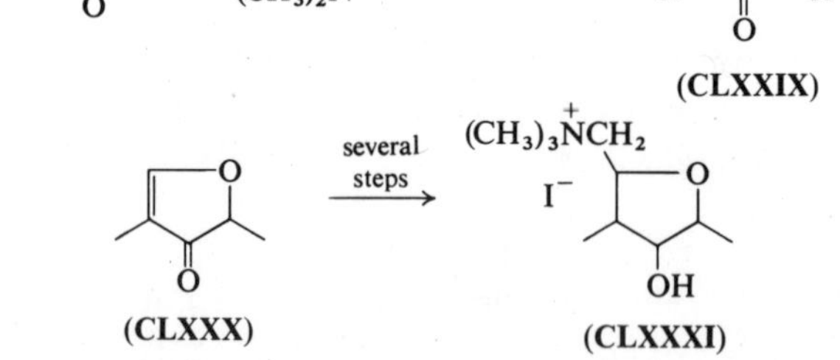

(CLXXIX)

(CLXXX)

(CLXXXI)

The reductive carbonylation of iodobenzene with triiron dodecacarbonyl in toluene gives benzophenone (Rhee *et al.*, 1967). Reaction of pentafluorophenyl iodide and nickel carbonyl in dimethylformamide or toluene gives perfluorobenzophenone and perfluorobiphenyl (Beckert and Lowe, 1967). Iodobenzene reacts with nickel carbonyl to give benzoic acid esters (in alcohols) or benzil (in tetrahydrofuran) (Bauld, 1963), possibly via a benzoylnickel intermediate. Reaction of this intermediate with enamines affords β-diketones in high yields (Seki *et al.*, 1975).

$$ArI + Ni(CO)_4 \longrightarrow [ArCONiI(CO)_n] \longrightarrow ArCOOR \text{ or } ArCOCOAr$$

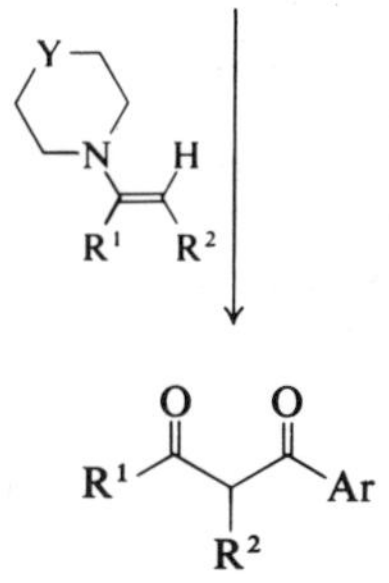

2. Anionic Metal Carbonyl Reagents

Certain transition metal carbonyls are converted to more electropositive anionic species, transition metal carbonylates, under the influence of anions. This tendency allows a base-catalyzed carboxylation of vinyl and aryl halides by nickel carbonyl in protic media (Corey and Hegedus, 1969a). Alkyl halides are much less reactive than halides attached to sp^2-hybridized carbon.

$$RX + Ni(CO)_4 \xrightarrow{CH_3O^- - CH_3OH} RCOOCH_3$$

However, a mixture of nickel carbonyl and potassium *tert*-butoxide in *tert*-butyl alcohol is a very powerful agent and can carboxylate even alkyl iodides. The carboxylation of *cis*- and *trans*-β-bromostyrene proceeds in a highly stereospecific manner with retention of double-bond configuration. The

$$RO^- + Ni(CO)_4 \longrightarrow [ROCONi(CO)_x]^-$$
$$[ROCONi(CO)_x]^- + R'{-}X \xrightarrow{-X^-} \{[ROCONi(CO)_x]^+ + R'^-\} \longrightarrow$$
$$ROCONiR'(CO)_x \xrightarrow{\text{CO or base}} R'COOR + Ni_z(CO)_y$$

reaction probably proceeds via an electron-transfer mechanism. This procedure has been applied to the synthesis of α-santalol **(CLXXXII)** (Corey *et al.*, 1970b).

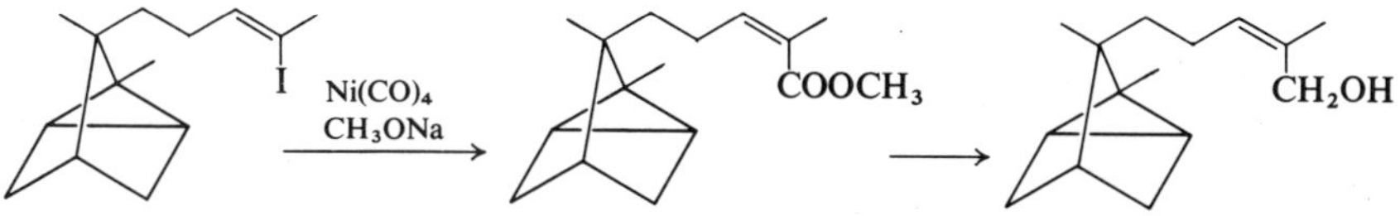

(CLXXXII)

The reaction of 1-iodo-5-decyne, nickel carbonyl, and potassium *tert*-butoxide leads to a mixture of three ester products **(CLXXXIII)–(CLXXXV)** (Crandall and Michaely, 1973).

$$C_4H_9C{\equiv}C(CH_2)_4I + Ni(CO)_4 + (CH_3)_3COK \longrightarrow [C_4H_9C{\equiv}C(CH_2)_4Ni(CO)_x]$$

$$\longrightarrow C_4H_9C{\equiv}C(CH_2)_4COOC(CH_3)_3 \quad \textbf{(CLXXXIII)}\ (30\%)$$

$$[C_4H_9C{\equiv}C(CH_2)_4Ni(CO)_x] \longrightarrow \left[(C_4H_9)(Ni(CO)_x)C{=}C_5H_8\right] \longrightarrow (C_4H_9)(COOC(CH_3)_3)C{=}C_5H_8\ \textbf{(CLXXXIV)}\ (30\%) + C_4H_9CH(COOC(CH_3)_3)\text{-}C_5H_7\ \textbf{(CLXXXV)}\ (17\%)$$

Addition of lithium dimethylamide to nickel carbonyl produces an air-sensitive anionic carbamoylnickel complex which is a powerful nucleophilic reagent toward a variety of organic halides (Fukuoka *et al.*, 1971).

$$LiN(CH_3)_2 + Ni(CO)_4 \longrightarrow Li[(CH_3)_2NCONi(CO)_3]$$

$$Li[(CH_3)_2NCONi(CO)_3] \xrightarrow{RX} RCON(CH_3)_2 \qquad Li[(CH_3)_2NCONi(CO)_3] \xrightarrow{R'COX} R'COCON(CH_3)_2$$

R = vinyl, aryl, allyl, benzyl

R′ = alkyl, phenyl

The nickel carbonylate **(CLXXXVI)** formed from nickel carbonyl and organolithium reagents reacts with benzyl bromide or acid chloride to give the adducts **CLXXXVII** and **CLXXXVIII**, respectively (Ryang, 1970).

$$\underset{\textbf{(CLXXXVI)}}{Li[RCONi(CO)_3]} \xrightarrow{C_6H_5CH_2Br} \underset{\textbf{(CLXXXVII)}}{RCOC(OH)(CH_2C_6H_5)R}$$

$$\underset{\textbf{(CLXXXVI)}}{Li[RCONi(CO)_3]} \xrightarrow{R'COCl} \underset{\textbf{(CLXXXVIII)}}{R'COOCR{=}CROCOR'}$$

Reaction of benzyl chloride and disodium iron tetracarbonyl **(CLXXXIX)** in tetrahydrofuran gives dibenzyl ketone. A dibenzyliron species may be the intermediate (Yoshisato and Tsutsumi, 1968a).

$$C_6H_5CH_2Cl + Na_2Fe(CO)_4 \longrightarrow (C_6H_5CH_2)_2Fe(CO)_n \longrightarrow C_6H_5CH_2COCH_2C_6H_5$$

(CLXXXIX)

Collman has demonstrated that the ferrate **(CLXXXIX)** is a highly versatile reagent for organic synthesis. The anionic complex undergoes nucleophilic displacement with aliphatic halides and tosylates to give alkyliron intermediates **(CXC)**, which, on reaction with certain donor ligands such as triphenylphosphine or carbon monoxide, are converted to the acyl complexes **(CXCI)**. Protonolysis of the latter gives aldehydes (Cooke, 1970). Reaction of **CLXXXIX** and acid halides leads to the acyl complex **(CXCI)** directly (Collman *et al.*, 1972a). The acyl- and alkyltetracarbonylferrates have been

$$Na_2Fe(CO)_4 + RX \xrightarrow{THF} Na[RFe(CO)_4] \;\textbf{(CXC)}$$

(CLXXXIX)

$$\textbf{(CXC)} \xrightarrow{L} Na[RCOFe(CO)_3L] \;\textbf{(CXCI)}$$

$$\textbf{(CLXXXIX)} + RCOCl \longrightarrow Na[RCOFe(CO)_3L] \;\textbf{(CXCI)}$$

$$\textbf{(CXCI)} \xrightarrow{H^+} RCHO; \quad \xrightarrow{R'X} RCOR'; \quad \xrightarrow{YH} RCOY$$

$YH = H_2O, R''OH, R''NH_2$

isolated and characterized. The acyl complex is trigonal bipyramid with the bulky RCO group occupying an apical position (Siegl and Collman, 1972). The alkyl–acyl migratory insertion is greatly assisted by alkali metal ions. Polar solvents such as hexamethylphosphoric amide or *N*-methylpyrrolidone show a marked inhibitory effect (Collman *et al.*, 1972b). The coupling reaction of **CXCI** with primary alkyl halides and tosylates gives unsymmetrical

$$\mathrm{OC{-}Fe(R)(CO)_2(CO\cdots Na^+)} \xrightarrow{L} \mathrm{L{-}Fe(CO)_3{-}C(=O\cdots Na^+){-}R}$$

ketones in good yields. (S)-(+)-2-Octyl tosylate is converted to (R)-(−)-3-methyl-**2**-nonanone stereospecifically, indicating that the initial step, **(CLXXXIX)** → **(CXC)**, proceeds with inversion at the chiral center. Lithium acyltetracarbonylferrates formed from organolithium reagents and iron carbonyl behave in a similar manner (Sawa *et al.*, 1970). Various hemifluorinated ketones are synthesized via this single-flask procedure (Collman and Hoffman, 1973). A variety of functional groups, such as esters or cyanides, are tolerable under these reaction conditions. Furthermore, with the aid of appropriate oxidants (oxygen, NaOCl, or halogen) and protic agents, the acyliron intermediates **(CXCI)** can be transformed selectively to carboxylic acids, esters, and amides in high yields (Collman *et al.*, 1973).

Tetracarbonylferrate reacts with phthaloyl dichloride to give dimeric biphthalidene in 23% yield (Mitsudo *et al.*, 1972). The ferrate complex is a versatile reagent for the synthesis of aldehydes (Watanabe *et al.*, 1971). Carboxylic acid anhydrides and carboxylic alkylcarbonic anhydrides can be reduced to the corresponding aldehydes (Watanabe *et al.*, 1973, 1975).

$$\mathrm{R(CO)_2O} + \mathrm{Na_2Fe(CO)_4} \longrightarrow \mathrm{R(CO{-}\bar{F}e(CO)_4)(COO^-)} \xrightarrow{H^+} \mathrm{R(CHO)(COOH)}$$

$$\mathrm{RCOOCOOC_2H_5} + \mathrm{Na_2Fe(CO)_4} \longrightarrow \mathrm{RCHO}$$

Organic halides or tosylates containing a carbon–carbon double bond at an appropriate position cyclize with the aid of the ferrate complex to give cyclic ketones (Mérour *et al.*, 1973).

$$\mathrm{Br(CH_2)_3CH{=}CH_2} + \mathrm{Fe(CO)_4{}^{2-}} \longrightarrow \mathrm{[(CO)_4Fe(CH_2)_3CH{=}CH_2]^-} \longrightarrow$$

$$\mathrm{[(CO)_4FeCO(CH_2)_3CH{=}CH_2]^-} \longrightarrow \left[\text{cyclohexanone-}\mathrm{Fe(CO)_3}\right]^- \longrightarrow \text{cyclohexanone}$$

(40%)

$$\text{(2-methylenecyclohexyl)ethyl-OTs} \xrightarrow{\mathrm{Fe(CO)_4{}^{2-}}} \text{2-decalone}$$

(65%)

Anionic hydridoiron species undergo selective addition to α,β-unsaturated carbonyl compounds in methanol to give reduction products (Noyori *et al.*, 1972c). Under an atmosphere of carbon monoxide, carbonylation takes place. Methylmalonate is selectively obtained from acrylate by treatment of the reaction mixture with alcoholic iodine (Masada *et al.*, 1970). The organo-iron complexes are readily acylated by treatment with alkyl iodides in an aprotic solvent (Mitsudo *et al.*, 1974).

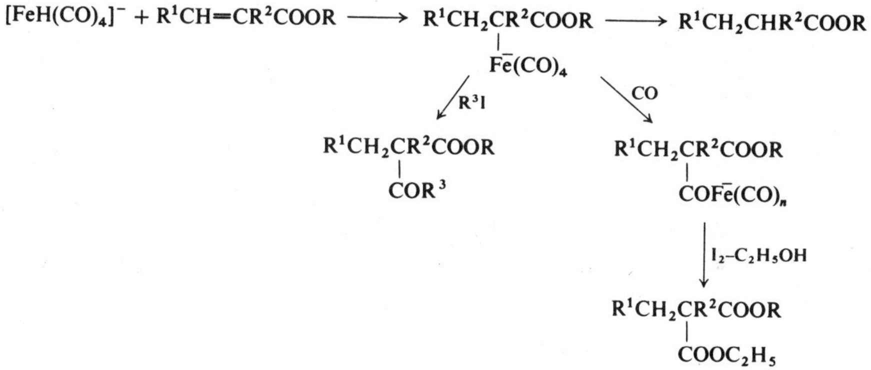

$$[FeH(CO)_4]^- + R^1CH{=}CR^2COOR \longrightarrow R^1CH_2CR^2COOR\,(Fe^-(CO)_4) \longrightarrow R^1CH_2CHR^2COOR$$

$$R^1CH_2CR^2COOR\,(Fe^-(CO)_4) \xrightarrow{R^3I} R^1CH_2CR^2COOR\,(COR^3)$$

$$R^1CH_2CR^2COOR\,(Fe^-(CO)_4) \xrightarrow{CO} R^1CH_2CR^2COOR\,(COFe^-(CO)_n) \xrightarrow{I_2-C_2H_5OH} R^1CH_2CR^2COOR\,(COOC_2H_5)$$

Thus, transition metal acyl species play an important role in various carbon–carbon bond-forming reactions. Organic synthesis via these intermediates has been reviewed by Heck (1968a).

A new ketone synthesis has been achieved by using the reaction of organomercuric halides and dicobalt octacarbonyl in tetrahydrofuran. This method is particularly useful for the preparation of symmetrical diaryl ketones (Seyferth and Spohn, 1968, 1969). The following mechanism has been postulated (Scheme 10).

$$Co_2(CO)_8 + THF \longrightarrow (THF)Co(CO)_4^+ + Co(CO)_4^-$$

$$RHgX + Co(CO)_4^- \longrightarrow RHgCo(CO)_4 + X^-$$

$$RHgCo(CO)_4 + (THF)Co(CO)_4^+ + Co(CO)_4^- \longrightarrow RCo(CO)_4 + Hg[Co(CO)_4]_2 + THF$$

$$2RHgCo(CO)_4 \longrightarrow R_2Hg + Hg[Co(CO)_4]_2$$

$$R_2Hg + (THF)Co(CO)_4^+ + Co(CO)_4^- \longrightarrow RCo(CO)_4 + RHgCo(CO)_4 + THF$$

$$RCo(CO)_4 \longrightarrow RCOCo(CO)_3$$

$$RCOCo(CO)_3 + RCo(CO)_4 \longrightarrow RCOR + Co_2(CO)_7$$

Scheme 10

Diaryl or dialkyl ketones are obtained in good yields by the reaction of nickel carbonyl with organomercuric halides in dipolar solvents (Hirota *et al.*, 1971).

III. NUCLEOPHILIC REACTIONS OF ORGANOTRANSITION METAL COMPLEXES WITH CARBONYL AND EPOXY COMPOUNDS

A. Reaction with Aldehydes and Ketones

Bis(π-allyl)nickel(0) complexes are nucleophilic and behave like Grignard reagents. Various homoallylic alcohols can be prepared from these complexes and carbonyl compounds. For instance, reaction of bis(π-allyl)nickel(0) (**X**) with benzaldehyde, followed by hydrolysis, affords the adduct **CXCII** (Heimbach *et al.*, 1970). The complex **XIII** reacts with acetaldehyde in a similar fashion. Coupling of the latter with acetyl chloride or allyl bromide gives products of increased chain length (Baker *et al.*, 1972b). The Grignard type reactions are also possible with active species formed from allylic halides

(X) + C_6H_5CHO ⟶ (CXCII)

and transition metals such as cobalt or nickel in water. The reaction may involve a π-allylic derivative which does not contain halogen (Agnès *et al.*, 1973).

π-Allylnickel(I) halides react with certain aldehydes, ketones, or epoxides to give the corresponding alcoholic adducts (Corey and Semmelhack, 1967; Hegedus and Stiverson, 1974). α-Diketones are the most reactive substrates. With conjugated ketones exclusive 1,2-attack results. The reaction of α-(2-carbethoxyallyl)nickel bromide with ketones and aldehydes leads to α-methylene-γ-butyrolactones (Hegedus *et al.*, 1975b).

(π-allyl)NiX dimer + RCOR′ $\xrightarrow{DMF}$ HO–C(R)(R′)–allyl

C_2H_5OOC–(π-allyl)NiBr dimer + RCOR′ $\xrightarrow{DMF}$ α-methylene-γ-butyrolactone (R, R′)

In general, organocopper and cuprate compounds are much less reactive toward the carbonyl group than organolithium or Grignard reagents (Kharasch *et al.*, 1941; Posner *et al.*, 1972). It should be noted, however, that the organocopper species formed from cyclopentadiene and a Cu_2O–isocyanide complex readily condenses with acetone or benzaldehyde (Saegusa *et al.*, 1971b). A novel type of cyclization, **(CXCIII)** → **(CXCIV)**, has been achieved with the aid of an organocuprate, the exact mechanism remaining uncertain (Corey and Kuwajima, 1970; Corey *et al.*, 1970a).

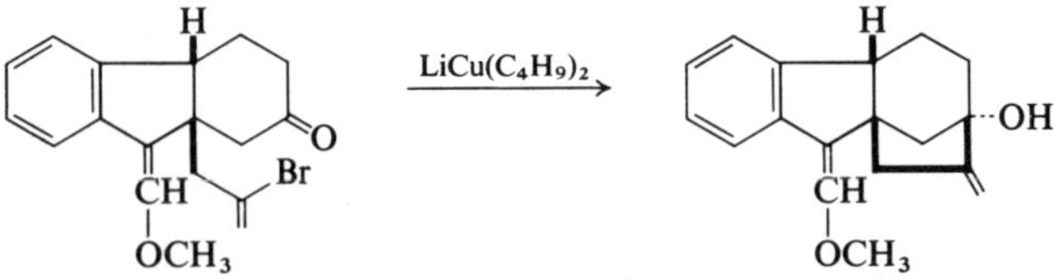

$LiCu(C_4H_9)_2$

(CXCIII) **(CXCIV)**

B. Reaction with Oxiranes

Ring opening of oxiranes with organocuprates is much more selective compared with that using organolithium or Grignard reagents (Herr *et al.*, 1970; Herr and Johnson, 1970; Anderson, 1970). For example, reaction of 1,2-epoxybutane and lithium dimethylcuprate affords 3-pentanol in 88% yield. α-Substituted malic acids can be obtained by this displacement reaction (Hill and Spencer, 1974). The ring-opening reaction of **CXCV** with lithium

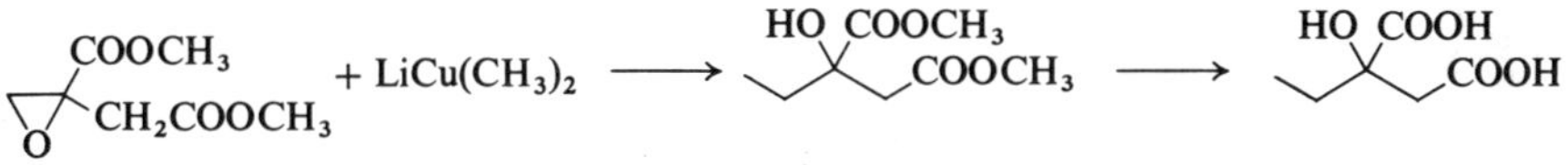

$COOCH_3$, CH_2COOCH_3 + $LiCu(CH_3)_2$ ⟶ HO $COOCH_3$, $COOCH_3$ ⟶ HO COOH, COOH

diallylcuprate has been used for the total synthesis of prostaglandin (Fried *et al.*, 1972). Furthermore, an elegant synthesis of *d,l-*C_{18} *Cecropia* juvenile

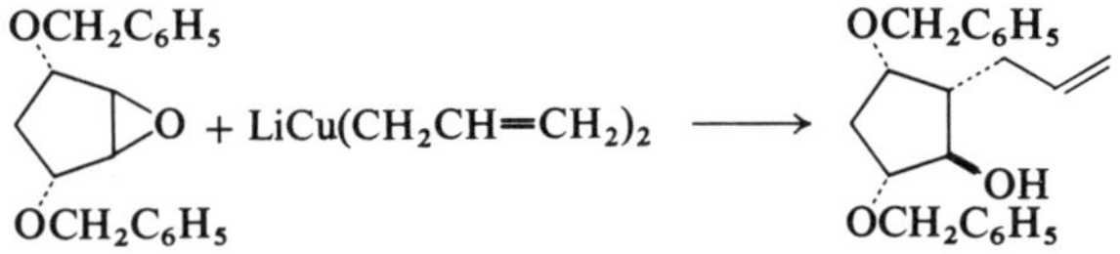

$OCH_2C_6H_5$, O, $OCH_2C_6H_5$ + $LiCu(CH_2CH{=}CH_2)_2$ ⟶ $OCH_2C_6H_5$, OH, $OCH_2C_6H_5$

(CXCV)

hormone **(LIX)** from farnesol has been achieved recently using lithium dimethylcuprate (Tanaka *et al.*, 1974). Very recently this method has been applied to the synthesis of methymycin, a macrolide antibiotic (Masamune *et al.*, 1975).

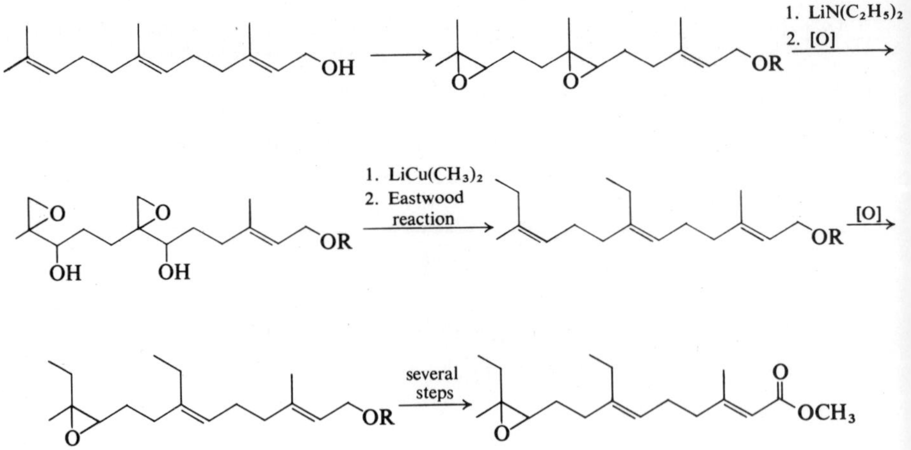
OH
O
O
OR
1. LiN(C₂H₅)₂
2. [O]
O
O
OH
OH
OR
1. LiCu(CH₃)₂
2. Eastwood reaction
OR
[O]
O
OR
several steps
O
O
OCH₃

(LIX)

A conjugate displacement has been observed with vinyloxiranes. Epoxides of 1,3- and 1,4-cyclohexanediene, **(CXCVI)** and **(CXCIX)**, behave in a somewhat different manner (Staroscik and Rickborn, 1971; Wieland and Johnson, 1971). The oxirane **(CXCVI)** undergoes both direct and conjugate addition to give **CXCVII** and **CXCVIII** with complete trans stereospecificity, whereas **CXCIX** gives the trans alcohol **(CC)** as only product.

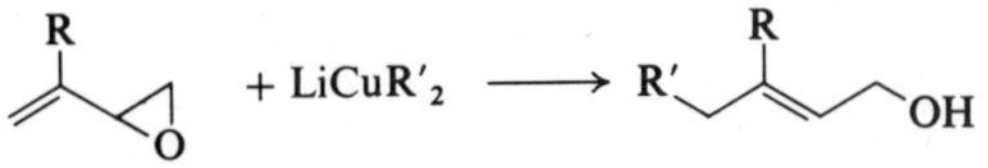
R
O
+ LiCuR'₂
R
R'
OH

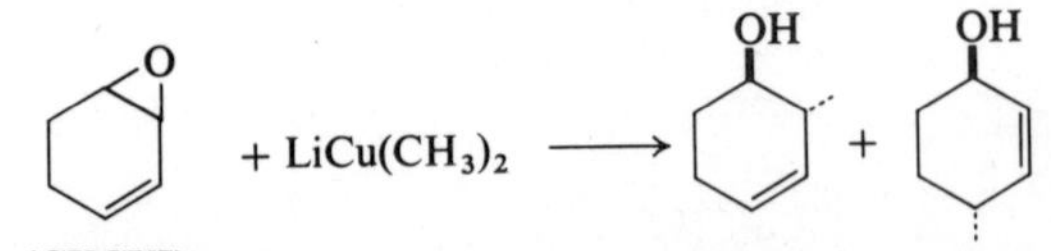
O
+ LiCu(CH₃)₂
OH
+
OH

(CXCVI) **(CXCVII)** **(CXCVIII)**

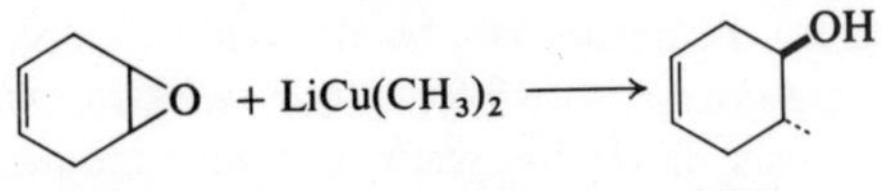
O
+ LiCu(CH₃)₂
OH

(CXCIX) **(CC)**

IV. ELECTROPHILIC REACTIONS OF ORGANOPALLADIUM COMPLEXES

Various nucleophilic reagents react with the π-allylpalladium species (Maitlis, 1971). Certain palladium and platinum catalysts promote coupling between 1,3-dienes and active methylene compounds such as β-diketones, β-keto esters, malonates, and esters having an electron-withdrawing group at the alpha position (Hata *et al.*, 1969, 1971). Nitroalkanes may also be used (Mitsuyasu *et al.*, 1971). The palladium-catalyzed reaction of isoprene and ethyl acetoacetate proceeds in high yield to give products of types **CCI–CCIII**. Here isoprene moieties are combined selectively in a tail-to-tail manner. In

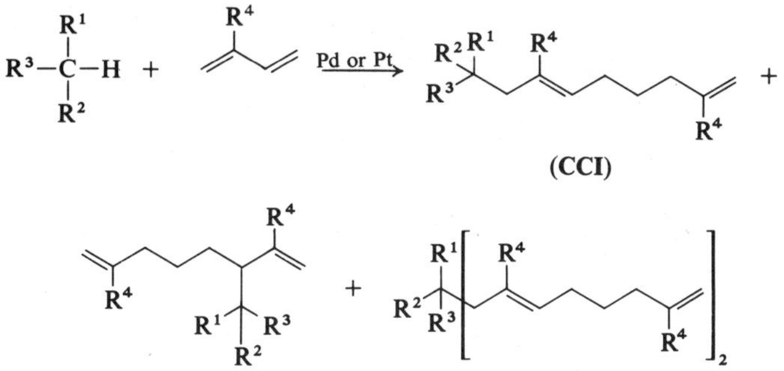

Pd or Pt

(CCI)

(CCII)

(CCIII)

$R^3 = H$

contrast, 1,3-pentadiene affords the corresponding head-to-tail adducts. Intermediacy of a four-coordinated π-allylpalladium complex of type **CCIV** (L = phosphine ligand) has been suggested. In fact, interaction of complex

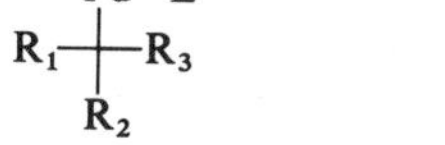

(CCIV)

CCV with carbon monoxide causes a ligand–ligand coupling reaction to produce the diketone **(CCVI)** (Takahashi *et al.*, 1967). Nickel(II) salts are also effective for the coupling between 1,3-dienes and active methylene

(CCV) (CCVI)

compounds, but the selectivity is moderate (Baker *et al.*, 1972a, 1974b).

When $PdBr_2[(C_6H_5)_2PCH_2CH_2P(C_6H_5)_2]$ is used as the catalyst, 1,3-diene–active methylene compound 1:1 adducts are produced selectively (Takahashi *et al.*, 1971, 1972). Primary and secondary amines may be used as

$R^1CH_2R^2$ + [butadiene] $\xrightarrow{Pd(II)}$ products

$R^1, R^2 = COCH_3, CO_2C_2H_5$

active methylene compounds. Allylic alcohols, esters, and amines in place of 1,3-dienes can be used as a source of the allylic moiety as well. They couple effectively with active methylene compounds in the presence of Pd catalysts (Atkins *et al.*, 1970).

The allylic position of olefins can be alkylated via activation through π-allylpalladium complexes. Thus, reaction of the complexes **CCVII** with stable carbanions C^-HZ_2 ($Z = COOR, SOCH_3, SO_2CH_3$, etc.) produces the desired alkylated derivatives **(CCVIII)** in good yield. The choice of

(CCVII) (CCVIII)

akylating species affects both the regioselectivity and stereoselectivity; the reaction of complex **CCIX** and the anion **CCX** gives selectively a single product **(CCXI)** in 80% yield (Trost and Fullerton, 1973). In order to determine whether bonding of the attacking reagent occurs initially at the metal with subsequent migration to carbon (Scheme 11, path *a*) or directly at carbon

(CCIX) (CCX) (CCXI)

(path *b*), the alkylation of 2-ethylidenenorpinane has been examined. The reaction of the palladium complex **(CCXII)** with the anion of dimethyl

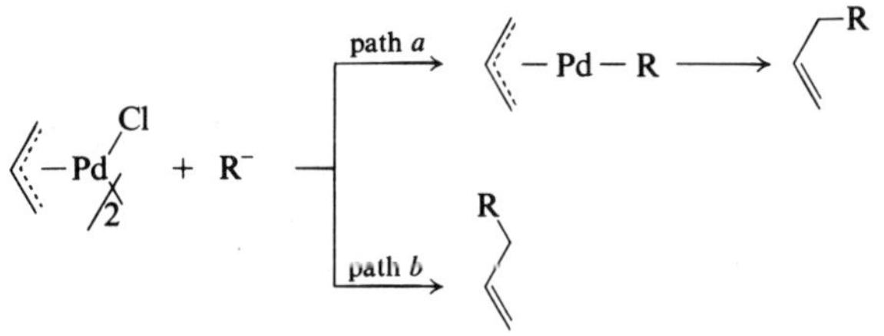
path a
Pd
R
Cl
Pd
2
+ R⁻
R
path b

Scheme 11

malonate in the presence of a bidentate phosphine ligand gives rise to a single diastereomer **(CCXIII)** in 69% yield, establishing that the alkylation occurs on the face of the π-allyl unit opposite to that of the palladium (Trost and Weber, 1975).

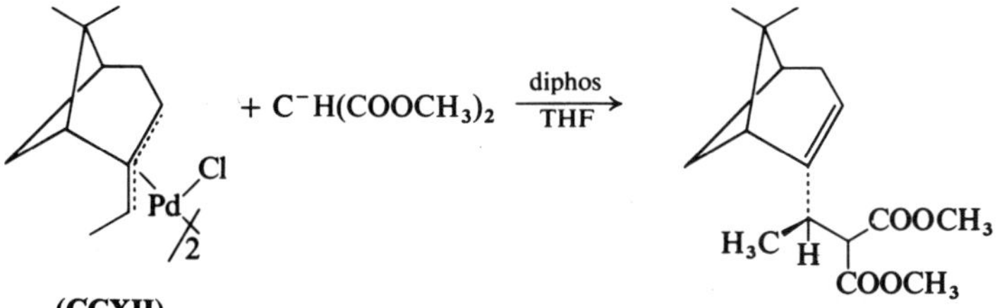
Cl
Pd
2
+ C⁻H(COOCH₃)₂
diphos
THF
COOCH₃
H₃C
H
COOCH₃

(CCXII) **(CCXIII)**

A remarkable regioselectivity and stereoselectivity has been observed in the allylic alkylation of the π-allyl complexes derived from methylenecyclohexane derivatives. Treatment of **CCXIV** (R_1 = H or *tert*-C_4H_9) with anions generated from methyl methylsulfonylacetate, methyl phenylsulfonylacetate, methyl phenylthioacetate, and methyl malonate in the presence of hexamethylphosphorus triamide leads to substitution at the primary carbon atom giving **CCXV**. On the other hand, utilization of a bulky activating ligand such as tri-*o*-tolylphosphine leads to predominate reaction at the secondary carbon atom, yielding the regioisomer **(CCXVI)**. Results of the reaction of CCXIV (R_1 = *tert*-C_4H_9), producing the adduct **(CCXVII)** as the major product, demonstrates that the allylic alkylation shows a ratio for axial vs. equatorial bond formation of 18 : 1 (Trost and Strege, 1975).

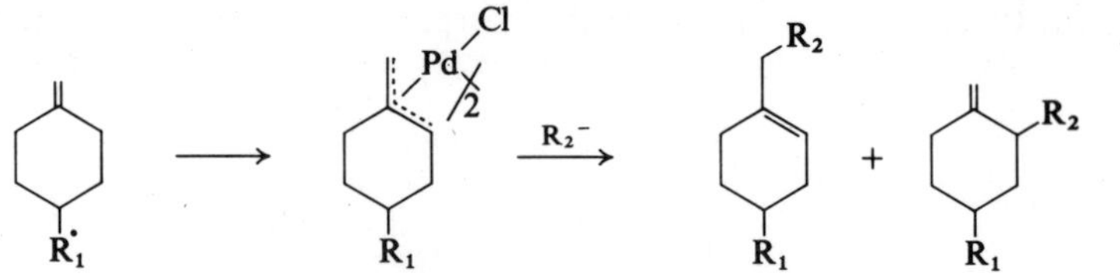

(CCXIV) (CCXV) (CCXVI)

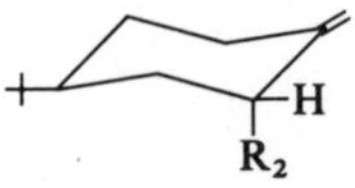

(CCXVII)

Selective alkylation of the methyl group of geranylacetone **(CCXVIII)** without carbonyl protection is possible via the intermediacy of the π-allylpalladium complexes **CCXIX** and **CCXX** (Scheme 12) (Trost *et al.*, 1973).

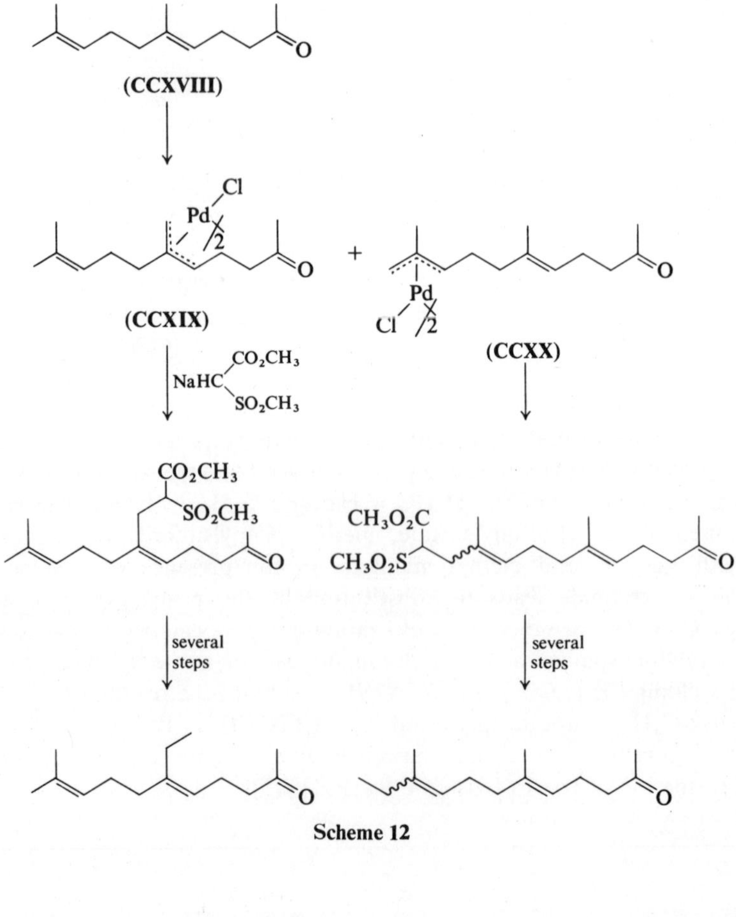

(CCXVIII)

(CCXIX) (CCXX)

$NaHC(CO_2CH_3)(SO_2CH_3)$

several steps

several steps

Scheme 12

When the reaction of **CCXXI** and diethyl sodiomalonate is carried out with an added chiral ligand such as (+)-2,3-*O*-isopropylidene-2,3-dihydroxy-1,4-bis(diphenylphosphino)butane, (+)-*o*-anisylcyclohexylmethylphosphine, (−)-dimethylisopropylphosphine, or (−)-spartein, the coupling product **(CCXXII)** is formed in up to 24% optical yield (Trost and Dietsche, 1973).

$$\text{(CCXXI)} + NaCH(COOC_2H_5)_2 \xrightarrow{L^*} \text{(CCXXII)}$$

(CCXXI) **(CCXXII)**

Reaction of π-allylpalladium complexes **(CCXXIII)** and the anions **(CCXXIV)** in the presence of triphenylphosphine in coordinating solvents provides the 1,3-dienes **(CCXXV)**; the coupling proceeds through alkylative elimination. The reaction occurs at the less substituted methylene carbon of the allylic moiety (Trost *et al.*, 1974).

$$\text{(CCXXIII)} + C_6H_5S(O)CH{=}C(OM)CH_3 \longrightarrow R\text{-CH=CH-CH=CH-}COOCH_3 \text{ (CCXXV)}$$

(CCXXIII) **(CCXXIV)** **(CCXXV)**

M = Na or Li

An allylic fragment can be introduced at the α-position of the ketones through the reaction of π-allylpalladium intermediates and enamines (Onoue *et al.*, 1973).

$$CH_2{=}CHCH_2OC_6H_5 + \text{1-piperidinocyclohexene} \xrightarrow[2.\ H_3O^+]{1.\ Pd(OAc)_2,\ P(C_6H_5)_3} \text{2-allylcyclohexanone } (70\%)$$

σ-Bonded organopalladium chloride dimers, unlike the π-allyl analogs, suffer nucleophilic attack at the central metal giving bisorganopalladium products. Recently a useful method for introducing an alkyl group at the ortho position of aromatic aldehydes has been found. The procedure consists of formation of the σ-bonded palladium complex **(CCXXVI)** from the Schiff base and palladium chloride, treatment with an alkyllithium, and acid hydrolysis of the coupling product **(CCXXVII)** (Murahashi *et al.*, 1974; Yamamura *et al.*, 1974).

$$\text{(CCXXVI)} \xrightarrow{\text{RLi}} \left[\text{Pd(R)(L) complex} \right] \xrightarrow{-\text{Pd(0)}} \text{(CCXXVII)} \xrightarrow{H_3O^+} o\text{-R}C_6H_4CHO$$

(CCXXVI) (CCXXVII)

Vinyl halides react with alkyllithiums with the aid of a palladium(0) complex to give olefins stereospecifically in good yields. When Grignard reagents are employed instead of alkyllithium compounds, these reactions can be carried out catalytically with the palladium complex (Yamamura *et al.*, 1975). *ortho*-Alkylated thiobenzophenones can be similarly prepared from *S*-donor ligand *ortho*-metalated complexes (H. Alper and J. Kamenof, unpublished results).

$$R^1R^2C{=}CR^3X + Pd[P(C_6H_5)_3]_4 \xrightarrow[\text{benzene}]{} R^1R^2C{=}CR^3PdX[P(C_6H_5)_3]_n \xrightarrow[\text{ether}]{\text{RLi}} R^1R^2C{=}CR^3R$$

V. ADDITION REACTIONS OF ORGANOMETALLIC COMPLEXES ACROSS UNSATURATED CARBON–CARBON BONDS

A. Electrophilic Additions to Olefinic Substrates

Many oligomerization and polymerization reactions of olefins and 1,3-dienes are considered to proceed through addition of σ-bonded organometallic intermediates across carbon–carbon double bonds. However, little is known about isolation or definite characterization of σ-bonded organometallic adducts formed from transition metal alkyls and simple olefins.

1. Vinyl Substitution Reactions

Addition of a σ-bonded organometallic compound across a carbon–carbon double bond, followed by elimination of a metal hydride species, results in net vinylic substitution. The arylpalladium complex **(CCXXVIII)**

$$R{-}M + HC{=}CH \longrightarrow {-}\underset{R}{\overset{H}{C}}{-}\underset{M}{\overset{H}{C}}{-} \longrightarrow RC{=}CH + H{-}M$$

undergoes the substitution with styrene to give **CCXXIX** (Tsuji, 1969b). Organopalladium species **CCXXX** produced from arylmercury compounds and palladium(II) salts or aryl halides and palladium metal react with

$$\left[\text{(2-}(CH_3)_2NCH_2C_6H_4)Pd(Cl)\right]_{2} + 2CH_2{=}CH{-}C_6H_5 \longrightarrow 2\ \text{2-}(CH_3)_2NCH_2C_6H_4{-}CH{=}CH{-}C_6H_5$$

(CCXXVIII) **(CCXXIX)**

olefinic substrates in a similar manner. This subject has been reviewed by Heck (1974).

$$ArHgOAc + Pd(OAc)_2 \xrightarrow[-Hg(OAc)_2]{L} ArPdL_2OAc \xrightarrow[-PdHL_2OAc]{CH_2{=}CHR} ArCH{=}CHR$$

(CCXXX)

Bis(triphenylphosphine)phenylnickel bromide undergoes a similar reaction with styrene. The reaction with methyl acrylate (probably nucleophilic) is more facile; in methanol, hydrocinnamate is obtained in 81% yield (Otsuka *et al.*, 1973).

$$(C_6H_5)_3P(Br)Ni(C_6H_5)P(C_6H_5)_3 \xrightarrow[80^\circ]{C_6H_5CH{=}CH_2} \underset{(85\%)}{C_6H_5CH{=}CHC_6H_5} + \underset{(4\%)}{C_6H_5CH_2CH_2C_6H_5}$$

$$(C_6H_5)_3P(Br)Ni(C_6H_5)P(C_6H_5)_3 \xrightarrow[20^\circ]{CH_2{=}CHCOOCH_3} \underset{(85\%)}{C_6H_5CH{=}CHCOOCH_3}$$

The sulfonylmethylpalladium **(CCXXXI)** adds across the double bond of simple alkenes. The organopalladium intermediate **(CCXXXII)** decomposes to give a formal vinylic substitution product **(CCXXXIII)** (Julia and Saussine, 1974).

$$\underset{\textbf{(CCXXXI)}}{(COD)Pd(Cl)CH_2SO_2C_6H_5} + R^1CH{=}CR^2R^3 \longrightarrow \underset{\textbf{(CCXXXII)}}{C_6H_5SO_2CH_2CHR^1C[PdCl(COD)]R^2R^3}$$

$$\longrightarrow \underset{\textbf{(CCXXXIII)}}{C_6H_5SO_2CH_2CR^1{=}CR^2R^3} + Pd(0) + HCl$$

2. Olefin Cyclopropanation

Transition metal salts or complexes are known to catalyze effectively the cyclopropanation of olefins with diazoalkanes. Asymmetric synthesis with chiral copper catalysts (Nozaki *et al.*, 1966, 1968; Noyori *et al.*, 1969; Moser, 1969), as well as a detailed kinetic study (Salomon and Kochi, 1973), has suggested the intervention of copper–carbene complexes as reactive intermediates. Recently synthesis of crysanthemic acid **(CCXXXIV)** (R = H) with high optical yield (60–70%) has been achieved by applying this asymmetric catalysis (Aratani *et al.*, 1975). The camphorglyoxime–cobalt(I) complex is also effective for the enantioselective reaction (Tatsuno *et al.*, 1974).

+ $N_2CHCOOR$ $\xrightarrow{Cu(L^*)_n}$ COOR + COOR

(CCXXXIVa) **(CCXXXIVb)**

B. Addition to 1,3-Dienes

Palladium complexes, in contrast to nickel(0) catalysts (Section I,B,1), promote *linear* oligomerization of 1,3-dienes. These reactions are considered

Pd catalyst

to involve electrophilic addition of π-allylpalladium species (for structural features, see Clarke, 1974) to 1,3-dienes (Baker, 1973).

Reaction of π-allylpalladium halides **(CCXXXV)** with 1,3-dienes **(CCXXXVI)** gives rise to new dimeric π-allylpalladium complexes **(CCXXXVII)** (Medema and van Helden, 1969, 1971; van Helden *et al.*, 1968; Medema *et al.*, 1969). Interestingly, the coupling reaction takes place at the

X Cl —Pd /2 + Y ⟶ X Y Pd Cl /2

(CCXXXV) **(CCXXXVI)** **(CCXXXVII)**

more substituted terminal carbon of the allylic moiety. Alkyl substituents on either the allyl group or diene moiety decelerate the reaction. Isoprene selectively gives the 1,1-disubstituted allylpalladium complexes of type **CCXXXVIII** (Takahashi *et al.*, 1969). A suggested mechanism involves an

(CCXXXVIII)

electrocyclization of the intermediates **CCXXXIX**, where the least substituted double bond of the diene and the carbon atom of the allylic moiety are coordinated to the central metal through π- and σ-interaction, respectively (Hughes and Powell, 1971, 1972).

(CCXXXIX)

Dimerization of isoprene with bis(triphenylphosphine)palladium maleic anhydride occurs exclusively in tail-to-tail fashion to give 2,7-dimethyl-1,3,7-octatriene (Josey, 1974). Dimerization of chloroprene (Guthrie and Nelson, 1972) and various terpenoids having a 1,3-diene unit (Dunne and McQuillin, 1970) is also a subject of recent interest.

C. Addition to Acetylenic Substrates

Carbon–carbon triple bonds are much more susceptible to attack by σ-bonded organometallics. Organocopper reagents, prepared from Grignard reagents and copper salts, react with acetylene and terminal alkynes in ethereal solvents to give cis addition products (Normant and Bourgain, 1971; Normant *et al.*, 1972, 1973, 1974). The resulting vinylcopper species of type **CCXL** are stereospecifically dimerized, halogenated, hydrolyzed, carbonated, and coupled with alkyl halides.

$$RC{\equiv}CH + R'CuMgX_2 \longrightarrow (R)(R')C{=}C(H)(Cu) \longrightarrow [(R)(R')C{=}C(H)]_{2}$$

(CCXL)

X_2: $(R)(R')C{=}C(H)(X)$; D_2O: $(R)(R')C{=}C(H)(D)$; CO_2: $(R)(R')C{=}C(H)(COOH)$; $R''X$: $(R)(R')C{=}C(H)(R'')$

Allylthioacetylenes are also alkylated by organocopper species (Vermeer *et al.*, 1974a).

$$RC{\equiv}CSCH_3 + R'MgX \xrightarrow{CuCl} \begin{matrix} R & & SCH_3 \\ & C{=}C & \\ H & & R' \end{matrix}$$

Organotransition metal species formed from methylmagnesium bromide and various transition metal salts add to carbon–carbon triple bonds (Light and Zeiss, 1970). In the presence of nickel(II) catalyst, the reaction leads to trisubstituted olefins in good yields (Duboudin and Jousseaume, 1972).

$$C_6H_5C{\equiv}CC_6H_5 + CH_3MgBr \xrightarrow[ether]{[(C_6H_5)_3P]_2NiCl_2} \begin{matrix} C_6H_5 & & C_6H_5 \\ & C{=}C & \\ H_3C & & H \end{matrix}$$

In the presence of palladium(II) chloride, diphenylacetylene is cis-dimethylated with methylmagnesium bromide. Dimethylpalladium may be involved as the active species (Garty and Michman, 1972).

$$C_6H_5C{\equiv}CC_6H_5 + 2CH_3MgBr + PdCl_2(C_6H_5CN)_2 \xrightarrow{60^\circ} \begin{matrix} H_5C_6 & & C_6H_5 \\ & C{=}C & \\ H_3C & & CH_3 \end{matrix} + Pd + 2C_6H_5CN$$

Certain monosubstituted acetylenes undergo facile linear dimerization in the presence of the Wilkinson catalyst **(CCXLI)** (Chini *et al.*, 1967; Singer and Wilkinson, 1968; Kern, 1968). Addition of the alkynylrhodium intermediate to the carbon–carbon triple bond can account for the dimerization.

$$C_6H_5C{\equiv}CH + RhCl[(C_6H_5)_3P]_3 \longrightarrow \underset{\textbf{(CCXLI)}}{HRh(C{\equiv}CC_6H_5)(Cl)[(C_6H_5)_3P]_3}$$

$$\xrightarrow{C_6H_5C{\equiv}CH} C_6H_5C{\equiv}CCH{=}CHC_6H_5$$

D. Nucleophilic Addition to Electron-Deficient Double and Triple Bonds

1. Organocopper Reagents

1,4-Addition to α,β-unsaturated carbonyl compounds is characteristic of organocopper and cuprate compounds, as alkyllithium and Grignard reagents react only with the carbonyl group of enones to give the 1,2-addition products. This synthetically useful reaction has been reviewed by Posner (1972).

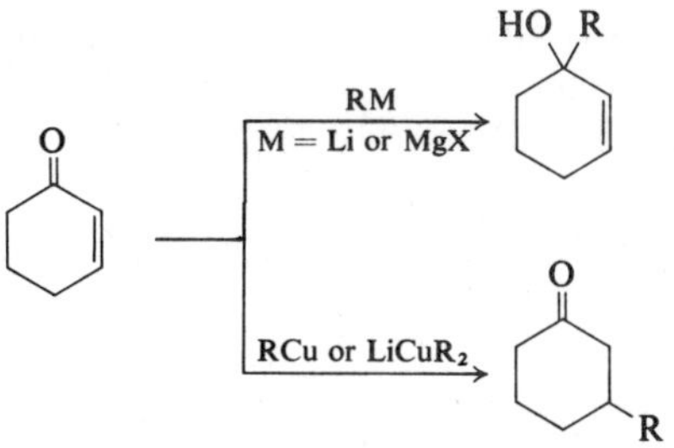

$$\text{2-cyclohexenone} \xrightarrow[M = \text{Li or MgX}]{\text{RM}} \text{1-R-2-cyclohexen-1-ol (HO, R)}$$

$$\text{2-cyclohexenone} \xrightarrow{\text{RCu or LiCuR}_2} \text{3-R-cyclohexanone}$$

Kharasch and Tawney (1941) reported that copper salts catalyze 1,4-addition of Grignard reagents to α,β-unsaturated ketones. Gilman *et al.* (1952) first discovered that phenylcopper reacts with benzalacetophenone in a 1,4-addition. Subsequently House and associates (1966) have revealed the scope of the conjugate addition of cuprate complexes. Now alkyl, vinyl, and aryl groups can be introduced specifically at the β position of α,β-unsaturated carbonyl compounds. Transfer of an allyl group from lithium diallylcuprate to 2-cyclohexenone is also known (House and Fischer, 1969). However, ethynyl, cyano, and hetero groups attached to the copper atom are difficult to transfer to electron-poor olefins.

Copper reagents complexed with lithium or magnesium halides are also employed in the reaction (House and Fischer, 1968; Luong-Thi and Rivière, 1968, 1970, 1971; Alexandre and Rouessac, 1974). Several mixed cuprate reagents useful for conjugate addition have been reported (Corey and Beams, 1972; Posner and Whitten, 1973; Posner *et al.*, 1973; House and Umen, 1973; Gorlier *et al.*, 1973).

A mechanism suggested by House and Umen (1972, 1973) is outlined in Scheme 13. The essential process is an initial electron transfer from the cuprate to the enone. Polarographic reduction potentials of the enones

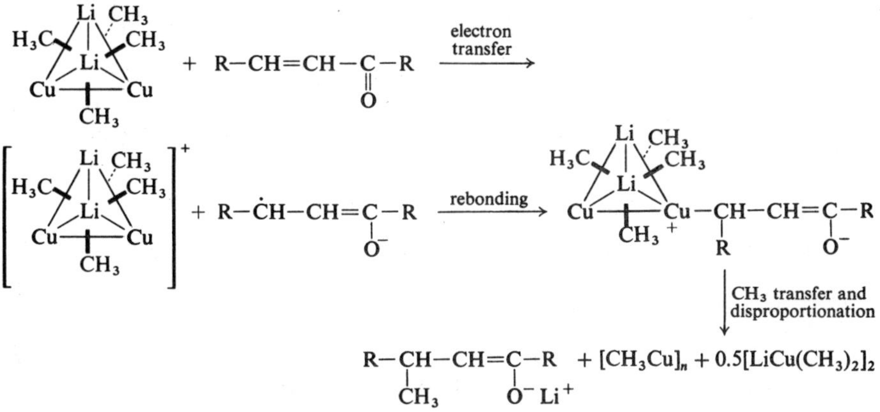

$$[LiCu(CH_3)_2]_2 + R{-}CH{=}CH{-}C({=}O){-}R \xrightarrow{\text{electron transfer}} [LiCu(CH_3)_2]_2^{+} + R{-}\dot{C}H{-}CH{=}C(O^-){-}R$$

$$\xrightarrow{\text{rebonding}} (CH_3)_3Cu_2Li_2{-}Cu^{+}{-}CH(R){-}CH{=}C(O^-){-}R$$

$$\xrightarrow{CH_3\text{ transfer and disproportionation}} R{-}CH(CH_3){-}CH{=}C(O^-Li^+){-}R + [CH_3Cu]_n + 0.5[LiCu(CH_3)_2]_2$$

Scheme 13

(House *et al.*, 1972) correlate well with the ease of conjugate addition. Failure of α-cyanoalkyl and α-ketoalkyl cuprates to undergo conjugate addition is ascribed to electron delocalization to the organic ligands. Recently, definitive chemical evidence for this electron-transfer mechanism has been presented (House and Weeks, 1975). Reaction of the enone tosylate **(CCXLII)** with lithium dimethylcuprate followed by hydrolytic workup affords **(CCXLIII)**, whereas treatment of the mixture with acetic anhydride gives the enol acetate **(CCXLIV)**. This may suggest a two-electron transfer from the cuprate to the enone substrate (Hannah and Smith, 1975).

OTs OTs
$LiCu(CH_3)_2$, −78°
MO MO
(CCXLII)
H_3O^+ $(CH_3CO)_2O$
AcO
(CCXLIII) **(CCXLIV)**

M = Li or Cu

Reaction of *endo*-2-norbornylcopper(I) reagent with mesityl oxide proceeds with high stereoselectivity to give an endo adduct (Whitesides and Kendall, 1972). Stereoselectivity in the conjugate addition of cuprates to 2-cyclohexenone, including octalones, has been examined in detail (Marshall *et al.*, 1966; Marshall and Roebke, 1968; House and Fischer, 1968).

Various naturally occurring materials have been synthesized using this procedure. For instance, *d,l*-fukinone **(CCXLV)** has been prepared through the conjugate addition of dimethylcuprate to an octalone (Marshall and Cohen, 1971). Similar reactions have been applied to the stereoselective

O H OAc
$LiCu(CH_3)_2$
O H OAc
several steps
H O
(CCXLV)

synthesis of *d,l*-elemophil-3,11-diene **(CCXLVI)** (Piers and Keziere, 1968, 1969). The cuprate conjugate addition is an important step in the total

CuCl

MgBr

several steps

$LiCu(CH_3)_2$

H

only cis

several steps

H

(CCXLVI)

syntheses of zizaene (Coates and Sowerby, 1972), eudesmol (Carlson and Zey, 1972), *d,l*-shionone (Ireland *et al.*, 1974), *d,l*-nootkatone (Pesaro *et al.*, 1968), and a boll weevil sex attractant (Babler and Mortell, 1972). A new, efficient route to *d,l*-muscone **(XXIII)** from cyclododecanone has been elaborated applying this method (Stork and Macdonald, 1975). Stereoselective

several steps

Cl

1. $LiCu(CH_3)_2$
2. Cr^{2+}

(XXIII)

addition of methylcuprate to the dienone **(CCXLVII)** leads to *β*-vetivone **(CCXLVIII)** (Bozzato *et al.*, 1974).

1. $LiCuCH_3Br–[(CH_3)_2CHCH_2]_2NH$
2. Δ

HCO

(CCXLVII)

(CCXLVIII)

Conjugate addition of lithium divinylcuprate to enones gives α,*β*-unsaturated ketones (Hooz and Layton, 1970). Addition of di-*cis*- or di-*trans*-propenylcuprate to 2-cyclohexenone proceeds in a completely stereospecific manner with retention of double-bond geometry (Casey and Boggs, 1971). Particularly important is the application to the synthesis of a variety of prostaglandin derivatives (Alvarez *et al.*, 1972). *l*-Prostaglandin E_1 **(CCLI)** has been prepared by the reaction of the cyclopentenone derivatives of type **(CCXLIX)** and the chiral copper reagent **(CCL)** (Sih *et al.*, 1972, 1973; Kluge *et al.*, 1972a). Similarly, 13-*cis*-15β-prostaglandins have been prepared in higher yields and with higher stereoselectivity (Kluge *et al.*, 1927b). The latter is readily

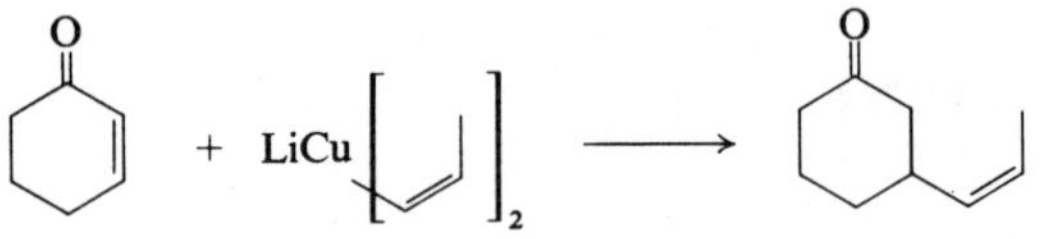

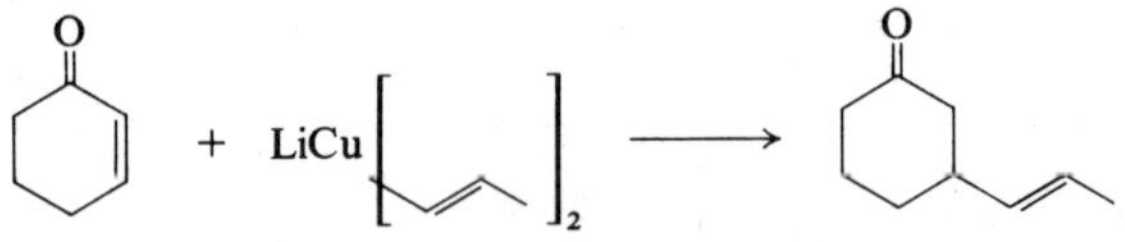

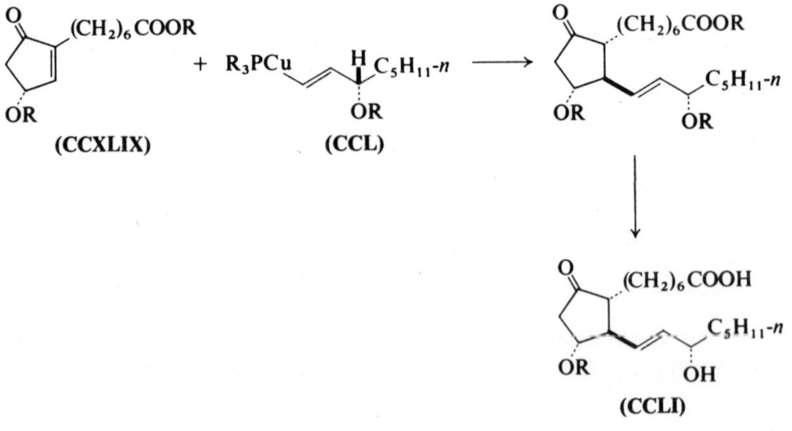

$(CH_2)_6COOR$ + R_3PCu ... C_5H_{11}-*n* → $(CH_2)_6COOR$ ... C_5H_{11}-*n*

(CCXLIX) **(CCL)**

$(CH_2)_6COOH$ ... C_5H_{11}-*n*

(CCLI)

convertible to the natural 13-*trans*-15α-prostaglandins (Miller *et al.*, 1974). Similarly, 11-deoxyprostaglandins and 11-deoxy-13-dehydroprostaglandins have been prepared (Patterson and Fried, 1974; Schaub and Weiss, 1973; Grudzinskas and Weiss, 1973).

An α-alkoxyvinyl group attached to the metal serves as an acyl anion equivalent. Reaction of di(α-methoxyvinyl)cuprate and α,β-unsaturated ketones followed by hydrolysis or ozonolysis affords 1,4-diketones and γ-keto esters, respectively (Chavdarian and Heathcock, 1975).

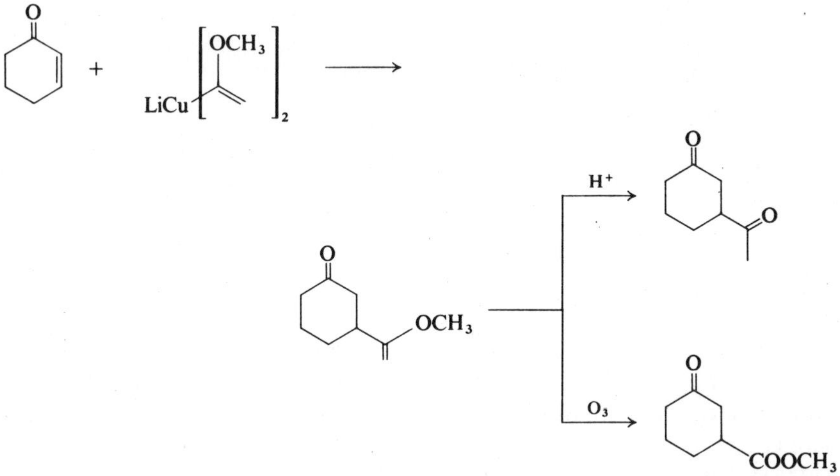

Transfer of an ethynyl group to the beta position of enones is precluded by the tenacity with which copper binds the ligand. Recently a nucleophilic ethynyl group equivalent **(CCLII)** has been demonstrated. The organotin moiety in the addition product is readily removed by the action of lead(IV) tetraacetate, thereby making possible functionalization of the angular position of the bicyclic system (Corey and Wollenberg, 1974).

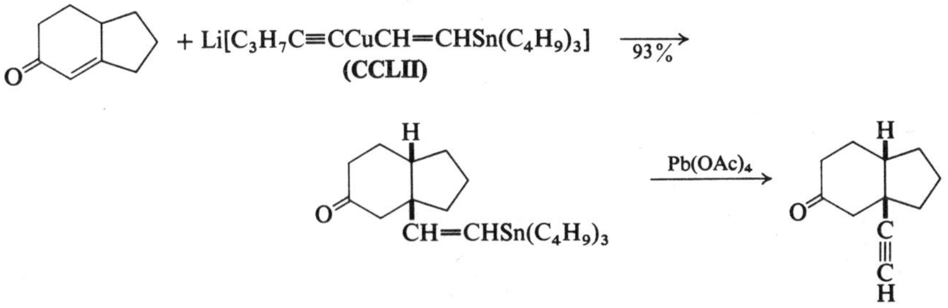

+ $Li[C_3H_7C{\equiv}CCuCH{=}CHSn(C_4H_9)_3]$ $\xrightarrow{93\%}$
(CCLII)

$CH{=}CHSn(C_4H_9)_3$ $\xrightarrow{Pb(OAc)_4}$

Electron-deficient unsaturated compounds other than enones also undergo conjugate addition of organocopper reagents. Synthesis of *d,l*-isolongifolene **(CCLIV)** has been achieved by reaction of the cyano ester **(CCLIII)** and dimethylcuprate (Sobti and Dev, 1967, 1970).

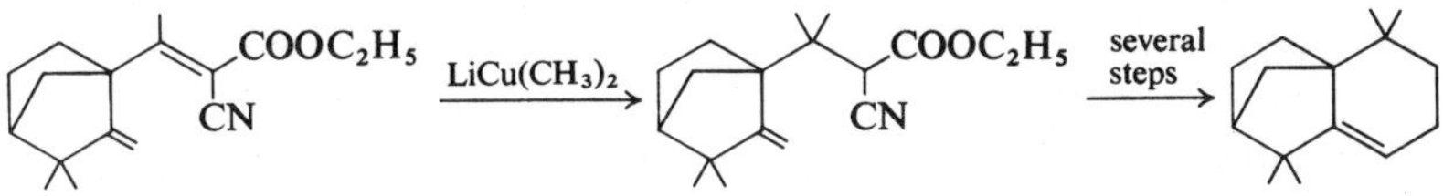

$COOC_2H_5$, CN $\xrightarrow{LiCu(CH_3)_2}$ $COOC_2H_5$, CN $\xrightarrow{\text{several steps}}$

(CCLIII) **(CCLIV)**

Dialkylcuprates react also with alkenyl sulfides and alkenyl sulfones to give the β-alkylated products (Posner and Brunelle, 1973a,b).

Electron-deficient triple bonds are highly reactive to copper reagents. Dialkylcuprates react with propiolate at −78° in a stereospecific manner to give trisubstituted olefins (Corey and Katzenellenbogen, 1969; Siddall *et al.*, 1969; Klein and Turk, 1969). At elevated temperatures, isomerization of the vinylic copper intermediate takes place to lead to nonstereospecific products.

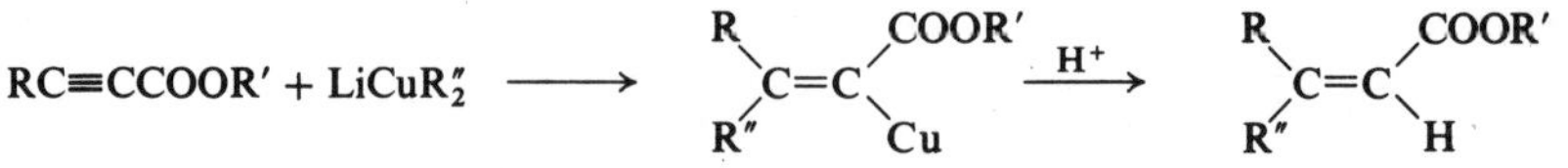

$$RC\equiv CCOOR' + LiCuR''_2 \longrightarrow (R)(R'')C=C(COOR')(Cu) \xrightarrow{H^+} (R)(R'')C=C(COOR')(H)$$

The codling moth sex pheromone **(CCLV)** has been synthesized by applying this cuprate addition procedure (Bowlus and Katzenellenbogen, 1973a,b; Cooke, 1973). The stereoselective conjugate addition has been utilized for the synthesis of juvenile hormone analogs as well (Anderson *et al.*, 1975).

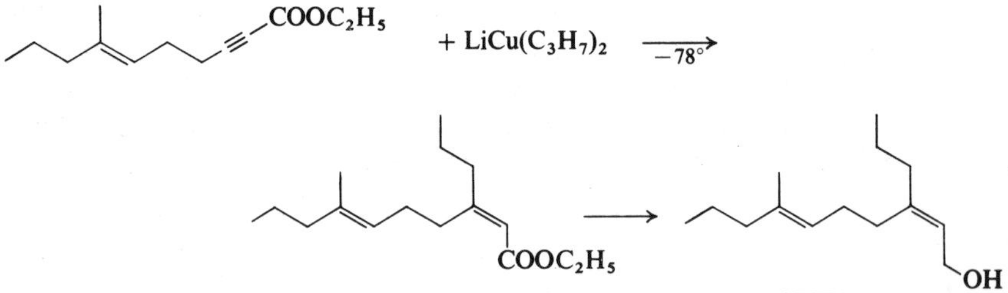

$$COOC_2H_5 + LiCu(C_3H_7)_2 \xrightarrow{-78^\circ}$$

$$COOC_2H_5 \longrightarrow OH$$

(CCLV)

When vinylcopper or allylcopper compounds are allowed to react with α,β-acetylenic esters, 1,3- and 1,4-dienes are synthesized stereospecifically (Näf and Degan, 1971; Corey *et al.*, 1972). The reaction of **CCLVI** and lithium divinylcuprate gives the addition product **(CCLVII)** in a purely cis form in >90% yield. The latter is converted readily to the prostaglandin intermediate **(CCLVIII)**. Vinylcopper may be a superior reagent for this purpose.

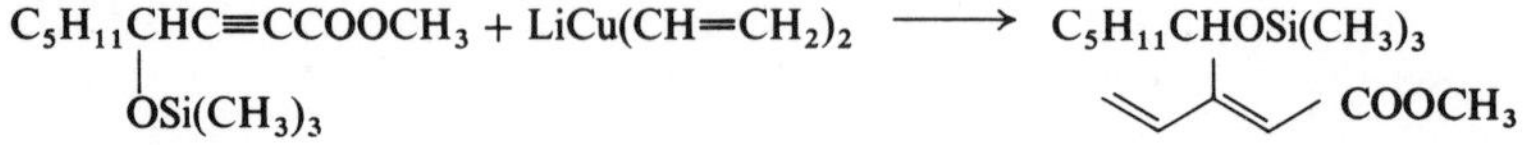

$$C_5H_{11}CH(OSi(CH_3)_3)C\equiv CCOOCH_3 + LiCu(CH{=}CH_2)_2 \longrightarrow C_5H_{11}CHOSi(CH_3)_3 \; COOCH_3$$

(CCLVI) **(CCLVII)**

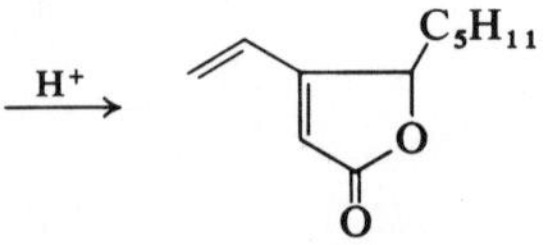

$$\xrightarrow{H^+} C_5H_{11} \; O \; O$$

(CCLVIII)

Monoalkylcopper reagents add to α,β-acetylenic sulfoxides with high cis stereospecificity to form β-alkylated α,β-ethylenic sulfoxides (Truce and Lusch, 1974; Vermeer *et al.*, 1974b).

Conjugate addition of organocuprates to α,β-unsaturated carbonyl compounds, followed by alkylation of the resulting enolates, leads to regiospecific α,β-dialkylation (Boeckman, 1973). This method can be applied to the synthesis of valerane **(CCLIX)** (Posner *et al.*, 1974, 1975a). The enolate intermediates are acylated as well. The method leads to a new synthesis of 7-oxoprostaglandins (Tanaka *et al.*, 1975).

$LiCuR_2$ → ; $R'X$ →

M = Li or Cu

$LiCu(CH_3)_2$, benzene, 0°–5° → ; HMPA 0° →

(25–30%) → **(CCLIX)**

The solution produced from the reaction of enones and dimethylcuprate undergoes aldol condensation with acetaldehyde in the presence of zinc chloride producing the β-methyl α-(1-hydroxyethyl) ketones in acceptable yields (Heng and Smith, 1975).

1. $LiCu(CH_3)_2$ 2. $CH_3CHO–ZnCl_2$ → ; H^+ →

Reaction of diorganocuprates with enol ethers or esters of β-dicarbonyl compounds gives β-alkylated α,β-unsaturated compounds via a conjugate addition–elimination mechanism (Casey *et al.*, 1973; Casey and Marten, 1974; Cacchi *et al.*, 1974). Similarly, β-alkylthio α,β-ethylenic ketones and esters can be converted to geminal dialkylated products (Coates and Sowerby, 1971; Posner and Brunelle, 1973c). Its modification makes α,β,β-trialkylation possible (Coates and Sandefur, 1974). Pulegone **(CCLX)** has been prepared by

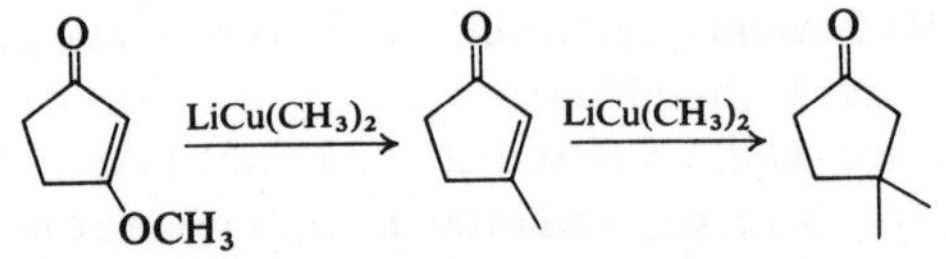

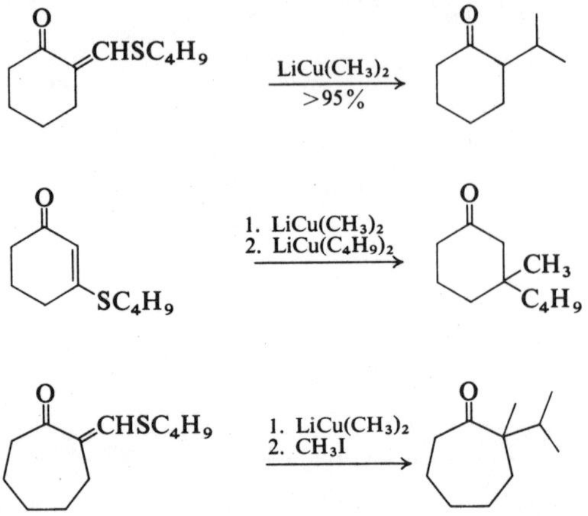

this method (Corey and Chen, 1973b). The stereospecific displacement can be

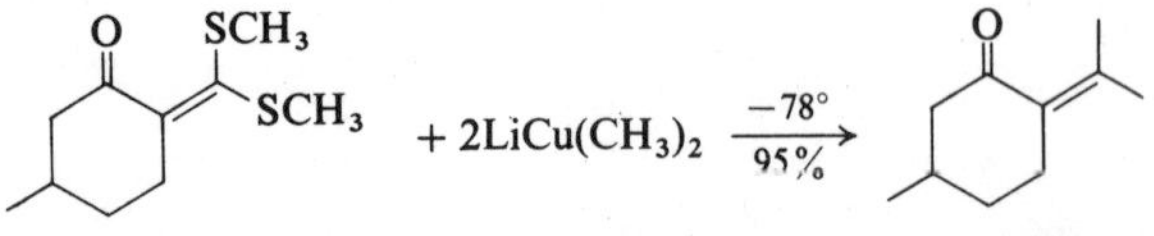

(CCLX)

used for the synthesis of ethyl geranate, methyl farnesoate, and precursors of *Cecropia* juvenile hormones (Kobayashi and Mukaiyama, 1974).

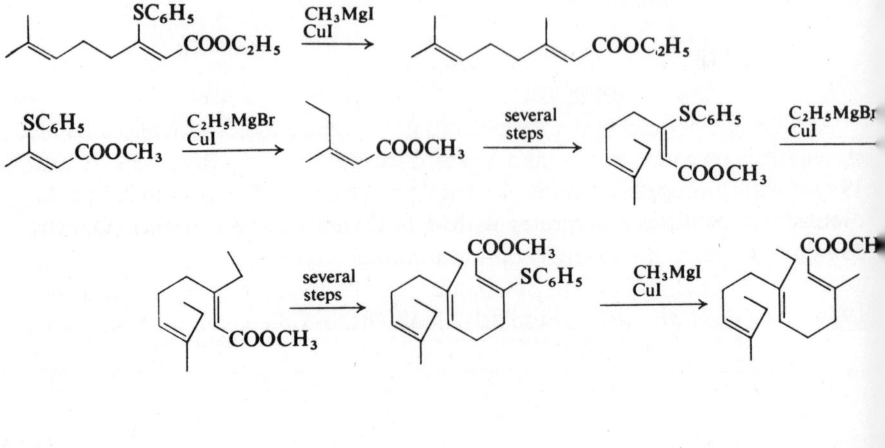

Reaction of dialkylcuprates with α-chlorotropone gives α-alkylated tropones (Cavazza and Pietra, 1974).

In certain cases, 1,4- and 1,6-conjugate additions take place competitively (Marshall *et al.*, 1971; Marshall and Ruden, 1971; Näf *et al.*, 1972; Daviaud and Miginiac, 1972). An instance of 1,7-addition is also known (Grieco and

+ $LiCu(CH_3)_2$ ⟶ +

(60 : 40)

+ $LiCu(CH_3)_2$ ⟶ +

Finkelhor, 1973). Homoconjugate addition of lithium divinylcuprate to the

$COOC_2H_5$ $COOC_2H_5$ + $LiCu(CH_3)_2$ ⟶ $COOC_2H_5$ $COOC_2H_5$

cyclopropane derivative **(CCLXI)** affords the adduct **(CCLXII)**. This process may be applicable to the synthesis of prostaglandins (Corey and Fuchs, 1972).

RO $COOCH_3$ + $LiCu(CH{=}CH_2)_2$ ⟶ RO $COOCH_3$

(CCLXI) **(CCLXII)**

Addition of organocopper reagents to $\Delta^{2,4}$-dienoic esters provides a highly stereoselective route to tri- and tetrasubstituted olefins (Corey and Chen, 1973a). 1,6-Conjugate addition of 1-trialkylsilylpropynylcoppers to 2,4-pentadienoates constitutes a simple route to 1,5-enlynes and 1,4,5-trienes (Ganem, 1974).

RCu + $COOCH_3$ ⟶ R $COOCH_3$

Organocopper species derived from copper(I)–isocyanide complexes and active methylene compounds are useful in organic synthesis. The intermediates undergo Michael-type addition to electron-deficient olefins; butenenitrile dimerizes in the presence of cuprous oxide–cyclohexyl isocyanide (Saegusa *et al.*, 1968, 1970a,b, 1972a, 1975). A new synthesis of pyrroline **(CCLXIII)** and oxazoline **(CCLXIV)** has been achieved by this method (Saegusa *et al.*, 1971c, 1972c). When the reaction is applied to allylic halides or α-halo-

$$\underset{H}{\overset{H_3C}{}}C{=}C\underset{CN}{\overset{H}{}} \rightleftharpoons \left[\begin{array}{c} CuCH_2CH{=}CHCN \\ \uparrow\downarrow \\ CH_2{=}CHCHCuCN \end{array}\right] \rightleftharpoons \underset{H}{\overset{H_3C}{}}C{=}C\underset{H}{\overset{CN}{}}$$

$$\downarrow CH_3CH{=}CHCN$$

$$\begin{array}{l} CH_2{=}CHCHCN \\ \quad\quad\;\; | \\ \quad\;\; CH_3CHCH_2CN \end{array}$$

$$R^1R^2CHN{\equiv}C \xrightarrow{Cu_2O} \begin{cases} \xrightarrow{CH_2{=}CHZ} \text{(CCLXIII): } R^1, R^2, N, Z \\ \xrightarrow{R_2C{=}O} \text{(CCLXIV): } R^1, R^2, N, O, R, R \end{cases}$$

(CCLXIII)

(CCLXIV)

carbonyl compounds and electron-deficient olefins, substituted cyclopropanes are obtained (Saegusa *et al.*, 1971a, 1972b, 1973; Ito *et al.*, 1974a). Cyclopentanecarboxylates are derived by coupling of 1,3-diiodopropane and α,β-unsaturated esters by a copper isocyanide complex (Ito *et al.*, 1974b; Saegusa and Ito, 1975).

$$ICH_2CH_2CH_2I + CH_2{=}CHCOOR \xrightarrow{Cu\text{-}RCN} \text{cyclopentane-}COOR$$

Reaction of diorganocuprates with α,α'-dibromo ketones provides a new method for the α-alkylation of a ketone; a nucleophilic attack of the cuprate on the intermediary cyclopropanones could explain the reaction (Posner and Sterling, 1973; Posner *et al.*, 1973). The cyclopropyl group can be introduced at the alpha position of ketones as well (Carlson and Mardis, 1975).

Br O Br $\xrightarrow{(t\text{-}C_4H_9O)RCuLi}$ O $\xrightarrow{(t\text{-}C_4H_9O)RCuLi}$ R OM $\xrightarrow{H^+}$ R O

M = Li or Cu

2. Organonickel Reagents

Organonickel intermediates formed from nickel(II) acetylacetonate and trimethylaluminum or tetramethylaluminate behave like cuprate reagents and undergo conjugate addition to α,β-unsaturated ketones (Ashby and Heinsohn, 1974).

A new synthesis of coenzyme Q_1 **(CXXVI)** and plastoquinone-1 **(CCLXV)** has been achieved by using the reaction of π-allylnickel(I) bromides and alkylated quinones in dimethylformamide or tetrahydrofuran (Hegedus *et al* 1972). Considerable amounts of hydroquinone derivatives are formed as a by-product. The reaction has been shown to proceed via an electron-transfer

Br –Ni /2 + CH_3O CH_3O O O $\xrightarrow{40\%}$ CH_3O CH_3O O O

Br –Ni /2 + O O $\xrightarrow{61\%}$ O O

(CCLXV)

mechanism. The reactivity of quinones parallels their reduction potentials, and the reaction site is correlated with spin density of the radical anions resulting from the electron-transfer process (Hegedus and Waterman, 1974).

π-Allylnickel halides, formed from allyl halides and nickel carbonyl, undergo addition to electron-deficient olefins such as acrylonitrile or methyl acrylate (Dubini *et al.*, 1965; Dubini and Montino, 1966). Reaction of benzyl chloride with triiron dodecacarbonyl gives dibenzyl ketone. Organoiron

$$CH_2{=}CHCH_2Cl + CH_2{=}CHCOOCH_3 + Ni(CO)_4 \xrightarrow{-[NiHCl]} CH_2{=}CHCH_2CH_2CH{=}CHCOOCH_3$$

species formed from the iron carbonyl and organic halides such as benzyl halide or iodobenzene react with electron-deficient olefins to afford the coupling products (Rhee *et al.*, 1967).

$$RX + Fe_3(CO)_{12} \longrightarrow RFeX(CO)_n \xrightarrow{CH_2=CHZ} \underset{FeX(CO)_n}{\overset{RCH_2CHZ}{|}} \longrightarrow RCH_2CH_2Z \text{ or } RCH{=}CHZ$$

Combination of organolithiums and nickel carbonyl provides a new method for generating reactive species synthetically equivalent to C-nucleophilic carbonyl groups (Corey and Hegedus, 1969b). The organonickel intermediates undergo conjugate addition to α,β-unsaturated carbonyl compounds to produce 1,4-dicarbonyl products in good yield. The reaction could be best explained by a mechanism involving an electron transfer from the initially generated $[RCONi(CO)_3]Li$ to the enone moieties. A reagent formed from

$$RLi + Ni(CO)_4 + {>}C{=}\underset{|}{C}{-}\overset{O}{\overset{\|}{C}}{-} \longrightarrow R{-}\overset{O}{\overset{\|}{C}}{-}\underset{|}{\overset{|}{C}}{-}\underset{|}{CH}{-}\overset{O}{\overset{\|}{C}}{-}$$

an aryllithium and nickel carbonyl reacts with terminal acetylenes to produce 2:1 adducts **(CLXXXVI)** (Sawa *et al.*, 1968). The aroylnickel species also

$$ArLi + Ni(CO)_4 + RC{\equiv}CH \xrightarrow[\text{ether}]{-78^\circ} \underset{\substack{(\sim 70\%) \\ \textbf{(CLXXXV)}}}{ArCOCHRCH_2COAr}$$

adds to activated olefins such as styrene, acrylonitrile, and ethyl acrylate to give the Michael-type adducts as the major products (Yoshisato *et al.*, 1969) Reaction with *N*-benzylidene alkylamine in dimethylformamide gives 1-alkyl-2-phenylindolin-3-one (Ryang *et al.*, 1973).

The reaction of potassium hexacyanodinickelate with organic halides forms unstable organonickel(II) complexes (Hashimoto *et al.*, 1969, 1970a,b). *trans*-β-Bromostyrene is particularly reactive and gives a mixture of the dimeric 1,3-diene **(CCLXVII)** and the substitution product **(CCLXVIII)**. The intermediate **(CCLXVI)** reacts with electron-deficient olefins to afford the chain-lengthened products **(CCLXIX)**. In addition, $K_4Ni_2(CN)_6$ serves as a

(CCLXVII)

+

CN

(CCLXVIII)

$K_2[Ni_2(CN)_6]$ + Br ⟶ K_2 [$Ni(CN)_3$]

(CCLXVI)

1. Z
2. H_2O

Z

(CCLXIX)

Z = CN, $COOC_2H_5$

useful agent for conversion of vinylic bromiaes to tne cyanides (Corey and Hegedus, 1969a). Nickel phosphine complexes, as well as palladium cyanide, catalyze substitution of aromatic halides with potassium cyanide (Cassar 1973; Takagi *et al.*, 1973).

3. Organopalladium Complexes

Iodobenzene and acrylic esters couple in the presence of a palladium(II) salt (or metallic palladium) and bases to give the substitution products, cinnamic esters. Phenylpalladium intermediates may be involved here (Mori *et al.*, 1973). The ligand triphenylphosphine, can facilitate the oxidative addition

$$C_6H_5I + CH_2{=}CHCOOR \xrightarrow[\text{base}]{\text{Pd}} C_6H_5CH{=}CHCOOR + HI$$

of organic halides to the palladium atom and allows the use of aryl and vinyl bromides in place of the iodides (Dieck and Heck, 1974). The mechanism is outlined in Scheme 14. The regeneration of the palladium complex makes the reaction catalytic.

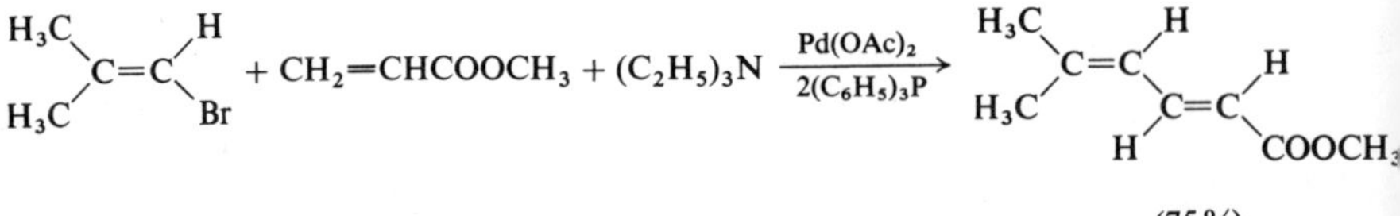

$$(H_3C)_2C{=}CHBr + CH_2{=}CHCOOCH_3 + (C_2H_5)_3N \xrightarrow[2(C_6H_5)_3P]{Pd(OAc)_2} (H_3C)_2C{=}CH{-}CH{=}CH{-}COOCH_3$$

(75%)

$$RX + PdL_n \longrightarrow \left[\begin{array}{c} R{-}Pd{-}X \\ | \\ L_n \end{array}\right] \xrightarrow{\ \text{H}{>}C{=}C{<}\ } R{-}\overset{H}{\underset{|}{\overset{|}{C}}}{-}\overset{H}{\underset{|}{\overset{|}{C}}}{-}\underset{L_n}{\underset{|}{Pd}}{-}X \longrightarrow \begin{array}{c} R \\ \end{array}{>}C{=}C{<} + \left[\begin{array}{c} H{-}Pd{-}X \\ | \\ L_n \end{array}\right]$$

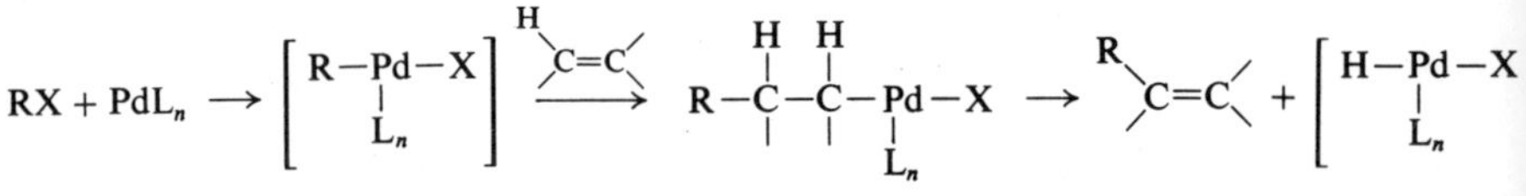

Scheme 14

VI. CARBONYLATION REACTIONS

Organic synthesis using carbon monoxide is now quite significant in both industry and laboratory. The transition metal-promoted processes through which carbon monoxide is incorporated in organic structures usually involve coupling reactions of σ- or π-bonded organometallic intermediates. This subject has been surveyed comprehensively (Falbe, 1970), and hence we will not discuss it here in detail.

Acrylic ester synthesis from acetylene, carbon monoxide, and alcohols with the aid of nickel carbonyl is known as the Reppe synthesis and is very important in industrial applications (Reppe, 1953). This carboxylation re-

$$HC{\equiv}CH + CO + ROH \xrightarrow{Ni(CO)_4} CH_2{=}CHCOOR$$

action has been utilized in a laboratory synthesis of sirenin **(LXXIV)** (Corey and Achiwa, 1970). The intramolecular version is used in the preparation of

$$\left[\begin{array}{c} H-Pd-X \\ | \\ L_n \end{array}\right] + R_3'N \longrightarrow PdL_n + R_3'\overset{+}{N}HX^-$$

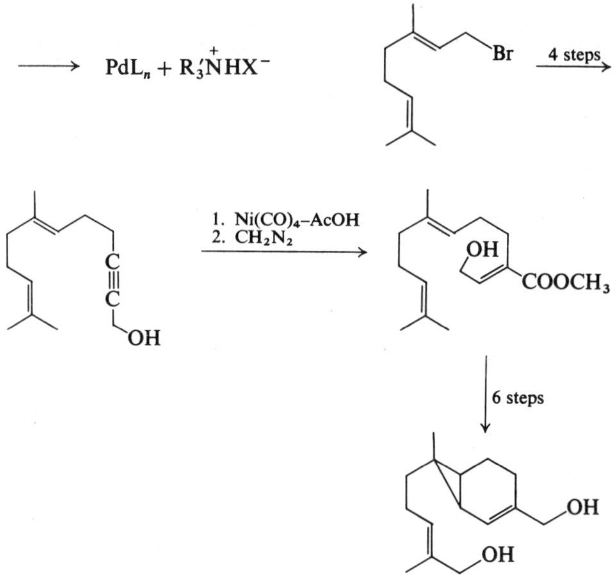

4 steps

1. $Ni(CO)_4$–AcOH
2. CH_2N_2

C≡C, OH, OH, $COOCH_3$, Br

6 steps

OH, OH

(LXXIV)

α-methylene-γ-butyrolactone (Jones *et al.*, 1950). The cyclization has also been achieved by use of a palladium(II) catalyst (Norton *et al.*, 1975).

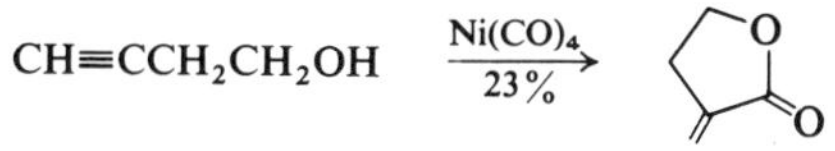

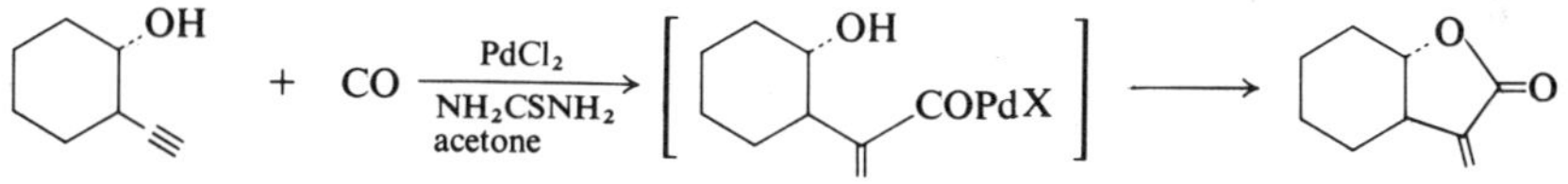

$$CH{\equiv}CCH_2CH_2OH \xrightarrow[23\%]{Ni(CO)_4}$$

OH + CO $\xrightarrow[\text{NH}_2\text{CSNH}_2,\ \text{acetone}]{\text{PdCl}_2}$ [OH, COPdX] ⟶ O, =O

(94%)

n-Butanol is prepared commercially by the iron carbonyl-promoted hydroxymethylation of propylene (Reppe and Vetter, 1953). Iron pentacarbonyl and a tertiary amine serve as a good catalyst system. The active species has been shown to be $HFe(CO)_4^-$ formed from the metal carbonyl and hydroxide ion (Wada and Matsuda, 1974).

$$CH_3CH{=}CH_2 + 3CO + 2H_2O \xrightarrow[\text{amine}]{Fe(CO)_5} CH_3CH_2CH_2CH_2OH + 2CO_2$$

Carbonylation of allylic halides is possible with nickel carbonyl. This process has been studied extensively by Chiusoli and reviews written by the same investigator have appeared (Chiusoli, 1969, 1971; Chiusoli and Cassar, 1967). In the presence of the nickel carbonyl catalyst, allylic halides undergo carbonylation in hydroxylic solvents under a carbon monoxide atmosphere (2–3 atm); the reaction proceeds by way of π-allylnickel intermediates.

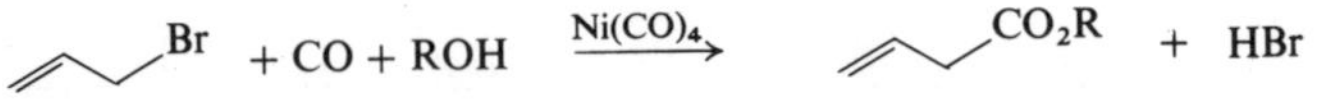

$$\text{CH}_2{=}\text{CHCH}_2\text{Br} + \text{CO} + \text{ROH} \xrightarrow{\text{Ni(CO)}_4} \text{CH}_2{=}\text{CHCH}_2\text{CO}_2\text{R} + \text{HBr}$$

When the reaction is carried out with acetylenes, 1,4-dienes are formed (Scheme 15). Methyl *trans*-chrysanthemate **(CCLXX)** has been synthesized by using this method as a key step (Corey and Jautelat, 1967).

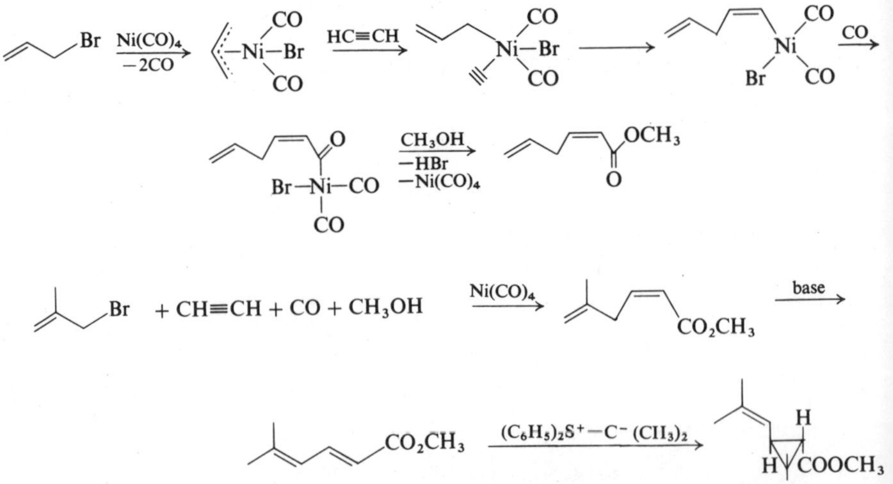

Br; $Ni(CO)_4$, $-2CO$; CO, Ni—Br, CO; HC≡CH; CO, Ni—Br, CO; CO, Ni, Br, CO; CO

O, Br—Ni—CO, CO; CH_3OH, $-HBr$, $-Ni(CO)_4$; OCH_3, O

Br + CH≡CH + CO + CH_3OH $\xrightarrow{Ni(CO)_4}$ CO_2CH_3 $\xrightarrow{\text{base}}$

CO_2CH_3 $\xrightarrow{(C_6H_5)_2S^+{-}C^-(CH_3)_2}$ H, H, $COOCH_3$

(CCLXX)

Scheme 15

Palladium-catalyzed carbonylation of olefins or dienes is a useful tool in organic synthesis. This topic has been reviewed by Tsuji (1969a,b). The carbonylation usually occurs at the less substituted terminal carbon of the

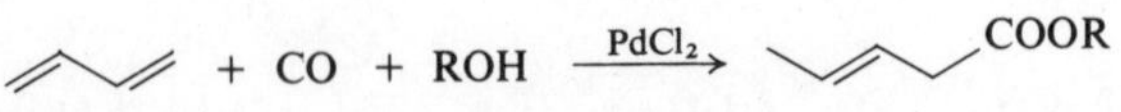

$$\text{CH}_2{=}\text{CHCH}{=}\text{CH}_2 + \text{CO} + \text{ROH} \xrightarrow{\text{PdCl}_2} \text{CH}_3\text{CH}{=}\text{CHCH}_2\text{COOR}$$

π-allylpalladium intermediates. Allylic halides, alcohols, ethers, esters, and anhydrides are carbonylated with palladium catalysts to afford the corresponding acid chloride. The products are readily converted to esters by the action of alcoholic media.

$$\text{R}^2\text{-allyl-X} \xrightarrow{PdCl_2} [\pi\text{-allyl-PdCl}]_{1/2} \xrightarrow{CO} \text{R}^2\text{-allyl-COCl} \xrightarrow{ROH} \text{R}^2\text{-allyl-COOR}$$

VII. MISCELLANEOUS

It should be noted that low-valent titanium complexes are powerful deoxygenating agents and, hence, can be used for various coupling reactions of organic molecules having oxygen functions.

The direct reductive coupling of benzyl and allyl alcohols with titanium-based complexes offers a convenient method for the synthesis of bibenzyl and 1,5-dienes. For example, when a mixture of titanium tetrachloride and sodium benzoxide is treated with potassium metal at 100°–140°, bibenzyl is produced in 51% yield. Allyl alcohol is converted to biallyl in 38% yield. Squalene

$$ROH \longrightarrow [(RO)_2Ti] \longrightarrow R{-}R + TiO_2$$

can be obtained in similar yield on carrying out the reduction sequence with farnesol. The reductive coupling of allylic alcohols is also effected without isolation of intermediates through the combined action of titanium trichloride and alkyl- or aryllithiums. Thus, geraniol dimerizes with a complex prepared from methyllithium and titanium trichloride (3:1 ratio) to provide hydrocarbons consisting of C-1–C-1′ : C-1–C-3′ coupled material in a ratio of 7:1 (71% yield). The cross-coupling of the allylic alcohols **(CCLXXI)** and **(CCLXXII)** (~10 *M* excess) gives the all-*trans*-acetal **(CCLXXIII)**. These coupling reactions have been postulated to proceed by a mechanism involving the production of titanium(II) alloxide or benzoxide, which decomposes

(CCLXXI) + **(CCLXXII)** ⟶ **(CCLXXIII)**

easily, providing a product by way of an allyl or benzyl radical route. (Schemes 16 and 17). Thus, coupling of pure *trans*-geraniol gives all possible (five)

$$RO^- + TiCl_4 \longrightarrow \{(RO)_2TiCl_2 \xrightarrow{e} [(RO)_2Ti]_n \rightleftharpoons n(RO)_2Ti \xrightarrow{\Delta} (TiO)_2 2R\cdot\} \longrightarrow R{-}R$$

Scheme 16

$$2ROH + TiCl_3 \xrightarrow{2R'Li} \{(RO)_2TiCl \xrightarrow{R'Li} (RO)_2TiR' \xrightarrow{-R'} (RO)_2Ti \xrightarrow{\Delta} (TiO_2)2R\cdot\} \longrightarrow R{-}R$$

Scheme 17

cis, trans, primary–primary, and primary–tertiary joined products (van Tamelen and Schwartz, 1965; Sharpless *et al.*, 1968; van Tamelen *et al.*, 1969).

A titanium(II) species formed from titanium trichloride and lithium aluminum hydride is a useful reagent for the reductive coupling of carbonyl compounds to olefins (McMurry, 1974; McMurry and Fleming, 1974). Both aliphatic and aromatic ketones can be converted to tetrasubstituted olefins in excellent yields. Reductive dimerization of retinal **(CCLXXIV)** affords β-carotene **(CCLXXV)** in 85% yield. The course of the reaction can be accounted for by assuming pinacol formation followed by loss of titanium dioxide.

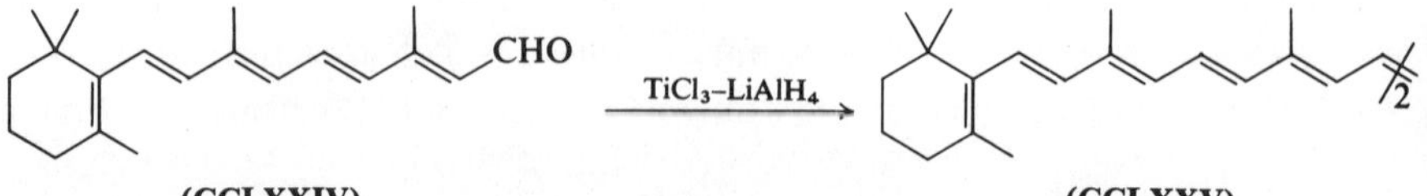

CHO $\xrightarrow{TiCl_3-LiAlH_4}$

(CCLXXIV) **(CCLXXV)**

$$R_2C{=}O \xrightarrow{Ti(II)} R_2C(O^-){-}C(O^-)R_2 \xrightarrow[-TiO_2]{Ti(II)} R_2C{=}CR_2$$

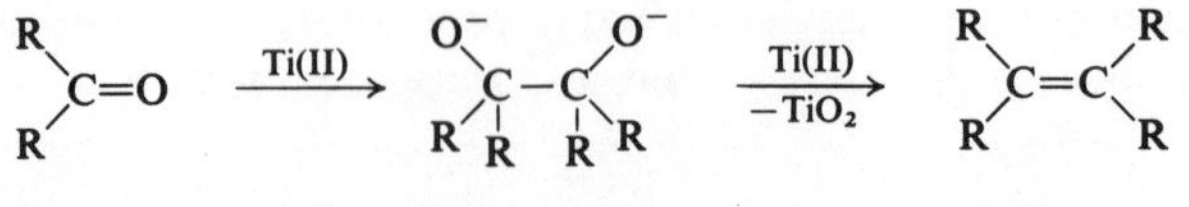

By this method, the sterically crowded hydrocarbon **(CCLXXVI)** with hindered rotation about a carbon–carbon single bond has been prepared (Langler and Tidwell, 1975; Bomse and Morton, 1975).

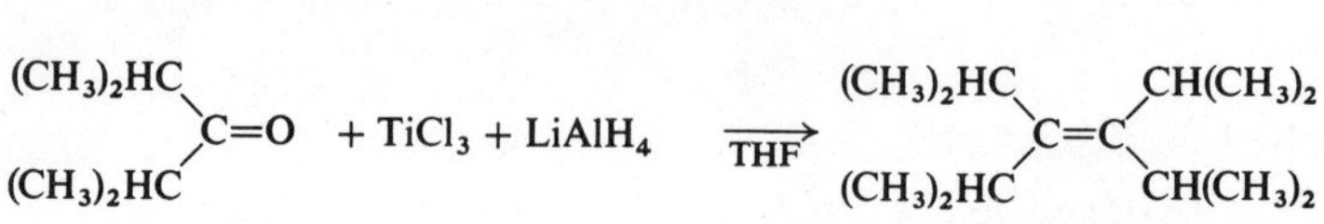

$$[(CH_3)_2HC]_2C{=}O + TiCl_3 + LiAlH_4 \xrightarrow{THF} [(CH_3)_2HC]_2C{=}C[CH(CH_3)_2]_2$$

(CCLXXVI)

Two related procedures for reductive coupling of aromatic ketones with titanium-based reagents have been reported (Tyrlik and Wolochowicz, 1973; Mukaiyama *et al.*, 1973). The deoxygenative dimerization of aromatic carbonyl compounds can be achieved with low-valent tungsten complexes formed from tungsten hexachloride and alkyllithium as well (Sharpless *et al.*, 1972).

REFERENCES

Agnès, G., Chiusoli, G. P., and Marraccini, A. (1973). *J. Organomet. Chem.* **49**, 239.

Alexandre, C., and Rouessac, F. (1974). *Bull. Soc. Chim. Belg.* **83**, 393.

Alper, H., and Keung, E. C. H. (1972). *J. Org. Chem.* **37**, 2566.

Alper, H., and Des Roches, D. (1976). *J. Org. Chem.* **41**, 806.

Alvarez, F. S., Wren, D., and Prince, A. (1972). *J. Am. Chem. Soc.* **94**, 7823.

Anderson, R. J. (1970). *J. Am. Chem. Soc.* **92**, 4978.

Anderson, R. J., Henrick, C. A., and Siddall, J. B. (1970). *J. Am. Chem. Soc.* **92**, 735.

Anderson, R. J., Henrick, C. A., Siddall, J. B., and Zurflüh, R. (1972). *J. Am. Chem. Soc.* **94**, 5379.

Anderson, R. J., Henrick, C. A., and Rosenblum, L. D. (1974). *J. Am. Chem. Soc.* **96**, 3654.

Anderson, R. J., Corbin, V. L., Cotterrell, G., Cox, G. R., Henrick, C. A., Schaub, F., and Siddall, J. B. (1975). *J. Am. Chem. Soc.* **97**, 1197.

Aratani, T., Yoneyoshi, Y., and Nagase, T. (1975). *Tetrahedron Lett.* p. 1707.

Ashby, E. C., and Heinsohn, G. (1974). *J. Org. Chem.* **39**, 3297.

Atkins, K. E., Walker, W. E., and Manyik, R. M. (1970). *Tetrahedron Lett.* p. 3821.

Atkinson, R. E., Curtis, R. F., Jones, D. M., and Taylor, J. A. (1967a). *Chem. Commun.* p. 718.

Atkinson, R. E., Curtis, R. F., and Taylor, J. A. (1967b). *J. Chem. Soc. C*, p. 578.

Atkinson, R. E., Curtis, R. F., and Phillips, G. T. (1967c). *J. Chem. Soc., C*, p. 2011.

Atkinson, R. E., Curtis, R. F., Jones, D. M., and Taylor, J. A. (1969). *J. Chem. Soc. C*, p. 2173.

Babler, J. H., and Mortell, T. R. (1972). *Tetrahedron Lett.* p. 669.

Bacon, R. G. R., and Hill, H. A. O. (1965). *Q. Rev., Chem. Soc.* **19**, 95.

Baddley, W. H., and Tupper, G. B. (1974). *J. Organomet. Chem.* **67**, C16.

Bähr, G., and Burba, P. (1970). *In* Houben-Weyl, "Methoden der organischen Chemie" (E. Müller, ed.), 4th ed., Vol. XIII, Part 1, p. 727. Thieme, Stuttgart.

Baker, R. (1973). *Chem. Rev.* **73**, 487.

Baker, R., Halliday, D. E., and Smith, T. N. (1972a). *J. Organomet. Chem.* **35**, C61.

Baker, R., Blackett, B. N., Cookson, R. C., Cross, R. C., and Madden, D. P. (1972b). *J. Chem. Soc., Chem. Commun.* p. 343.

Baker, R., Blackett, B. N., and Cookson, R. C. (1972c). *J. Chem. Soc., Chem. Commun.* p. 802.

Baker, R., Cookson, R. C., and Vinson, J. R. (1974a). *J. Chem. Soc., Chem. Commun.* p. 515.

Baker, R., Cook, A. H., and Smith, T. N. (1974b). *J. Chem. Soc., Perkin Trans. 2* p. 1517.

Barnett, B., Büssemeier, B., Heimbach, P., Jolly, P. W., Krüger, C., Tkatchenko, I., and Wilke, G. (1972). *Tetrahedron Lett.* p. 1457.

Bauld, N. L. (1962). *Tetrahedron Lett.* p. 859.

Bauld, N. L. (1963). *Tetrahedron Lett.* p. 1841.
Beckert, W. F., and Lowe, J. U., Jr. (1967). *J. Org. Chem.* **32**, 1215.
Bergbreiter, D. E., and Whitesides, G. M. (1974). *J. Am. Chem. Soc.* **96**, 4937.
Bergbreiter, D. E., and Whitesides, G. M. (1975). *J. Org. Chem.* **40**, 779.
Bertelo, C. A., and Schwartz, J. (1975). *J. Am. Chem. Soc.* **97**, 228.
Billups, W. E., Cross, J. H., and Smith, C. V. (1973). *J. Am. Chem. Soc.* **95**, 3438.
Binger, P., and McMeeking, J. (1974). *Angew. Chem.* **86**, 518.
Bird, C. W. (1967). "Transition Metal Intermediates in Organic Synthesis." Academic Press, New York.
Björklund, C., Nilsson, M., and Wennerström, O. (1970). *Acta Chem. Scand.* **24**, 3599.
Blomquist, A. T., and Maitlis, P. M. (1962). *J. Am. Chem. Soc.* **84**, 2329.
Boeckman, R. K., Jr. (1973). *J. Org. Chem.* **38**, 4450.
Bogdanović, B. (1973). *Angew. Chem., Int. Ed. Engl.* **12**, 954.
Bogdanović, B., Heimbach, P., Kroner, M., Wilke, G., Hoffman, E. G., and Brandt, J. (1969). *Justus Liebigs Ann. Chem.* **727**, 143.
Bomse, D. S., and Morton, T. H. (1975). *Tetrahedron Lett.* p. 781.
Bönnemann, H. (1973). *Angew. Chem., Int. Ed. Engl.* **12**, 964.
Bönnemann, H., Bogdanović, B., and Wilke, G. (1967). *Angew. Chem., Int. Ed. Engl.* **6**, 804.
Bowlus, S. B., and Katzenellenbogen, J. A. (1973a). *J. Org. Chem.* **38**, 2733.
Bowlus, S. B., and Katzenellenbogen, J. A. (1973b). *Tetrahedron Lett.* p. 1277.
Bozzato, G., Bachmann, J.-P., and Pesaro, M. (1974). *J. Chem. Soc., Chem. Commun.* p. 1005.
Braterman, P. S., and Cross, R. J. (1973). *Chem. Soc. Rev.* **2**, 271.
Braünling, H., Binnig, F., and Staab, H. A. (1967). *Chem. Ber.* **100**, 880.
Breil, H., and Wilke, G. (1970). *Angew. Chem., Int. Ed. Engl.* **9**, 367.
Brenner, W., Heimbach, P., Hey, H., Muller, E., and Wilke, G. (1969). *Justus Liebigs Ann. Chem.* **727**, 161.
Brown, M. P., Puddephatt, R. J., and Upton, C. E. E. (1973). *J. Organomet. Chem.* **49**, C61.
Bruggink, A., and McKillop, A. (1974). *Angew. Chem.* **86**, 349.
Büchi, G., and Carlson, J. A. (1968). *J. Am. Chem. Soc.* **90**, 5356.
Burdon, J., Coe, P. L., Marsh, C. R., and Tatlow J. C. (1967). *Chem. Commun.* p. 1259.
Burdon, J., Coe, P. L., Marsh, C. R., and Tatlow, J. C. (1972a). *J. Chem. Soc., Perkin Trans. 1* p. 639.
Burdon, J., Coe, P. L., Marsh, C. R., and Tatlow, J. C. (1972b). *J. Chem. Soc., Perkin Trans. 1* p. 763.
Büssemeier, B., Jolly, P. W., and Wilke, G. (1974). *J. Am. Chem. Soc.* **96**, 4726.
Cacchi, S., Caputo, A., and Misiti, D. (1974). *Indian J. Chem.* **12**, 325.
Cairncross, A., and Sheppard, W. A. (1971). *J. Am. Chem. Soc.* **93**, 247.
Cairncross, A., Roland, J. R., Henderson, R. M., and Sheppard, W. A. (1970). *J. Am. Chem. Soc.* **92**, 3187.
Cairncross, A., Omura, H., and Sheppard, W. A. (1971). *J. Am. Chem. Soc.* **93**, 248.
Candlin, J. P., Taylor, K. A., and Thompson, D. T. (1968). "Reactions of Transition-Metal Complexes." Elsevier, Amsterdam.
Carlson, R. G., and Mardis, W. S. (1975). *J. Org. Chem.* **40**, 817.
Carlson, R. G., and Zey, E. G. (1972). *J. Org. Chem.* **37**, 2468.
Casey, C. P., and Boggs, R. A. (1971). *Tetrahedron Lett.* p. 2455.
Casey, C. P., and Marten, D. F. (1974). *Tetrahedron Lett.* p. 925.
Casey, C. P., Marten, D. F., and Boggs, R. A. (1973). *Tetrahedron Lett.* p. 2071.

Cassar, L. (1973). *J. Organomet. Chem.* **54**, C57.
Castro, C. E., Gaughan, E. J., and Owsley, D. C. (1966). *J. Org. Chem.* **31**, 4071.
Castro, C. E., Havlin, R., Honwad, V. K., Malte, A., and Mojé, S. (1969). *J. Am. Chem. Soc.* **91**, 6464.
Cavazza, M., and Pietra, F. (1974). *J. Chem. Soc., Chem. Commun.* p. 501.
Chavdarian, C. G., and Heathcock, C. H. (1975). *J. Am. Chem. Soc.* **97**, 3822.
Chini, P., Santambrogio, A., and Palladino, N. (1967). *J. Chem. Soc. C* p. 830.
Chiusoli, G. P. (1969). *Bull. Soc. Chim. Fr.* p. 1139.
Chiusoli, G. P. (1971). *Proc. Int. Congr. Pure Appl. Chem.* Vol. 6, 169-199.
Chiusoli, G. P., and Cassar, L. (1967). *Angew. Chem., Int. Ed. Engl.* **6**, 124.
Clarke, H. L. (1974). *J. Organomet. Chem.* **80**, 155.
Coates, G. E. (1968). "Organometallic Compounds," 2nd ed. Methuen, London.
Coates, R. M., and Sandefur, L. O. (1974). *J. Org. Chem.* **39**, 275.
Coates, R. M., and Sowerby, R. L. (1971). *J. Am. Chem. Soc.* **93**, 1027.
Coates, R. M., and Sowerby, R. L. (1972). *J. Am. Chem. Soc.* **94**, 5386.
Coe, P. L., and Milner, N. E. (1972). *J. Organomet. Chem.* **39**, 395.
Coffey, C. E. (1961). *J. Am. Chem. Soc.* **83**, 1623.
Cohen, T., and Poeth, T. (1972). *J. Am. Chem. Soc.* **94**, 4363.
Collman, J. P., and Hoffman, N. W. (1973). *J. Am. Chem. Soc.* **95**, 2689.
Collman, J. P., and Kang, J. W. (1967). *J. Am. Chem. Soc.* **89**, 844.
Collman, J. P., Kang, J. W., Little, W. F., and Sullivan, M. F. (1968). *Inorg. Chem.* **7**, 1298.
Collman, J. P., Winter, S. R., and Clark, D. R. (1972a). *J. Am. Chem. Soc.* **94**, 1788.
Collman, J. P., Cawse, J. N., and Brauman, J. I. (1972b). *J. Am. Chem. Soc.* **94**, 5905.
Collman, J. P., Winter, S. R., and Komoto, R. G. (1973). *J. Am. Chem. Soc.* **95**, 249.
Consiglio, G., and Botteghi, C. (1973). *Helv. Chim. Acta* **56**, 460.
Cooke, M. P., Jr. (1970). *J. Am. Chem. Soc.* **92**, 6080.
Cooke, M. P., Jr. (1973). *Tetrahedron Lett.* p. 1281.
Corey, E. J., and Achiwa, K. (1970). *Tetrahedron Lett.* p. 2245.
Corey, E. J., and Beams, D. J. (1972). *J. Am. Chem. Soc.* **94**, 7210.
Corey, E. J., and Broger, E. A. (1969). *Tetrahedron Lett.* p. 1779.
Corey, E. J., and Chen, R. H. K. (1973a). *Tetrahedron Lett.* p. 1611.
Corey, E. J., and Chen, R. H. K. (1973b). *Tetrahedron Lett.* p. 3817.
Corey, E. J., and Fuchs, P. L. (1972). *J. Am. Chem. Soc.* **94**, 4014.
Corey, E. J., and Hamanaka, E. (1967). *J. Am. Chem. Soc.* **89**, 2758.
Corey, E. J., and Hegedus, L. S. (1969a). *J. Am. Khem. Soc.* **91**, 1233.
Corey, E. J., and Hegedus, L. S. (1969b). *J. Am. Chem. Soc.* **91**, 4926.
Corey, E. J., and Jautelat, M. (1967). *J. Am. Chem. Soc.* **89**, 3912.
Corey, E. J., and Jautelat, M. (1968). *Tetrahedron Lett.* p. 5787.
Corey, E. J., and Katzenellenbogen, J. A. (1969). *J. Am. Chem. Soc.* **91**, 1851.
Corey, E. J., and Kirst, H. A. (1972). *J. Am. Chem. Soc.* **94**, 667.
Corey, E. J., and Kuwajima, I. (1970). *J. Am. Chem. Soc.* **92**, 395.
Corey, E. J., and Kuwajima, I. (1972). *Tetrahedron Lett.* p. 487.
Corey, E. J., and Mann, J. (1973). *J. Am. Chem. Soc.* **95**, 6832.
Corey, E. J., and Posner, G. H. (1967). *J. Am. Chem. Soc.* **89**, 3911.
Corey, E. J., and Posner, G. H. (1968). *J. Am. Chem. Soc.* **90**, 5615.
Corey, E. J., and Posner, G. H. (1970). *Tetrahedron Lett.* p. 315.
Corey, E. J., and Semmelhack, M. F. (1966). *Tetrahedron Lett.* p. 6237.
Corey, E. J., and Semmelhack, M. F. (1967). *J. Am. Chem. Soc.* **89**, 2755.
Corey, E. J., and Wat, E. K. W. (1967). *J. Am. Chem. Soc.* **89**, 2757.
Corey, E. J., and Wollenberg, R. H. (1974). *J. Am. Chem. Soc.* **96**, 5581.

Corey, E. J., Katzenellenbogen, J. A., and Posner, G. H. (1967). *J. Am. Chem. Soc.* **89**, 4245.
Corey, E. J., Semmelhack, M. F., and Hegedus, L. S. (1968a). *J. Am. Chem. Soc.* **90**, 2416.
Corey, E. J., Hegedus, L. S., and Semmelhack, M. F. (1968b). *J. Am. Chem. Soc.* **90**, 2417.
Corey, E. J., Katzenellenbogen, J. A., Gilman, N. W., Roman, S. A., and Erickson, B. W. (1968c). *J. Am. Chem. Soc.* **90**, 5618.
Corey, E. J., Achiwa, K., and Katzenellenbogen, J. A. (1969). *J. Am. Chem. Soc.* **91**, 4318.
Corey, E. J., Narisada, M., Hiraoka, T., and Ellison, R. A. (1970a). *J. Am. Chem. Soc.* **92**, 396.
Corey, E. J., Kirst, H. A., and Katzenellenbogen, J. A. (1970b). *J. Am. Chem. Soc.* **92**, 6314.
Corey, E. J., Kim, C. U., Chen, R. H. K., and Takada, M. (1972). *J. Am. Chem. Soc.* **94**, 4395.
Corriu, R. J. P., and Masse, J. P. (1972). *J. Chem. Soc., Chem. Commun.* p. 144.
Crandall, J. K., and Michaely, W. J. (1973). *J. Organometal. Chem.* **51**, 375.
Curtis, R. F., and Taylor, J. A. (1966). *J. Chem. Soc., C* p. 1813.
Danehy, J. P., Killian, D. B., and Niewuland, J. A. (1936). *J. Am. Chem. Soc.* **58**, 611.
Das Gupta, S. K., Crump, D. R., and Gut, M. (1974). *J. Org. Chem.* **39**, 1658.
Dauben, W. G., Beasley, G. H., Broadhurst, M. D., Muller, B., Peppard, D. J., Pesnele, P., and Suter, C. (1974). *J. Am. Chem. Soc.* **97**, 4724.
Daviaud, G., and Miginiac, P. (1972). *Tetrahedron Lett.* p. 997.
DePasquale, R. J., and Tamborski, C. (1969). *J. Org. Chem.* **34**, 1736.
Descoins, C., Henrick, C. A., and Siddall, J. B. (1972). *Tetrahedron Lett.* p. 3777.
Dieck, H. A., and Heck, R. F. (1974). *J. Am. Chem. Soc.* **96**, 1133.
Dietrich, H., and Uttech, R. (1963). *Naturwissenschaften* **50**, 613.
Dietrich, H., and Uttech, R. (1965). *Z. Kristallogr., Kristallgeom., Kristallphys., Kristallchem.* **122**, 60.
Dubini, M., and Montino, F. (1966). *J. Organomet. Chem.* **6**, 188.
Dubini, M., Montino, F., and Chiusoli, G. P. (1965). *Chim. Ind. (Milan)* **47**, 839.
Dubois, J.-E., and Boussu, M. (1969). *C. R. Hebd. Seances Acad. Sci., Ser. C* **268**, 1603.
Dubois, J.-E., and Boussu, M. (1970). *Tetrahedron Lett.* p. 2523.
Dubois, J.-E., and Boussu, M. (1971). *C. R. Hebd. Seances Acad. Sci., Ser. C* **273**, 1101.
Dubois, J.-E., Leheup, B., Hennequin, F., and Bauer, P. (1967a). *Bull. Soc. Chim. Fr.* p. 1150.
Dubois, J.-E., Chastrette, M., and Létoquart, C. (1967b). *C. R. Hebd. Seances Acad. Sci., Ser. C* **264**, 1124.
Dubois, J.-E., Hennequin, F., and Boussu, M. (1969). *Bull. Soc. Chim. Fr.* p. 3615.
Dubois, J.-E., Boussu, M., and Lion, C. (1971). *Tetrahedron Lett.* p. 829.
Duboudin, J.-G., and Jousseaume, B. (1972). *J. Organomet. Chem.* **44**, C1.
Dunne, K., and McQuillin, F. J. (1970). *J. Chem. Soc. C* pp. 2196, 2200, and 2203.
Eglinton, G., and McCrae, W. (1963). *Adv. Org. Chem.* **4**, 225.
Falbe, J. (1970). "Carbon Monoxide in Organic Synthesis." Springer-Verlag, Berlin and New York.
Fanta, P. E. (1964). *Chem. Rev.* **64**, 613.
Fanta, P. E. (1974). *Synthesis* p. 9.
Felkin, H., and Swierczewski, G. (1972). *Tetrahedron Lett.* p. 1433.
Fischer, E. O., and Bürger, G. (1961). *Z. Naturforsch.* **16B**, 77.
Fischer, K., Jonas, K., Misbach, P., Stabba, R., and Wilke, G. (1973). *Angew. Chem., Int. Ed. Engl.* **12**, 943.
Fouquet, G., and Schlosser, M. (1974). *Angew. Chem., Int. Ed. Engl.* **13**, 82.

Fraser, A. R., Bird, P. H., Bezman, S. A., Shapley, J. R., White, R., and Osborn, J. A. (1973). *J. Am. Chem. Soc.* **95**, 597.
Fried, J., Lin, C. H., Sih, J. C., Delven, P., and Cooper, G. F. (1972). *J. Am. Chem. Soc.* **94**, 4342.
Friedman, L., and Shani, A. (1974). *J. Am. Chem. Soc.* **96**, 7101.
Fryer, R. I., Gilman, N. W., and Holland, B. C. (1975). *J. Org. Chem.* **40**, 348.
Fukuoka, S., Ryang, M., and Tsutsumi, S. (1971). *J. Org. Chem.* **36**, 2721.
Ganem, B. (1974). *Tetrahedron Lett.* p. 4467.
Gardner, S. A., Andrews, P. S., and Rausch, M. D. (1973). *Inorg. Chem.* **12**, 2396.
Garratt, P. J., and Wyatt, M. (1974). *J. Chem. Soc., Chem. Commun.* p. 251.
Garty, N., and Michman, M. (1972). *J. Organomet. Chem.* **36**, 391.
Gilman, H., and Kirby, J. E. (1929). *Recl. Trav. Chim. Pays-Bas* **48**, 155.
Gilman, H., Jones, R. G., and Woods, L. A. (1952). *J. Org. Chem.* **17**, 1630.
Gorlier, J.-P., Hamon, L., Levisalles, J., and Wagnon, J. (1973). *J. Chem. Soc., Chem. Commun.* p. 88.
Grandjean, J., Laszlo, P., and Stockis, A. (1974). *J. Am. Chem. Soc.* **96**, 1622.
Grevels, F.-W., Schulz, D., and Koerner von Gustorf, E. (1974). *Angew. Chem., Int. Ed. Engl.* **13**, 534.
Grieco, P. A., and Finkelhor, R. (1973). *J. Org. Chem.* **38**, 2100.
Grudzinskas, C. V., and Weiss, M. J. (1973). *Tetrahedron Lett.* p. 141.
Guerrieri, F., and Chiusoli, G. P. (1969). *Chim. Ind.* (*Milan*) **51**, 1252.
Guerrieri, F., Chiusoli, G. P., and Merzoni, S. (1974). *Gazz. Chim. Ital.* **104**, 557.
Gump, K., Moje, S. W., and Castro, C. E. (1967). *J. Am. Chem. Soc.* **89**, 6770.
Guthrie, D. J. S., and Nelson, S. M. (1972). *Coord. Chem. Rev.* **8**, 139.
Hannah, D. J., and Smith, R. A. J. (1975). *Tetrahedron Lett.* p. 187.
Harrison, I. T., Kimura, E., Bohme, E., and Fried, J. H. (1969). *Tetrahedron Lett.* p. 1589.
Hart, D. W., and Schwartz, J. (1974). *J. Am. Chem. Soc.* **96**, 8115.
Hashimoto, H., and Nakano, T. (1966). *J. Org. Chem.* **31**, 891.
Hashimoto, I., Ryang, M., and Tsutsumi, S. (1969). *Tetrahedron Lett.* p. 3291.
Hashimoto, I., Tsuruta, N., Ryang, M., and Tsutsumi, S. (1970a). *J. Org. Chem.* **35**, 3748.
Hashimoto, I., Ryang, M., and Tsutsumi, S. (1970b). *Tetrahedron Lett.* p. 4567.
Hata, G., Takahashi, K., and Miyake, A. (1969). *Chem. Ind.* (*London*) p. 1836.
Hata, G., Takahashi, K., and Miyake, A. (1971). *J. Org. Chem.* **36**, 2116.
Hayakawa, Y., Sakai, M., and Noyori, R. (1975). *Chem. Lett.* p. 509.
Heck, R. F. (1968a). *In* "Organic Syntheses via Metal Carbonyls" I. Wender and P. Pino, eds.), Vol. 1, p. 373. Wiley (Interscience), New York.
Heck, R. F. (1968b). *J. Am. Chem. Soc.* **90**, 5518 and 5546.
Heck, R. F. (1974). "Organotransition Metal Chemistry. A Mechanistic Approach," Academic Press, New York.
Hegedus, L. S., and Stiverson, R. K. (1974). *J. Am. Chem. Soc.* **96**, 3250.
Hegedus, L. S., and Miller, L. L. (1975). *J. Am. Chem. Soc.* **97**, 459.
Hegedus, L. S., and Waterman, E. L. (1974). *J. Am. Chem. Soc.* **96**, 6789.
Hegedus, L. S., Waterman, E. L., and Catlin, J. (1972). *J. Am. Chem. Soc.* **94**, 7155.
Hegedus, L. S., Lo, S. M., and Bloss, D. E. (1973). *J. Am. Chem. Soc.* **95**, 3040.
Hegedus, L. S., Kendall, P. M., Lo, S. M., and Sheats, J. R. (1975a). *J. Am. Chem. Soc.* **97**, 5448.
Hegedus, L. S., Wagner, S. D., Waterman, E. L., and Siirala-Hansen, K. (1975b). *J. Org. Chem.* **40**, 593.
Heimbach, P. (1973). *Angew. Chem., Int. Ed. Engl.* **12**, 975.
Heimbach, P., and Wilke, G. (1969). *Justus Liebigs Ann. Chem.* **727**, 183.

Heimbach, P., Jolly, P. W., and Wilke, G. (1970). *Adv. Organomet. Chem.* **8**, 29.
Heng, K. H., and Smith, R. A. J. (1975). *Tetrahedron Lett.* p. 589.
Henry, P. M. (1968). *Tetrahedron Lett.* p. 2285.
Herr, R. W., and Johnson, C. R. (1970). *J. Am. Chem. Soc.* **92**, 4979.
Herr, R. W., Wieland, D. M., and Johnson, C. R. (1970). *J. Am. Chem. Soc.* **92**, 3813.
Hill, R. K., and Spencer, H. K. (1974). *168th Am. Chem. Soc. Natl. Meet.*, 1974 ORGN 87.
Hirota, Y., Ryang, M., and Tsutsumi, S. (1971). *Tetrahedron Lett.* p. 1531.
Hodgson, G. L., MacSweeney, D. F., Mills, R. W., and Money, T. (1973). *J. Chem. Soc., Chem. Commun.* p. 235.
Hoogzand, C., and Hübel, W. (1968). *In* "Organic Syntheses via Metal Carbonyls" (I. Wender and P. Pino, eds.), p. 349. Wiley (Interscience), New York.
Hooz, J., and Layton, R. B. (1970). *Can. J. Chem.* **48**, 1626.
House, H. O., and Fischer, W. F., Jr. (1968). *J. Org. Chem.* **33**, 949.
House, H. O., and Fischer, W. F., Jr. (1969). *J. Org. Chem.* **34**, 3615.
House, H. O., and Umen, M. J. (1972). *J. Am. Chem. Soc.* **94**, 5495.
House, H. O., and Umen, M. J. (1973). *J. Org. Chem.* **38**, 3893.
House, H. O., and Weeks, P. D. (1975). *J. Am. Chem. Soc.* **97**, 2770, 2778.
House, H. O., Respess, W. L., and Whitesides, G. M. (1966). *J. Org. Chem.* **31**, 3128.
House, H. O., Huber, L. E., and Umen, M. J. (1972). *J. Am. Chem. Soc.* **94**, 8471.
Hughes, R. P., and Powell, J. (1971). *Chem. Commun.* p. 275.
Hughes, R. P., and Powell, J. (1972). *J. Am. Chem. Soc.* **94**, 7723.
Iataaki, H., and Yoshimoto, H. (1973). *J. Org. Chem.* **38**, 76.
Inoue, S., Saito, K., Kato, K., Nozaki, S., and Sato, K. (1974). *J. Chem. Soc., Perkin Trans. 1* p. 2097.
Ireland, R. E., Lipinski, C. A., Kowalski, C. J., Tilley, J. W., and Walba, D. M. (1974). *J. Am. Chem. Soc.* **96**, 3333.
Ito, Y., Yonezawa, K., and Saegusa, T. (1974a). *J. Org. Chem.* **39**, 1763.
Ito, Y., Nakayama, K., Yonezawa, K., and Saegusa, T. (1974b). *J. Org. Chem.* **39**, 3273.
Jallabert, C., Luong-Thi, N.-T., and Rivière, H. (1970). *Bull. Soc. Chim. Fr.* p. 797.
James, B. R. (1973). "Homogeneous Hydrogenation." Wiley, New York.
Johnson, C. R., and Dutra, G. A. (1973). *J. Am. Chem. Soc.* **95**, 7777 and 7783.
Jolly, P. W., and Wilke, G. (1974–1975). "The Organic Chemistry of Nickel," Vols. I and II. Academic Press, New York.
Jones, E. R. H., Shen, T. Y., and Whiting, M. C. (1950). *J. Chem. Soc.* p. 230.
Josey, A. D. (1974). *J. Org. Chem.* **39**, 139.
Jukes, A. E. (1974). *Adv. Organomet. Chem.* **12**, 215.
Jukes, A. E., Dua, S. S., and Gilman, H. (1968). *J. Organomet. Chem.* **12**, P44.
Jukes, A. E., Dua, S. S., and Gilman, H. (1970a). *J. Organomet. Chem.* **21**, 241.
Jukes, A. E., Dua, S. S., and Gilman, H. (1970b). *J. Organomet. Chem.* **24**, 791.
Julia, M., and Saussine, L. (1974). *Tetrahedron Lett.* p. 3443.
Kalli, M., Landor, P. D., and Landor, S. R. (1972). *J. Chem. Soc., Chem. Commun.* p. 593.
Katzenellenbogen, J. A., and Corey, E. J. (1972). *J. Org. Chem.* **37**, 1441.
Katzenellenbogen, J. A., and Utawanit, T. (1974). *J. Am. Chem. Soc.* **96**, 6153.
Kauffmann, T. (1974). *Angew. Chem., Int., Ed. Engl.* **13**, 291.
Kauffmann, T., and Sahm, W. (1967). *Angew. Chem., Int. Ed. Engl.* **6**, 85.
Kauffmann, T., Jackisch, J., Woltermann, A., and Röwemeier, P. (1972a). *Angew. Chem., Int. Ed. Engl.* **11**, 844.
Kauffmann, T., Legler, J., Ludorff, E., and Fischer, H. (1972b). *Angew. Chem., Int. Ed. Engl.* **11**, 846.
Kende, A. S., Liebeskind, L. S., and Braitsch, D. M. (1975). *Tetrahedron Lett.* p. 3375.

Kern, R. J. (1968). *Chem. Commun.* p. 706.
Kharasch, M. S., and Tawney, P. O. (1941). *J. Am. Chem. Soc.* **63**, 2308.
Kharasch, M. S., Kleiger, S. C., Martin, J. A., and Mayo, F. R. (1941). *J. Am. Chem. Soc.* **63**, 2305.
Kiso, Y., Tamao, K., and Kumada, M. (1973). *J. Organomet. Chem.* **50**, C12.
Kiso, Y., Tamao, K., Miyake, N., Yamamoto, K., and Kumada, M. (1974). *Tetrahedron Lett.* p. 3.
Klein, J., and Levene, R. (1972). *J. Am. Chem. Soc.* **94**, 2520.
Klein, J., and Turk, R. M. (1969). *J. Am. Chem. Soc.* **91**, 6186.
Kluge, A. F., Untch, K. G., and Fried, J. H. (1972a). *J. Am. Chem. Soc.* **94**, 7827.
Kluge, A. F., Untch, K. G., and Fried, J. H. (1972b). *J. Am. Chem. Soc.* **94**, 9256.
Knight, D. W., and Pettenden, G. (1974). *J. Chem. Soc., Chem. Commun.* p. 188.
Kobayashi, S., and Mukaiyama, T. (1974). *Chem. Lett.* pp. 705 and 1425.
Kochi, J. K. (1974). *Acc. Chem. Res.* **7**, 351.
Kricka, K. J., and Ledwith, A. (1974). *Synthesis* p. 539.
Kuwajima, I., and Doi, Y. (1972). *Tetrahedron Lett.* p. 1163.
Langler, R. F., and Tidwell, T. T. (1975). *Tetrahedron Lett.* p. 777.
Light, J. R. C., and Zeiss, H. H. (1970). *J. Organomet. Chem.* **21**, 517.
Luong-Thi, N.-T., and Rivière, H. (1968). *C. R. Hebd. Seances Acad. Sci., Ser. C* **267**, 776.
Luong-Thi, N.-T., and Rivière, H. (1970). *Tetrahedron Lett.* pp. 1579 and 1583.
Luong-Thi, N.-T., and Rivière, H. (1971). *Tetrahedron Lett.* p. 587.
Luong-Thi, N.-T., Rivière, H., Bégué, J.-P., and Forestier, C. (1971). *Tetrahedron Lett.* p. 2113.
McDermott, J. X., and Whitesides, G. M. (1974). *J. Am. Chem. Soc.* **96**, 947.
McLoughlin, V. C. R., and Thrower, J. (1969). *Tetrahedron* **25**, 5921.
McMurry, J. E. (1974). *Acc. Chem. Res.* **9**, 281.
McMurry, J. E., and Fleming, M. P. (1974). *J. Am. Chem. Soc.* **96**, 4708.
MacPhee, J. A., and Dubois, J.-E. (1972). *Tetrahedron Lett.* p. 467.
Maitlis, P. M. (1971). "The Organic Chemistry of Palladium," Vols. 1 and 2. Academic Press, New York.
Maitlis, P. M. (1973). *Pure Appl. Chem.* **33**, 489.
Malte, A. M., and Castro, C. E. (1967). *J. Am. Chem. Soc.* **89**, 6770.
Mango, F. D., and Schachtschneider, J. H. (1971). *In* "Transition Metals in Homogeneous Catalysis" (G. N. Schrauzer, ed.), Chapter 6, p. 223. Dekker, New York.
Mantzaris, J., and Weissberger, E. (1974). *J. Am. Chem. Soc.* **96**, 1873, 1880.
Marino, J. P., and Floyd, D. M. (1974). *J. Am. Chem. Soc.* **96**, 7138.
Marshall, J. A., and Cohen, G. M. (1971). *J. Org. Chem.* **36**, 877.
Marshall, J. A., and Roebke, H. (1968). *J. Org. Chem.* **33**, 840.
Marshall, J. A., and Ruden, R. A. (1971). *Tetrahedron Lett.* p. 2875.
Marshall, J. A., and Ruth, J. A. (1974). *J. Org. Chem.* **39**, 1971.
Marshall, J. A., Fanta, W. I., and Roebke, H. (1966). *J. Org. Chem.* **31**, 1016.
Marshall, J. A., Ruden, R. A., Hirsch, L. K., and Phillippe, M. (1971). *Tetrahedron Lett.* p. 3795.
Masada, H., Mizuno, M., Suga, S., Watanabe, Y., and Takegami, Y. (1970). *Bull. Chem. Soc. Jpn.* **43**, 3824.
Masamune, S., Kim, C. U., Wilson, K. E., Spessard, G. O., Georghiou, P. E., and Bates, G. S. (1975). *J. Am. Chem. Soc.* **97**, 3512.
Medema, D., and van Helden, R. (1969). *Prepr., Div. Pet. Chem., Am. Chem. Soc.* **14**, B92.
Medema, D., and van Helden, R. (1971). *Recl. Trav. Chim. Pays-Bas* **90**, 304.
Medema, D., van Helden, R., and Kohll, C. F. (1969). *Inorg. Chem. Acta* **3**, 2550.

Meriwether, L. S., Leto, M. F., Colthup, E. C., and Kennerly, G. W. (1962). *J. Org. Chem.* **27**, 3930.
Mérour, J. Y., Roustan, J. L., Charrier, C., and Collin, J. (1973). *J. Organomet. Chem.* **51**, C24.
Miller, J. G., Kurz, W., Untch, K. G., and Stork, G. (1974). *J. Am. Chem. Soc.* **96**, 6774.
Miller, R. G., Fahey, D. R., and Kuhlman, D. P. (1968). *J. Am. Chem. Soc.* **90**, 6248.
Mitsudo, T., Watanabe, Y., Tanaka, M., Yamamoto, K., and Takegami, Y. (1972). *Bull. Chem. Soc. Jpn.* **45**, 305.
Mitsudo, T., Watanabe, Y., Yamashita, M., and Takegami, Y. (1974). *Chem. Lett.* p. 1385.
Mitsuyasu, T., Hara, M., and Tsuji, J. (1971). *Chem. Commun.* p. 345.
Miyake, A., Kondo, H., and Nishino, M. (1971a). *Angew. Chem., Int. Ed. Engl.* **10**, 802.
Miyake, A., Kondo, H., Nishino, M., and Tokizane, S. (1971b). *Proc. Int. Congr. Pure Appl. Chem.* **23**, Vol. 6, 201-211.
Mladenović, S. A., and Castro, C. E. (1968). *J. Heterocycl. Chem.* **5**, 227.
Mori, K., Mizoroki, T., and Ozaki, A. (1973). *Bull. Chem. Soc. Jpn.* **46**, 1505.
Moseley, K., and Maitlis, P. M. (1974). *J. Chem. Soc., Dalton Trans.* p. 169.
Moser, W. R. (1969). *J. Am. Chem. Soc.* **91**, 1135.
Mukaiyama, T., Sato, T., and Hanna, J. (1973). *Chem. Lett.* p. 1041.
Müller, E. (1974). *Synthesis* p. 761.
Müller, E., and Beissner, C. (1973). *Chem.-Ztg.* **97**, 207.
Murahashi, S.-I., Tanba, Y., Yamamura, M., and Moritani, I. (1974). *Tetrahedron Lett.* p. 3749.
Näf, P., and Degen, P. (1971). *Helv. Chim. Acta* **54**, 1939.
Näf, P., Degen, P., and Ohloff, G. (1972). *Helv. Chim. Acta* **55**, 82.
Neumann, S. M., and Kochi, J. K. (1975). *J. Org. Chem.* **40**, 599.
Nilsson, M., and Ullenius, C. (1970). *Acta Chem. Scand.* **24**, 2379.
Nilsson, M., and Ullenius, C. (1971). *Acta Chem. Scand.* **25**, 2428.
Nilsson, M., and Wahren, R. (1969). *J. Organomet. Chem.* **16**, 515.
Nilsson, M., and Wennerström, O. (1969). *Tetrahedron Lett.* p. 3307.
Nilsson, M., and Wennerström, O. (1970). *Acta Chem. Scand.* **24**, 482.
Nilsson, M., Wahren, R., and Wennerström, O. (1970). *Tetrahedron Lett.* p. 4583.
Normant, J. F. (1972). *Synthesis* p. 63.
Normant, J. F., and Bourgain, M. (1970). *Tetrahedron Lett.* p. 2659.
Normant, J. F., and Bourgain, M. (1971). *Tetrahedron Lett.* p. 2583.
Normant, J. F., Cahiez, G., Chuit, C., and Alexakis, A. (1972). *J. Organomet. Chem.* **40**, C49.
Normant, J. F., Cahiez, G., and Chuit, C. (1973). *J. Organomet. Chem.* **54**, C53.
Normant, J. F., Cahiez, G., Chuit, C., and Villieras, J. (1974). *J. Organomet. Chem.* **77**, 269, 281.
Norton, J. R., Shenton, K. E., and Schwartz, J. (1975). *Tetrahedron Lett.* p. 51.
Noyori, R. (1973). *Tetrahedron Lett.* p. 1691.
Noyori, R. (1975). *In* "Organotransition-Metal Chemistry" (Y. Ishii and M. Tsutsui, eds.), pp. 231-241. Plenum, New York.
Noyori, R., Takaya, H., Nakanisi, Y., and Nozaki, H. (1969). *Can. J. Chem.* **47**, 1242.
Noyori, R., Odagi, T., and Takaya, H. (1970). *J. Am. Chem. Soc.* **92**, 5780.
Noyori, R., Makino, S., and Takaya, H. (1971a). *J. Am. Chem. Soc.* **93**, 1272.
Noyori, R., Suzuki, T., Kumagai, Y., and Takaya, H. (1971b). *J. Am. Chem. Soc.* **93**, 5894.
Noyori, R., Suzuki, T., and Takaya H. (1971c). *J. Am. Chem. Soc.* **93**, 5896.

Noyori, R., Yokoyama, K., Makino, S., and Hayakawa, Y. (1972a). *J. Am. Chem. Soc.* **94**, 1772.
Noyori, R., Kumagai, Y., Umeda, I., and Takaya, H. (1972b). *J. Am. Chem. Soc.* **94**, 4018.
Noyori, R., Umeda, I., and Ishigami, T. (1972c). *J. Org. Chem.* **37**, 1542.
Noyori, R., Hayakawa, Y., Funakura, M., Takaya, H., Murai, S., Kobayashi, R., and Tsutsumi, S. (1972d). *J. Am. Chem. Soc.* **94**, 7202.
Noyori, R., Hayakawa, Y., Makino, S., and Takaya, H. (1973a). *Chem. Lett.* p. 3.
Noyori, R., Baba, Y., Makino, S , and Takaya, H. (1973b). *Tetrahedron Lett.* p. 1741.
Noyori, R., Makino, S., and Takaya, H. (1937c). *Tetrahedron Lett.* p. 1745.
Noyori, R., Yokoyama, K., and Hayakawa, Y. (1973d). *J. Am. Chem. Soc.* **95**, 2722.
Noyori, R., Hayakawa, Y., Makino, S., Hayakawa, N., and Takaya, H. (1973e). *J. Am. Chem. Soc.* **95**, 4103.
Noyori, R., Kumagai, Y., and Takaya, H. (1974a). *J. Am. Chem. Soc.* **96**, 634.
Noyori, R., Baba, Y., and Hayakawa, Y. (1974b). *J. Am. Chem. Soc.* **96**, 3336.
Noyori, R., Makino, S., Baba, Y., and Hayakawa, Y. (1974c). *Tetrahedron Lett.* p. 1049.
Noyori, R., Kawauchi, H., and Takaya, H. (1974d). *Tetrahedron Lett.* p. 1749.
Noyori, R., Umeda, I., Kawauchi, H., and Takaya, H. (1975a). *J. Am. Chem. Soc.* **97**, 812.
Noyori, R., Makino, S., Okita, T., and Hayakawa, Y. (1975b). *J. Org. Chem.* **40**, 806.
Noyori, R., Souchi, T., and Hayakawa, Y. (1975c). *J. Org. Chem.* **40**, 2681.
Nozaki, H., Moriuti, S., Takaya, H., and Noyori, R. (1966). *Tetrahedron Lett.* p. 5239.
Nozaki, H., Takaya, H., Moriuti, S., and Noyori, R. (1968). *Tetrahedron* **24**, 3655.
Ohbe, Y., and Matsuda, T. (1973). *Tetrahedron* **29**, 2989.
Onoue, H., Moritani, I., and Murahashi, S.-I. (1973). *Tetrahedron Lett.* p. 121.
Onuma, K., and Hashimoto, H. (1972). *Bull. Chem. Soc. Jpn.* **45**, 2582.
Oshima, K., Yamamoto, H., and Nozaki, H. (1973). *J. Am. Chem. Soc.* **95**, 7926.
Otsuka, S., Nakamura, A., Yoshikatsu, T., Naruto, M., and Ataka, K. (1973). *J. Am. Chem. Soc.* **95**, 3180.
Owsley, D. C., and Castro, C. E. (1972). *Org. Synth.* **52**, 128.
Patterson, J. W., Jr., and Fried, J. H. (1974). *J. Org. Chem.* **39**, 2506.
Pesaro, M., Bozzato, G., and Schudel, P. (1968). *Chem. Commun.* p. 1152.
Piers, E., and Keziere, R. J. (1968). *Tetrahedron Lett.* p. 583.
Piers, E., and Keziere, R. J. (1969). *Can. J. Chem.* **47**, 137.
Posner, G. H. (1972). *Org. React.* **19**, 1.
Posner, G. H. (1975). *Org. React.* **22**, 253.
Posner, G. H., and Brunelle, D. J. (1972). *Tetrahedron Lett.* p. 293.
Posner, G. H., and Brunelle, D. J. (1973a). *Tetrahedron Lett.* p. 935.
Posner, G. H., and Brunelle, D. J. (1973b). *J. Org. Chem.* **38**, 2747.
Posner, G. H., and Brunelle, D. J. (1973c). *J. Chem. Soc., Chem. Commun.* p. 907.
Posner, G. H., and Sterling, J. J. (1973). *J. Am. Chem. Soc.* **95**, 3076.
Posner, G. H., and Ting, J.-S. (1974). *Tetrahedron Lett.* p. 683.
Posner, G. H., and Whitten, C. E. (1973). *Tetrahedron Lett.* p. 1815.
Posner, G. H., Whitten, C. E., and McFarland, P. E. (1970). *Tetrahedron Lett.* p. 4647.
Posner, G. H., Whitten, C. E., and McFarland, P. E. (1972). *J. Am. Chem. Soc.* **94**, 5106.
Posner, G. H., Whitten, C. E., and Sterling, J. J. (1973). *J. Am. Chem. Soc.* **95**, 7788.
Posner, G. H., Whitten, C. E., Sterling, J. J., and Brunelle, D. J. (1974). *Tetrahedron Lett.* p. 2591.
Posner, G. H., Sterling, J. J., Whitten, C. E., Lentz, C. M., and Brunelle, D. J. (1975a). *J. Am. Chem. Soc.* **97**, 107.
Posner, G. H., Loomis, G. L., and Sawaya, H. S. (1975b). *Tetrahedron Lett.* p. 1373.
Rapson, W. S., Shuttleworth, R. G., and van Niekerk, J. N. (1943). *J. Chem. Soc.* p. 326.

Rausch, M. D., Siegel, A., and Klemann, L. P. (1966). *J. Org. Chem.* **31**, 2703.
Rausch, M. D., Siegel, A., and Klemann, L. P. (1969). *J. Org. Chem.* **34**, 468.
Reich, R. (1923). *C. R. Hebd. Seances Acad. Sci.* **177**, 322.
Reppe, W. (1953). *Justus Liebigs Ann. Chem.* **582**, 1.
Reppe, W., and Vetter, H. (1953). *Justus Liebigs Ann. Chem.* **582**, 133.
Rhee, I., Ryang, M., and Tsutsumi, S. (1967). *J. Organomet. Chem.* **9**, 331.
Rhee, I., Mizuta, N., Ryang, M., and Tsutsumi, S. (1968). *Bull. Chem. Soc. Jpn.* **41**, 1417.
Rhee, I., Ryang, M., and Tsutsumi, S. (1969). *Tetrahedron Lett.* p. 4593.
Roe, D. M., Calvo, C., Krishnamachari, N., and Maitlis, P. M. (1975). *J. Chem. Soc., Dalton Trans.* p. 125.
Rona, P., and Crabbé, P. (1968). *J. Am. Chem. Soc.* **90**, 4733.
Rona, P., and Crabbé, P. (1969). *J. Am. Chem. Soc.* **91**, 3289.
Ryang, M. (1970). *Organomet. Chem. Rev. Sect. A* **5**, 67.
Ryang, M., and Tsutsumi, S. (1971). *Synthesis* p. 55.
Ryang, M., Song, K. M., Sawa, Y., and Tsutsumi, S. (1966). *J. Organomet. Chem.* **5**, 305.
Ryang, M., Toyoda, Y., Murai, S., Sonoda, N., and Tsutsumi, S. (1973). *J. Org. Chem.* **38**, 62.
Rylander, P. N. (1973). "Organic Syntheses with Noble Metal Catalysts." Academic Press, New York.
Saegusa, T., and Ito, Y. (1975). *Synthesis* p. 291.
Saegusa, T., Ito, Y., Kobayashi, S., and Tomita, S. (1968). *Chem. Commun.* p. 273.
Saegusa, T., Ito, Y., Kinoshita, H., and Tomita, S. (1970a). *Bull. Chem. Soc. Jpn.* **43**, 877.
Saegusa, T., Ito, Y., Tomita, S., and Kinoshita, H. (1970b). *J. Org. Chem.* **35**, 670.
Saegusa, T., Ito, Y., Yonezawa, K., Inubushi, Y., and Tomita, S. (1971a). *J. Am. Chem. Soc.* **93**, 4049.
Saegusa, T., Ito, Y., and Tomita, S. (1971b). *J. Am. Chem. Soc.* **93**, 5656.
Saegusa, T., Ito, Y., Kinoshita, H., and Tomita, S. (1971c). *J. Org. Chem.* **36**, 3316.
Saegusa, T., Ito, Y., Tomita, S., and Kinoshita, H. (1972a). *Bull. Chem. Soc. Jpn.* **45**, 496.
Saegusa, T., Yonezawa, K., and Ito, Y. (1972b). *Synth. Commun.* **2**, 431.
Saegusa, T., Murase, I., and Ito, Y. (1927c). *Bull. Chem. Soc. Jpn.* **45**, 830.
Saegusa, T., Ito, Y., Yonezawa, K., Konike, T., Tomita, S., and Ito, Y. (1973). *J. Org. Chem.* **38**, 2319.
Salomon, R. G., and Kochi, J. K. (1973). *J. Am. Chem. Soc.* **95**, 3300.
Sato, K., Inoue, S., Ota, S., and Fujita, Y. (1972a). *J. Org. Chem.* **37**, 462.
Sato, K., Inoue, S., and Yamaguchi, R. (1972b). *J. Org. Chem.* **37**, 1889.
Sato, K., Inoue, S., and Saito, K. (1972c). *J. Chem. Soc., Chem. Commun.* p. 953.
Sato, K., Inoue, S., and Saito, K. (1973). *J. Chem. Soc., Perkin Trans. 1* p. 2289.
Sawa, Y., Hashimoto, I., Ryang, M., and Tsutsumi, S. (1968). *J. Org. Chem.* **33**, 2159.
Sawa, Y., Ryang, M., and Tsutsumi, S. (1970). *J. Org. Chem.* **35**, 4183.
Schaeffer, D. J., and Zieger, H. E. (1969). *J. Org. Chem.* **34**, 3958.
Schaub, R. E., and Weiss, M. J. (1973). *Tetrahedron Lett.* p. 129.
Schlosser, M. (1974). *Angew. Chem., Int. Ed. Engl.* **13**, 701.
Schrauzer, G. N. (1964). *Angew. Chem.* **76**, 28.
Schwartz, J., and Cannon, J. B. (1974). *J. Am. Chem. Soc.* **96**, 4721.
Schwartz, J., Hart, D. W., and Holden, J. L. (1972). *J. Am. Chem. Soc.* **94**, 9269.
Seitz, L. M., and Madl, R. (1972). *J. Organomet. Chem.* **34**, 415.
Seki, Y., Murai, S., Ryang, M., and Sonoda, N. (1975). *J. Chem. Soc., Chem. Commun.* p. 528.
Semmelhack, M. F. (1967). Ph.D. Thesis, Harvard University, Cambridge, Massachusetts.
Semmelhack, M. F. (1972). *Org. React.* **19**, 115.

Semmelhack, M. F., and Ryono, L. S. (1973). *Tetrahedron Lett.* p. 2967.
Semmelhack, M. F., and Ryono, L. S. (1975). *J. Am. Chem. Soc.* **97**, 3873.
Semmelhack, M. F., Helquist, P. M., and Jones, L. D. (1971). *J. Am. Chem. Soc.* **93**, 5908.
Semmelhack, M. F., Helquist, P. M., and Gorzynski, J. D. (1972). *J. Am. Chem. Soc.* **94**, 9234.
Semmelhack, M. F., Stauffer, R. D., and Rogerson, T. D. (1973). *Tetrahedron Lett.* p. 4519.
Semmelhack, M. F., Chong, B. P., Stauffer, R. D., Rogerson, T. D., Chong, A., and Jones, L. D. (1975). *J. Am. Chem. Soc.* **97**, 2507.
Seyferth, D., and Millar, M. D. (1972). *J. Organomet. Chem.* **38**, 373.
Seyferth, D., and Spohn, R. J. (1968). *J. Am. Chem. Soc.* **90**, 540.
Seyferth, D., and Spohn, R. J. (1969). *J. Am. Chem. Soc.* **91**, 3037.
Sharpless, K. B., Hanzlik, R. P., and van Tamelen, E. E. (1968). *J. Am. Chem. Soc.* **90**, 209.
Sharpless, K. B., Umbreit, M. A., Nieh, M. T., and Flood, T. C. (1972). *J. Am. Chem. Soc.* **94**, 6538.
Sheppard, W. A. (1970). *J. Am. Chem. Soc.* **92**, 5419.
Siddall, J. B., Biskup, M., and Fried, J. H. (1969). *J. Am. Chem. Soc.* **91**, 1853.
Siegl, W. O., and Collman, J. P. (1972). *J. Am. Chem. Soc.* **94**, 2516.
Sih, C. J., Price, P., Sood, R., Salomon, R. G., Peruzzotti, G., and Casey, M. (1972). *J. Am. Chem. Soc.* **94**, 3643.
Sih, C. J., Heather, J. B., Peruzzotti, G. P., Price, P., Sood, R., and Lee, L.-F. H. (1973). *J. Am. Chem. Soc.* **95**, 1676.
Singer, H., and Wilkinson, G. (1968). *J. Chem. Soc. A* p. 849.
Sladkov, A. M., and Ukhin, L. Yu. (1968). *Russ. Chem. Rev.* (*Engl. Transl.*) **37**, 748.
Smith, M. R., Jr., and Gilman, H. (1972). *J. Organomet. Chem.* **42**, 1.
Smith, M. R., Jr., Rahman, M. T., and Gilman, H. (1971). *Organomet. Chem. Synth.* **1**, 295.
Sobti, R. R., and Dev, S. (1967). *Tetrahedron Lett.* p. 2893.
Sobti, R. R., and Dev, S. (1970). *Tetrahedron* **26**, 649.
Soloski, E. J., Ward, W. E., and Tamborski, C. (1973). *J. Fluorine Chem.* **2**, 361.
Staab, H. A., and Binnig, F. (1967a). *Chem. Ber.* **100**, 293.
Staab, H. A., and Binnig, F. (1967b). *Chem. Ber.* **100**, 889.
Staroscik, J., and Rickborn, B. (1971). *J. Am. Chem. Soc.* **93**, 3046.
Stephens, R. D., and Castro, C. E. (1963). *J. Org. Chem.* **28**, 3313.
Stork, G., and Macdonald, T. L. (1975). *J. Am. Chem. Soc.* **97**, 1264.
Takagi, K., Okamoto, T., Sakakibara, Y., and Oka, S. (1973). *Chem. Lett.* p. 471.
Takahashi, K., Miyake, A., and Hata, G. (1971). *Chem. Ind.* (*London*) p. 488.
Takahashi, K., Miyake, A., and Hata, G. (1972). *Bull. Chem. Soc. Jpn.* **45**, 1183.
Takahashi, S., Suzuki, Y., and Hagihara, N. (1974). *Chem. Lett.* p. 1363.
Takahashi, Y., Sakai, S., and Ishii, Y. (1967). *Chem. Commun.* p. 1092.
Takahashi, Y., Sakai, S., and Ishii, Y. (1969). *J. Organomet. Chem.* **16**, 177.
Tamaki, A., and Kochi, J. K. (1973). *J. Organomet. Chem.* **61**, 441.
Tamaki, A., Magennis, S. A., and Kochi, J. K. (1973). *J. Am. Chem. Soc.* **95**, 6487.
Tamao, K., Sumitani, K., and Kumada, M. (1972a). *J. Am. Chem. Soc.* **94**, 4374.
Tamao, K., Kiso, Y., Sumitani, K., and Kumada, M. (1972b). *J. Am. Chem. Soc.* **94**, 9268.
Tamao, K., Zembayashi, M., Kiso, Y., and Kumada, M. (1973). *J. Organomet. Chem.* **55**, C91.
Tamao, K., Minato, A., Miyake, N., Matsuda, T., Kiso, Y., and Kumada, M. (1975). *Chem. Lett.* p. 133.
Tamura, M., and Kochi, J. (1971a). *J. Am. Chem. Soc.* **93**, 1485.
Tamura, M., and Kochi, J. (1971b). *Bull. Chem. Soc. Jpn.* **44**, 3063.
Tamura, M., and Kochi, J. (1971c). *Synthesis* p. 303.

Tamura, M., and Kochi, J. (1972). *J. Organomet. Chem.* **92**, 205.
Tanaka, S., Yamamoto, H., Nozaki, H., Sharpless, K. B., Michaelson, R. C., and Cutting, J. D. (1974). *J. Am. Chem. Soc.* **96**, 5254.
Tanaka, T., Kurozumi, S., Toru, T., Kobayashi, M., Miura, S., and Ishimoto, S. (1975). *Tetrahedron Lett.* p. 1535.
Tatsuno, Y., Konishi, A., Nakamura, A., Otsuka, S. (1974). *J. Chem. Soc., Chem. Commun.*, p. 588.
Trost, B. M., and Dietsche, T. J. (1973). *J. Am. Chem. Soc.* **95**, 8200.
Trost, B. M., and Fullerton, T. J. (1973). *J. Am. Chem. Soc.* **95**, 292.
Trost, B. M., and Strege, P. E. (1975). *J. Am. Chem. Soc.* **97**, 2534.
Trost, B. M., and Weber, L. (1975). *J. Am. Chem. Soc.* **97**, 1611.
Trost, B. M., Dietsche, T. J., and Fullerton, T. J. (1973). *J. Org. Chem.* **39**, 737.
Trost, B. M., Conway, W. P., Strege, P. E., and Dietsche, T. J. (1974). *J. Am. Chem. Soc.* **96**, 7165.
Truce, W. E., and Lusch, M. J. (1974). *J. Org. Chem.* **39**, 3174.
Tsuji, J. (1969a). *Adv. Org. Chem.* **6**, 109.
Tsuji, J. (1969b). *Acc. Chem. Res.* **2**, 144.
Tsuji, J. (1975). "Organic Synthesis by Means of Transition Metal Complexes," Springer-Verlag, Berlin and New York.
Tyrlik, S., and Wolochowicz, I. (1973). *Bull. Soc. Chim. Fr.* p. 2147.
van Helden, R., and Verberg, G. (1965). *Recl. Trav. Chim. Pays-Bas* **84**, 1263.
van Helden, R., Kohll, C. F., Medema, D., Verberg, G., and Jonkhoff. T. (1968). *Recl. Trav. Chim. Pays-Bas* **87**, 961.
van Koten, G., and Noltes, J. G. (1972). *J. Chem. Soc., Chem. Commun.* p. 940.
van Koten, G., Leusink, A. J., and Noltes, J. G. (1970). *Chem. Commun.*, p. 1107.
van Koten, G., Leusink, A. J., and Noltes, J. G. (1971). *Inorg. Nucl. Chem. Lett.* **7**, 227.
van Koten, G., Leusink, A. J., and Noltes, J. G. (1975a). *J. Organomet. Chem.* **84**, 117.
van Koten, G., Leusink, A. J., and Noltes, J. G. (1975b). *J. Organomet. Chem.* **85**, 105.
van Tamelen, E. E., and McCormick, J. P. (1970). *J. Am. Chem. Soc.* **92**, 737.
van Tamelen, E. E., and Schwartz, M. A. (1965). *J. Am. Chem. Soc.* **87**, 3277.
van Tamelen, E. E., Åkermark, B., and Sharpless, K. B. (1969). *J. Am. Chem. Soc.* **91**, 1552.
Vermeer, P., DeGraaf, C., and Meijer, J. (1974a). *Recl. Trav. Chim. Pays-Bas*, **93**, 24.
Vermeer, P., Meijer, J., and Eylander, C. (1974b). *Recl. Trav. Chim. Pays-Bas* **93**, 240.
Vig, O. P., Kapur, J. C., and Sharma, S. D. (1968a). *J. Indian Chem. Soc.* **45**, 734 and 1026.
Vig, O. P., Matta, K. L., Kapur, J. C., and Vig, B. (1968b). *J. Indian Chem. Soc.* **45**, 973.
Vig, O. P., Sharma, S. D., and Kapur, J. C. (1969). *J. Indian Chem. Soc.* **46**, 167.
Vollhardt, K. P. C., and Bergman, R. G. (1974). *J. Am. Chem. Soc.* **96**, 4996.
Wada, F., and Matsuda, T. (1974). *Chem. Lett.* p. 197.
Wagner, F., and Meier, H. (1974). *Tetrahedron* **30**, 773.
Wagner, F., and Meier, H. (1975). *Synthesis* p. 324.
Wakatsuki, Y., and Yamazaki, H. (1973a). *Tetrahedron Lett.* p. 3383.
Wakatsuki, Y., and Yamazaki, H. (1973b). *J. Chem. Soc., Chem. Commun.* p. 280.
Wakatsuki, Y., and Yamazaki H. (1976). *Synthesis*, p. 26.
Wakatsuki, Y., Akoi, K., and Yamazaki, H. (1974a). *J. Am. Chem. Soc.* **96**, 5284.
Wakatsuki, Y., Kuramitsu, T., and Yamazaki, H. (1974b). *Tetrahedron Lett.* p. 4549.
Walter, D., and Wilke, G. (1966). *Angew. Chem., Int. Ed. Engl.* **5**, 897.
Watanabe, Y., Mitsudo, T., Tanaka, M., Yamamoto, K., Okajima, T., and Takegami, Y. (1971). *Bull. Chem. Soc. Jpn.* **44**, 2569.

Watanabe, Y., Yamashita, M., Mitsudo, T., Tanaka, M., and Takegami, Y. (1973). *Tetrahedron Lett.* p. 3535.
Watanabe, Y., Yamashita, M., Mitsudo, T., Igami, M., Tomi, K., and Takegami, Y. (1975). *Tetrahedron Lett.* p. 1063.
Webb, I. D., and Borcherdt, G. T. (1951). *J. Am. Chem. Soc.* **73**, 2654.
Wender, I., and Pino, P., eds. (1968). "Organic Syntheses via Metal Carbonyls." Wiley (Interscience), New York.
Whitesides, G. M., and Casey, C. P. (1966). *J. Am. Chem. Soc.* **88**, 4541.
Whitesides, G. M., and Ehmann, W. J. (1969). *J. Am. Chem. Soc.* **91**, 3800.
Whitesides, G. M., and Kendall, P. E. (1972). *J. Org. Chem.* **37**, 3718.
Whitesides, G. M., San Filippo, J., Jr., Casey, C. P., and Panek, E. J. (1967). *J. Am. Chem. Soc.* **89**, 5302.
Whitesides, G. M., Fischer, W. F., Jr., San Filippo, J., Jr., Bashe, R. W., and House, H. O. (1969a). *J. Am. Chem. Soc.* **91**, 4871.
Whitesides, G. M., San Filippo, J. Jr., Stedronsky, E. R., and Casey, C. P. (1969b). *J. Am. Chem. Soc.* **91**, 6542.
Whitesides, G. M., Casey, C. P., and Krieger, J. K. (1971). *J. Am. Chem. Soc.* **93**, 1379.
Whitesides, G. M., Panek, E. J., and Stedronsky, E. R. (1972). *J. Am. Chem. Soc.* **94**, 232.
Whitesides, G. M., Bergbreiter, D. E., and Kendall, P. E. (1974). *J. Am. Chem. Soc.* **96**, 2806.
Wieland, D. M., and Johnson, C. R. (1971). *J. Am. Chem. Soc.* **93**, 3047.
Wilke, G. (1963). *Angew. Chem., Int. Ed. Engl.* **2**, 105.
Wilke, G., and Bogdanović, G. (1961). *Angew. Chem.* **73**, 756.
Wilke, G., Bogdanović, B., Hardt, P., Heimbach, P., Keim, W., Kroner, M., Oberkirch, W., Tanaka, K., Steinrucke, E., Walter, D., and Zimmerman, H. (1966). *Angew. Chem., Int. Ed. Engl.* **5**, 151.
Wittig, G., and Klar, G. (1967). *Justus Liebigs Ann. Chem.* **704**, 91.
Woo, E. P., and Sondheimer, F. (1970). *Tetrahedron* **26**, 2933.
Worm, A. T., and Brewster, J. H. (1970). *J. Org. Chem.* **35**, 1715.
Yamamoto, T., Yamamoto, A., and Ikeda, S. (1971). *J. Am. Chem. Soc.* **93**, 3350 and 3360.
Yamamura, M., Moritani, I., and Murahashi, S.-I. (1974). *Chem. Lett.* p. 1423.
Yamamura, M., Moritani, I., and Murahashi, S.-I. (1975). *J. Organomet. Chem.* **91**, C39.
Yamazaki, H., and Hagihara, N. (1970). *J. Organomet. Chem.* **21**, 431.
Yoshisato, E., and Tsutsumi, S. (1968a). *J. Org. Chem.* **33**, 869.
Yoshisato, E., and Tsutsumi, S. (1968b). *J. Am. Chem. Soc.* **90**, 4488.
Yoshisato, E., and Tsutsumi, S. (1968c). *Chem. Commun.* p. 33.
Yoshisato, E., Ryang, M., and Tsutsumi, S. (1969). *J. Org. Chem.* **34**, 1500.
Zweifel, G., and Miller, R. L. (1970). *J. Am. Chem. Soc.* **92**, 6678.

3

METAL–CARBENE COMPLEXES IN ORGANIC SYNTHESIS

CHARLES P. CASEY

I. INTRODUCTION

The first stable transition metal–carbene complex was synthesized and characterized by E. O. Fischer's group in 1964. Fischer and Maasböl (1964)

$$W(CO)_6 + C_6H_5Li \longrightarrow (CO)_5W^{-}{-}C(=O)C_6H_5 \longrightarrow (CO)_5W{=}C(OCH_3)C_6H_5$$

found that metal carbonyls react with organolithium reagents to produce acyl metal anions which could subsequently be alkylated on oxygen to give transition metal–carbene complexes. Since then, transition metal–carbene complexes have attracted a great deal of attention as a novel class of compounds, as model compounds for the study of catalytic cyclopropanation and as intermediates for organic synthesis. This chapter will review the chemistry of transition metal–carbene complexes with a view toward their potential use in organic synthesis. Several reviews of transition metal–carbene complexes have appeared (Cardin *et al.*, 1972a, 1973; Cotton and Lukehart, 1972). In addition, Fischer has summarized the work of his group several times, most recently in his Nobel Prize Address (Fischer, 1970, 1972, 1974).

Initially, carbene complexes excited a great deal of attention as potential sources of free carbenes. However, in many of the reactions of carbene complexes there is ample evidence against the intervention of a free carbene and, in no instance, has a compelling case for the intervention of a free carbene been made. This should not be overly surprising since a free carbene is a high-energy species (in contrast to alkenes, CO, phosphines, etc.) and should be reluctant to leave the coordination sphere of the metal. Thus, from the synthetic viewpoint, the metal–carbene complex is most profitably viewed not as a source of free carbenes, but as a functional group which can be synthesized, modified, and ultimately converted into other functional groups. Therefore, after brief considerations of the structure and bonding of metal–carbene complexes, this chapter will deal with (1) the synthesis of new metal–carbene complexes, (2) the modification of existing metal–carbene complexes, and (3) the conversion of metal–carbene complexes into useful organic products.

II. STRUCTURE, BONDING, AND SPECTRA OF METAL–CARBENE COMPLEXES

The structure and properties of metal–carbene complexes will be discussed using $(CO)_5CrC(OCH_3)C_6H_5$ and $(CO)_5CrC(OCH_3)CH_3$ as examples of typical metal–carbene complexes. Both the red phenyl-substituted carbene complex and the yellow methyl-substituted carbene complex are thermally stable to well over 100°. The solids are air-stable, but solutions are slowly air-oxidized. The carbene complexes are very soluble in organic solvents and insoluble in water; consequently, the isolation of metal–carbene complexes often involves a workup using ether–water mixtures. Metal–carbene complexes are not decomposed by dilute aqueous acids and bases. Although $(CO)_5CrC(OCH_3)C_6H_5$ has a relatively high dipole moment of 4.08 D (Fischer and Maasböl, 1967), it behaves as a nonpolar compound which is eluted rapidly with hexane on silica gel chromatography.

There are three important resonance structures for alkoxy-substituted carbene complexes (**Ia-c**) which are a very useful aid to the understanding of their structure and reactivity. The bonding of the carbene ligand to a transi-

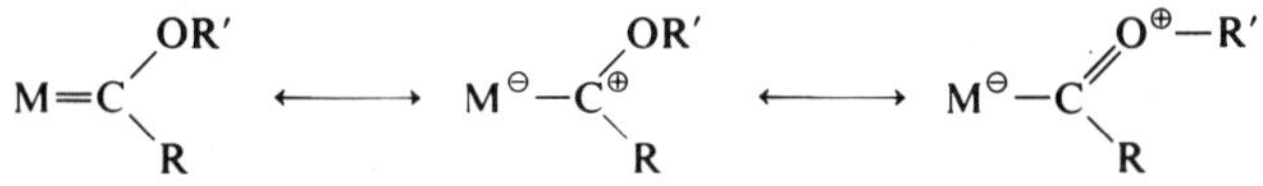

$$M{=}C(OR')R \longleftrightarrow M^{\ominus}{-}C^{\oplus}(OR')R \longleftrightarrow M^{\ominus}{-}C({=}O^{\oplus}{-}R')R$$

(Ia) **(Ib)** **(Ic)**

tion metal is conveniently described in terms of the donation of the sp^2-hybridized lone-pair electrons of the carbene carbon to the metal atom together with a concomitant acceptance of electrons from the filled d orbitals of the transition metal into the empty p_z atomic orbital of the carbene. The extent to which these separate processes occur determines the amount of double-bonding present and the metal–carbon bond length. In addition, electron donation from the lone pair of adjacent heteroatoms into the empty p_z orbital of the carbene carbon atom generates double-bond character in the carbon–heteroatom bond.

The X-ray determination of the crystal structures of several carbene complexes have confirmed the bonding scheme outlined above (Cardin *et al.*, 1972a, 1973; Cotton and Lukehart, 1972). In the crystal structure of $(CO)_5CrC(OCH_3)C_6H_5$ (Mills and Redhouse, 1968), the carbene carbon atom and the three atoms (Cr, C, O) attached to it are coplanar. The chromium–carbene distance of 2.04 Å is shorter than the predicted single-bond distance of 2.21 Å and indicates partial double-bond character of the chromium–carbene bond. The amount of double-bond character in the chromium carbene bond is, however, less than the double-bond character of the Cr–CO

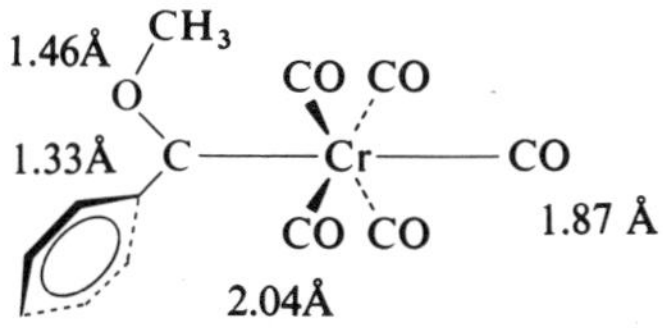

CH_3 (1.46Å) — O — (1.33Å) C(Ph) — Cr(CO)$_5$; C–Cr 2.04Å; Cr–CO 1.87 Å

bonds in the molecule which are considerably shorter (1.88 Å). The carbene carbon–oxygen bond length of 1.33 Å is considerably shorter than the methyl carbon–oxygen bond length of 1.46 Å and indicates partial double-bond character of the carbene carbon–oxygen bond. The methyl group is coplanar with the carbene group and the C–O–CH_3 angle is 121° in agreement with a carbene–oxygen partial double bond.

Nuclear magnetic resonance (nmr) studies have indicated the partial double-bond character of the carbon–heteroatom bond in the metal–carbene complexes as depicted in resonance structure **Ic**. The barrier to rotation about the carbon–nitrogen bond in **II** was found to be greater than 25 kcal $mole^{-1}$

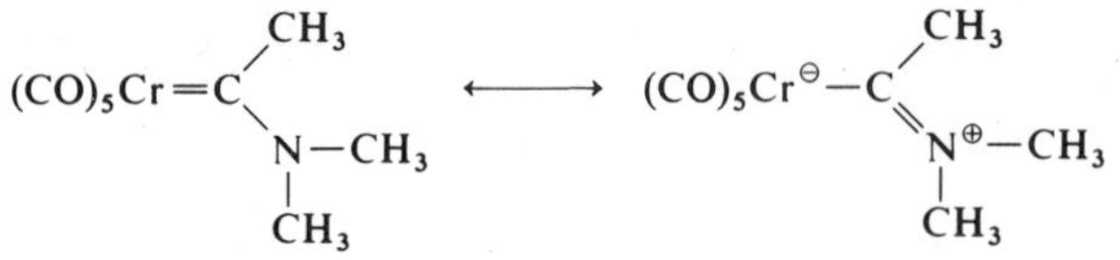

$$(CO)_5Cr{=}C(CH_3)N(CH_3)CH_3 \longleftrightarrow (CO)_5Cr^{\ominus}{-}C(CH_3){=}N^{\oplus}(CH_3)CH_3$$

(II)

(Fischer and Moser, 1968b). Similarly, the activation energy for the interconversion of the two geometrical isomers **IIIa** and **IIIb** was found to be 12.4 kcal $mole^{-1}$ (Fischer and Kreiter, 1969).

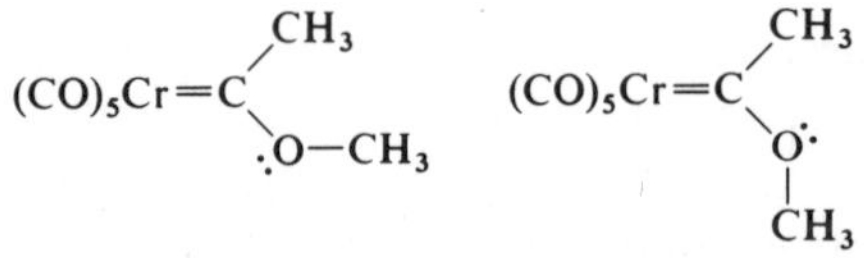

$$(CO)_5Cr{=}C(CH_3){-}\ddot{O}{-}CH_3 \qquad (CO)_5Cr{=}C(CH_3){-}\ddot{O}{-}CH_3$$

(IIIa) **(IIIb)**

Rotation about the carbene carbon–metal partial double bond is very rapid. This is expected since the carbon *p* orbital can interact with either of two orthogonal *d* orbitals on the metal and the net *p–d* overlap does not vary with rotation. The infrared spectrum of $(C_6H_6)Cr(CO)_2C(OCH_3)C_6H_5$ (**IV**) exhibits four CO bands indicative of two rotational isomers in solution (Beck *et al.*, 1971). The nmr spectrum of **IV** has only a single methyl resonance at low temperature which provides evidence for the rapid interconversion of these rotamers.

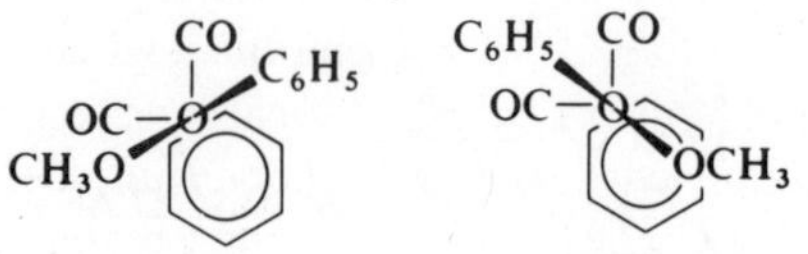

CO, OC, CH_3O, C_6H_5 **(IVa)**

C_6H_5, CO, OC, OCH_3 **(IVb)**

The ^{13}C nmr spectra of carbene complexes are characterized by marked deshielding of the carbene carbon atom (Formacek and Kreiter, 1972). The chemical shift of the carbene carbon atom in $(CO)_5CrC(OCH_3)CH_3$ occurs 360 ppm downfield from tetramethylsilane (TMS). The enormous downfield shifts of the carbene carbon atom are useful in characterizing the complexes, but are not readily interpreted. For example, downfield ^{13}C chemical shifts are thought to reflect the electropositive character of the carbon atoms. Other more important effects must be involved for metal–carbene complexes since $(CO)_5W^{13}C(C_6H_5)_2$ has a chemical shift of 358 ppm (T. J. Burkhardt and C. P. Casey, unpublished observations, 1974), whereas the related carbonium ion $^{13}C^{\oplus}(C_6H_5)_3$ has a chemical shift of only 212 ppm.

The ir spectrum of $(CO)_5CrC(OCH_3)C_6H_5$ in the CO region consists of three bands at 2066, 1992, and 1953 cm^{-1} (Fischer and Massböl, 1967). The A_1 absorption due to the *trans*-CO of $(CO)_5CrC(OCH_3)C_6H_5$ at 1953 cm^{-1} is shifted to much lower energy than the A_{1g} band of $Cr(CO)_6$ which is found at 2108 cm^{-1}. This indicates that the carbene ligand is a strong σ donor and a weak π acceptor.

The mass spectrum of $(CO)_5CrC(OCH_3)CH_3$ indicates that the carbene ligand is bound more tightly to the metal than are the CO ligands. The spectrum is dominated by a group of intense peaks due to the $(CO)_xCrC(OCH_3)$-CH_3^+ ions; the base peak in the spectrum is $CrC(OCH_3)CH_3^+$ (Connor and Müller, 1969).

III. DIRECT SYNTHESIS OF METAL–CARBENE COMPLEXES

A. Nucleophilic Attack on Metal Carbonyls

Although a variety of different methods for the synthesis of metal–carbene complexes have appeared in the last 10 years, Fischer's original preparation is probably still the most useful and general procedure for the direct synthesis of carbene complexes from noncarbene complex precursors (see Scheme 1). Phenyllithium reacts with $Cr(CO)_6$ to produce an anionic acyl complex. In many cases, the anionic acyl complexes have been isolated and characterized as stable tetraalkylammonium salts (Fischer and Maasböl, 1967). In his initial preparation of $(CO)_5WC(C_6H_5)OCH_3$, Fischer reacted the anionic acyl complex with diazomethane in acidic solution to give a 55% yield of the carbene complex. In acid solution, the anionic acyl complex is known to be protonated and several hydroxy-substituted carbene complexes have now been reported (Fischer and Riedel, 1968; Fischer *et al.*, 1973a). It is somewhat surprising that $(CO)_5WC(C_6H_5)OCH_3$ can be obtained in good yield using CH_2N_2 as an alkylating agent since CH_2N_2 reacts rapidly with metal–carbene complexes at 0° to give vinyl ethers (Casey *et al.*, 1973a). The preferred alkylating agents for the synthesis of alkoxy-substituted metal–carbene complexes are trialkyloxonium salts (Aumann and Fischer, 1969) and CH_3OSO_2F (Casey *et al.*, 1973b) which give yields in excess of 80%.

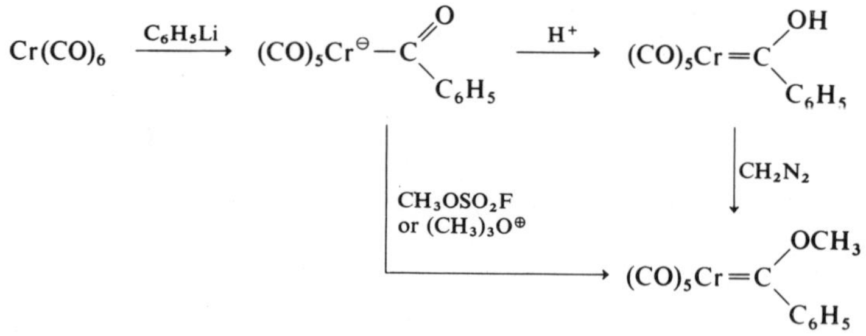

$$Cr(CO)_6 \xrightarrow{C_6H_5Li} (CO)_5Cr^{\ominus}-C(=O)C_6H_5 \xrightarrow{H^+} (CO)_5Cr=C(OH)C_6H_5 \xrightarrow{CH_2N_2} (CO)_5Cr=C(OCH_3)C_6H_5$$

$$(CO)_5Cr^{\ominus}-C(=O)C_6H_5 \xrightarrow{CH_3OSO_2F \text{ or } (CH_3)_3O^{\oplus}} (CO)_5Cr=C(OCH_3)C_6H_5$$

Scheme 1

Related acetoxy-substituted carbene complexes have been prepared by acylation of anionic metal acyl complexes with acetyl chloride (Connor and Jones, 1971a). The silylation of anionic acyl complexes was used to prepare $(CO)_5CrC[OSi(CH_3)_3]CH_3$ (Fischer and Moser, 1968a). Phenoxy-substituted carbene complexes have been obtained from the reaction of anionic acyl com-

$$\text{(2-furyl)}-Li + Cr(CO)_6 \longrightarrow \xrightarrow{CH_3COCl} (CO)_5Cr=C(OCOCH_3)(\text{2-furyl})$$

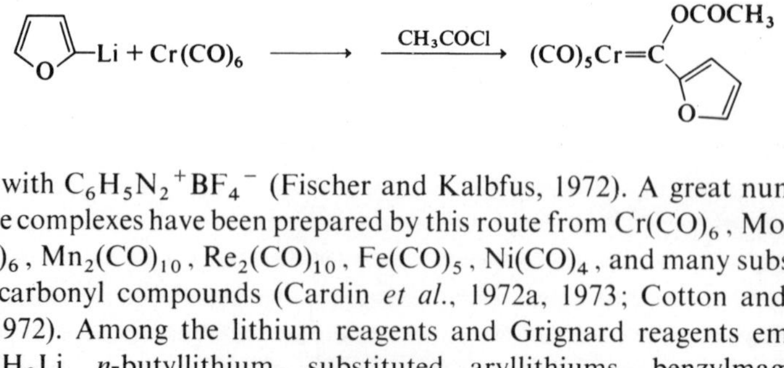

plexes with $C_6H_5N_2{}^+BF_4{}^-$ (Fischer and Kalbfus, 1972). A great number of carbene complexes have been prepared by this route from $Cr(CO)_6$, $Mo(CO)_6$, $W(CO)_6$, $Mn_2(CO)_{10}$, $Re_2(CO)_{10}$, $Fe(CO)_5$, $Ni(CO)_4$, and many substituted metal carbonyl compounds (Cardin *et al.*, 1972a, 1973; Cotton and Lukehart, 1972). Among the lithium reagents and Grignard reagents employed are CH_3Li, *n*-butyllithium, substituted aryllithiums, benzylmagnesium bromide, furyllithium, and ferrocenyllithium. Amides (Fischer *et al.*, 1971, 1972b) can also be used to prepare metal–carbene complexes from metal carbonyls. An interesting biscarbene complex was prepared from $(CH_3)_2P^{\ominus}$ (Fischer *et al.*, 1972c).

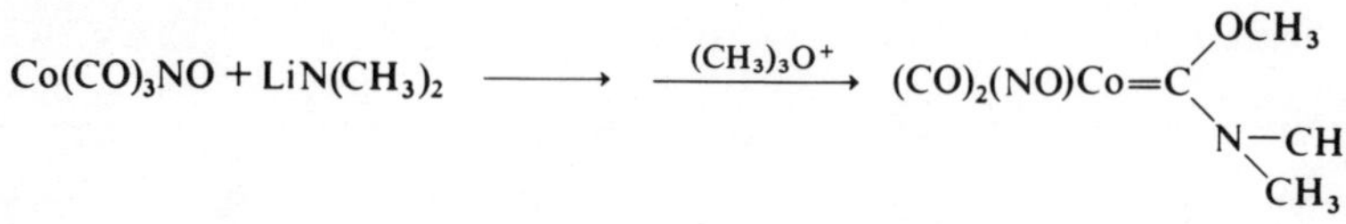

$$Co(CO)_3NO + LiN(CH_3)_2 \longrightarrow \xrightarrow{(CH_3)_3O^+} (CO)_2(NO)Co=C(OCH_3)N(CH_3)_2$$

$$W(CO)_6 + LiP(CH_3)_2 \longrightarrow \xrightarrow{Et_3O^+} cis\text{-}(CO)_4W{=}C(P(CH_3)_2)(OCH_2CH_3)\ \text{and}\ {=}C(P(CH_3)_2)(OCH_2CH_3)$$

B. Nucleophilic Attack on Isocyanide Complexes

The reactions of isocyanide complexes with nucleophiles parallel the reactions of metal carbonyls. Alcohols were found to add across the carbon–nitrogen bond of complexed isocyanides (Badley *et al.*, 1969, 1971). Amines

$$C_6H_5{-}N{\equiv}C{-}PtCl_2(PEt_3) \xrightarrow{EtOH} (CH_3CH_2O)(C_6H_5{-}N{-}H)C{=}PtCl_2(PEt_3)$$

also readily add to complexed isocyanides (Crociani *et al.*, 1972). In com-

$$C_6H_5N{\equiv}C{-}PdCl_2(P\phi_3) + CH_3C_6H_4NH_2 \longrightarrow (CH_3C_6H_4HN)(C_6H_5HN)C{=}PdCl_2(P\phi_3)$$

plexes containing several isocyanide ligands, addition of a primary amine across two different isocyanides can occur (Miller *et al.*, 1971). Alternatively,

$$Fe(CNCH_3)_6^{2+} + CH_3NH_2 \longrightarrow \left[(CH_3NC)_4Fe\langle C(NHCH_3){-}N(CH_3){-}C(NHCH_3)\rangle\right]^{2+}$$

addition of two or even three molecules of amine can lead to bis- and triscarbene complexes (Chatt *et al.*, 1973). Addition of hydrazine to a polyiso-

$$Os(CNCH_3)_6^{2+} + CH_3NH_2 \longrightarrow (CH_3NC)_4Os[C(NHCH_3)_2]_2^{2+} + (CH_3NC)_3Os[C(NHCH_3)_2]_3^{2+}$$

cyanide complex can lead to cyclic biscarbene complexes (Balch and Miller, 1972).

$$Fe(CNCH_3)_6^{2+} + NH_2NH_2 \longrightarrow (CH_3NC)_4Fe\begin{matrix} =C(NHCH_3)-NH \\ | \\ =C(NHCH_3)-NH \end{matrix}\ ^{2+}$$

C. Diazoalkane Precursors

The reaction of diazo compounds with metal complexes has recently been reported to give carbene complexes (Herrmann, 1974). This method is particularly noteworthy since complexes of electron-poor carbenes can be prepared via this route. Presumably, this transformation occurs by dissociation

$$\underset{\textbf{(V)}}{(CH_3C_5H_4)Mn(CO)_2(THF)} + C_6H_5COCN_2C_6H_5 \longrightarrow \underset{\textbf{(VI)}}{\underset{(37\%)}{(CH_3C_5H_4)(CO)_2Mn{=}C}}\begin{matrix} C_6H_5 \\ C(=O)C_6H_5 \end{matrix}$$

of tetrahydrofuran (THF) from **V**, followed by reaction of the coordinatively unsaturated intermediate with the diazoalkane to give carbene complex **VI**. Carbene complex **VI** is interesting in relation to the Wolff rearrangement of diazoketones. Upon photolysis at 10°, **VI** is converted to a ketene complex (Herrmann, 1975). A diphenylcarbene complex has been prepared by reaction of **V** with diphenyldiazomethane (Herrmann, 1975).

$$\textbf{(VI)} \xrightarrow{h\nu} (CH_3C_5H_4)(CO)_2Mn-\begin{matrix} O{=}C \\ \| \\ C(C_6H_5)_2 \end{matrix}$$

D. Alkylation of Acyl Complexes

The O-alkylation of acyl mercurials produces cationic carbene complexes (Gerhart and Schöllkopf, 1967). Similar procedures have been used to prepare

$$Et_2N-\overset{O}{\overset{\|}{C}}-Hg-\overset{O}{\overset{\|}{C}}-NEt_2 \xrightarrow{(CH_3)_3O^+} \left[\begin{matrix} CH_3O \\ Et_2N \end{matrix} C{=}Hg{=}C \begin{matrix} OCH_3 \\ NEt_2 \end{matrix}\right]^{2+}$$

complexes of iron, molybdenum, and ruthenium (Green *et al.*, 1971).

$$C_5H_5FeCO(P\phi_3)COCH_3 \xrightarrow{Et_3O^+} C_5H_5FeCO(P\phi_3)[C(OCH_2CH_3)CH_3]^+$$

The reaction of $NaMn(CO)_5$ with 1,3-dibromopropane produces the cyclic carbene complex **(VII)** (Casey, 1970; King, 1963) presumably via intramolecular alkylation of an intermediate acyl complex (see Scheme 2).

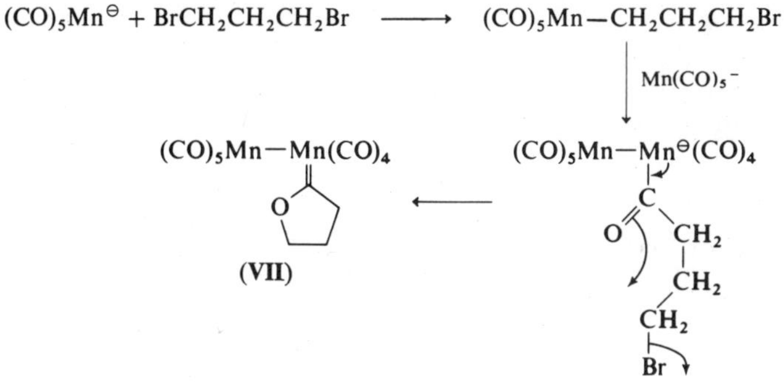

$$(CO)_5Mn^{\ominus} + BrCH_2CH_2CH_2Br \longrightarrow (CO)_5Mn-CH_2CH_2CH_2Br$$

$$\xrightarrow{Mn(CO)_5^-} (CO)_5Mn-Mn^{\ominus}(CO)_4[C(=O)CH_2CH_2CH_2Br] \longrightarrow (CO)_5Mn-Mn(CO)_4(=\text{cyclic } C_4H_6O)$$

(VII)

Scheme 2

Further studies on the reaction of $CH_3Mn(CO)_5$ with $Mn(CO)_5^-$ indicated that these species are in equilibrium with an anionic acyl complex which can be trapped as a carbene complex by O-alkylation (Anderson and Casey, 1971). The corresponding reaction of $Re(CO)_5^{\ominus}$ with $CH_3Mn(CO)_5$ did not produce the expected manganese–carbene complex **(VIII)**, but gave the rhenium–

$$Mn(CO)_5^- + CH_3Mn(CO)_5 \rightleftharpoons (CO)_5Mn-\overset{\ominus}{Mn}(CO)_4[C(=O)CH_3]$$

$$\xrightarrow{(CH_3)_3O^+} (CO)_5Mn-Mn(CO)_4=C(OCH_3)CH_3$$

carbene complex **(IX)** (Casey *et al.*, 1975). Evidently the initially formed manganese–carbene complex **(VIII)** rearranges to the rhenium complex **(IX)** possibly via bridged intermediate **(X)**. The formation of cyclic carbene

$$(CO)_4Mn{-}Re(CO)_5 \text{ (=C(CH}_3\text{)OCH}_3\text{ on Mn) (VIII)} \longrightarrow (CO)_4Mn(\mu\text{-CO})(\mu\text{-C(CH}_3\text{)OCH}_3)Re(CO)_4 \text{ (X)} \longrightarrow (CO)_5Mn{-}Re(CO)_4 \text{ (=C(CH}_3\text{)OCH}_3\text{ on Re) (IX)}$$

complexes via a similar route has been reported by Cotton and Lukehart (1971).

$$C_5H_5(CO)_4MoCH_2CH_2CH_2Br + P\phi_3 \longrightarrow C_5H_5(CO)_2(P\phi_3)Mo{=}C\langle\text{cyclic }OCH_2CH_2CH_2\rangle^{\oplus}$$

E. Reactions of Metal Carbonyl Anions

The reaction of $Na_2Cr(CO)_5$ with 1,2-diphenyl-3,3-dichlorocyclopropene gives the novel carbene complex (**XI**) (Öfele, 1968). The high stability of **XI** is undoubtedly related to the aromaticity of the diphenylcyclopropenium cation. A similar palladium complex has been prepared from palladium metal (Öfele, 1970).

$$Na_2Cr(CO)_5 + Cl_2C_3\phi_2 \longrightarrow (CO)_5Cr^{\ominus}{-}C_3\phi_2^{\oplus} \quad \textbf{(XI)}$$

The reaction of $NaFe(CO)_4H$ with the 1,3-dimethylimidazolium ion gives a carbon-bonded imidazole complex (Kreiter and Öfele, 1972).

$$HFe^{\ominus}(CO)_4 + \text{1,3-dimethylimidazolium}^{\oplus}(2\text{-H}) \longrightarrow (CO)_4Fe{=}\text{C(1,3-dimethylimidazol-2-ylidene)}$$

Related N-bonded imidazole complexes have been prepared and shown to rearrange to C-bonded imidazoles (Sundberg *et al.*, 1974).

$$Ru(H_2O)(NH_3)_5^{2+} + \text{(1-methylimidazole)} \longrightarrow (NH_3)_5Ru{-}N(\text{imidazole})N{-}CH_3 \;\; ^{2+}$$

$$\xrightarrow{H^+} (NH_3)_5Ru{=}C(N(CH_3)CH{=}CHNH) \;\; ^{2+}$$

The reaction of $Cr(CO)_5{}^{2-}$ with $(CH_3)_2N{=}CHCl^+$ gives rise to the secondary carbene complex **(XII)** (Cetinkaya *et al.*, 1972). In related reactions,

$$(CO)_5Cr^{2-} + (CH_3)_2N{=}CHCl^+ \longrightarrow (CO)_5Cr{=}C(N(CH_3)CH_3)(H)$$

(XII)

imidoyl chlorides have been converted to metal–carbene complexes (Lappert and Oliver, 1972).

$$\phi(Cl)C{=}N{-}CH_3 + (CH_3\phi_2P)_2Rh(CO)Cl \longrightarrow Cl_2(CO)(P\phi_2CH_3)_2Rh{=}C(C_6H_5)(NHCH_3)$$

F. Alkene Scission Reactions

Electron-rich alkenes are capable of reacting with metal complexes to split the alkene and give carbene complexes (Cardin *et al.*, 1972b). A bis-

$$[(Et_3P)PtCl_2]_2 + \text{(bis(1,3-diphenylimidazolidin-2-ylidene))} \longrightarrow (Et_3P)Cl_2Pt{=}C(N(C_6H_5)CH_2CH_2N(C_6H_5))$$

carbene complex of chromium was prepared by a similar route (Cetinkaya *et al.*, 1974). These reactions are particularly interesting in relation to olefin

$$Cr(CO)_4(NCCH_3)_2 + [(CH_3NCH_2CH_2NCH_3)C{=}C(CH_3NCH_2CH_2NCH_3)] \longrightarrow cis\text{-}(CO)_4Cr[{=}C(NCH_3CH_2CH_2NCH_3)]_2$$

metathesis as a rhodium–carbene complex has been found to metathesize tetraaminoethylenes (Cardin *et al.*, 1972c).

G. Carbene Transfer Reactions

Light-catalyzed transfer of a carbene ligand from molybdenum to iron has been observed (Beck and Fischer, 1971). It is not known whether light

$$(C_5H_5)Mo(CO)(NO){=}C(OCH_3)C_6H_5 + Fe(CO)_5 \xrightarrow{h\nu} (CO)_4Fe{=}C(C_6H_5)(OCH_3) + (C_5H_5)Mo(CO)_2NO$$

activates the iron or the molybdenum complex to catalyze this reaction. A similar reaction has provided a synthesis of a biscarbene complex of chromium (Herberhold and Öfele, 1970).

$$(CH_3NCH{=}CHNCH_3)C{=}Cr(CO)_5 \xrightarrow{h\nu} (CO)_4Cr[{=}C(NCH_3CH{=}CHNCH_3)]_2 \quad (8\%)$$

We recently noted the thermally activated transfer of a carbene ligand from chromium to tungsten (Anderson and Casey, 1975). The reaction proceeds at a

$$(CO)_5Cr{=}\overline{CCH_2CH_2CH_2O} + W(CO)_6 \underset{}{\overset{140^\circ}{\rightleftharpoons}} (CO)_6Cr + (CO)_5W{=}\overline{CCH_2CH_2CH_2O} \qquad (K_{eq} = 3)$$

rate similar to that for CO exchange of the carbene complex and for the thermal decomposition of the carbene complex. A possible mechanism is shown in Scheme 3.

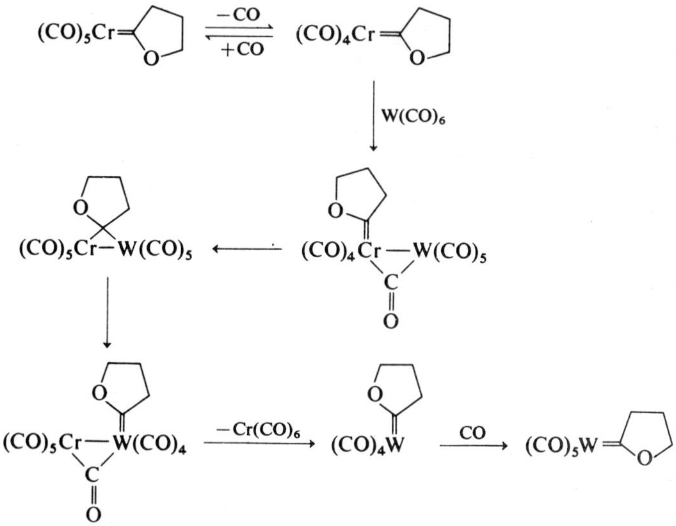

Scheme 3

H. α-Elimination Reactions

Reaction of $[(CH_3)_3CCH_2]_3TaCl_2$ with $(CH_3)_3CCH_2Li$ gives a tantalum–carbene complex (Schrock, 1974). Labeling studies indicate that the reaction proceeds via α-elimination from an intermediate R_5Ta compound. Carbene complexes have been considered as possible intermediates in olefin metathesis

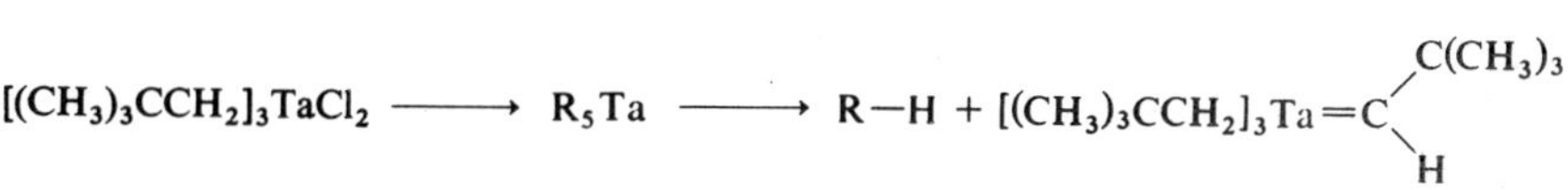

$$[(CH_3)_3CCH_2]_3TaCl_2 \longrightarrow R_5Ta \longrightarrow R{-}H + [(CH_3)_3CCH_2]_3Ta{=}C\begin{matrix} C(CH_3)_3 \\ H \end{matrix}$$

catalyzed by WCl_6 and alkylaluminum or alkyllithium compounds. α-Elimination reactions offer a possible route to carbene complexes under these reaction conditions.

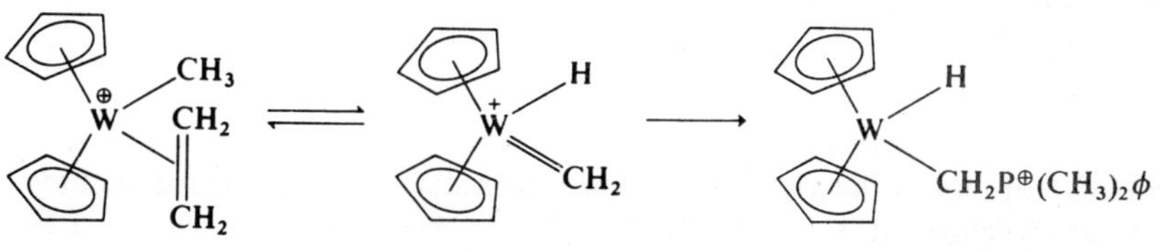

Reversible α-elimination of hydrogen from a methyltungsten compound to give a tungsten–methylene hydride derivative has been proposed as a key step in the reactions of $(C_5H_5)_2W(CH_2{=}CH_2)CH_3^+$ with $P(CH_3)_2C_6H_5$ (Cooper and Green, 1974). Green has suggested that the reversible insertion of a metal into the C–H bond of a bound alkyl group may be a much more general process than previously believed.

I. Miscellaneous Methods

Hydride abstraction from an alkyliron compound has been used to prepare a nonheteroatom-stabilized carbene complex (Sanders *et al.*, 1973).

$$C_5H_5(CO)_2Fe\text{–}C(H)(\text{benzocyclobutenyl}) + \phi_3C^+ \longrightarrow \phi_3CH + C_5H_5(CO)_2Fe^{\oplus}{=}C(\text{benzocyclobutenylidene})$$

The generation of coordinatively unsaturated platinum complexes in the presence of terminal acetylenes and alcohols gives alkoxy-substituted carbene complexes (Chisholm and Clark, 1971). Cyclic carbene complexes are

$$trans\text{-}CH_3PtCl[P\phi(CH_3)_2]_2 + AgPF_6 + HC{\equiv}CH + CH_3OH \longrightarrow trans\text{-}CH_3[(CH_3)_2\phi P]_2Pt{=}C(OCH_3)(CH_3)^+ \; PF_6^-$$

formed by intramolecular cyclization of acetylenic alcohols. The mechanism of these reactions has not been established, but vinyl ether complexes, which

$$trans\text{-}CH_3PtCl[P\phi(CH_3)_2]_2 + H{-}C{\equiv}C{-}CH_2CH_2OH + Ag^+PF_6^- \longrightarrow trans\text{-}CH_3[(CH_3)_2\phi P]_2Pt^+{=}C(\text{2-oxacyclopentylidene}) \; PF_6^-$$

were originally thought to be intermediates, were shown to be stable under the reaction conditions (Chisholm *et al.*, 1971).

IV. MODIFICATION AND ELABORATION OF THE STRUCTURE OF METAL–CARBENE COMPLEXES

In addition to the syntheses of metal–carbene complexes from compounds not containing the carbene ligand, a variety of methods for modifying the structure of carbene complexes have been developed. Three different reactive centers in a transition metal–carbene complex can be utilized in designing a synthesis of a new carbene complex (see Scheme 4). For example, $(CO)_5$-$CrC(OCH_3)CH_3$ can be converted into a wide range of carbene complexes by (*a*) nucleophilic attack at the carbene carbon atom followed by loss of methoxide, (*b*) ligand substitution reactions in which CO is replaced by a new ligand, and (*c*) reactions involving anions generated alpha to the carbene carbon atom. Each of these three pathways will be discussed in relation to the elaboration of metal–carbene complexes.

$$(CO)_5Cr{=}C(OCH_3)CH_3 \xrightarrow[HN(CH_3)_2]{(a)} \left[(CO)_5Cr^{\ominus}{-}C(OCH_3)(CH_3)\overset{\oplus}{N}H(CH_3)_2\right] \xrightarrow{-OCH_3^-} (CO)_5Cr{=}C(CH_3)N(CH_3)CH_3$$

$$(CO)_5Cr{=}C(OCH_3)CH_3 \xrightarrow[-CO]{(b)} \left[(CO)_4Cr{=}C(OCH_3)CH_3\right] \xrightarrow{P\phi_3} (CO)_4(P\phi_3)Cr{=}C(OCH_3)CH_3$$

$$(CO)_5Cr{=}C(OCH_3)CH_3 \xrightarrow[base]{(c)} (CO)_5Cr^{\ominus}{-}C(OCH_3){=}CH_2 \xrightarrow{CH_3X} (CO)_5Cr{=}C(OCH_3)CH_2CH_3$$

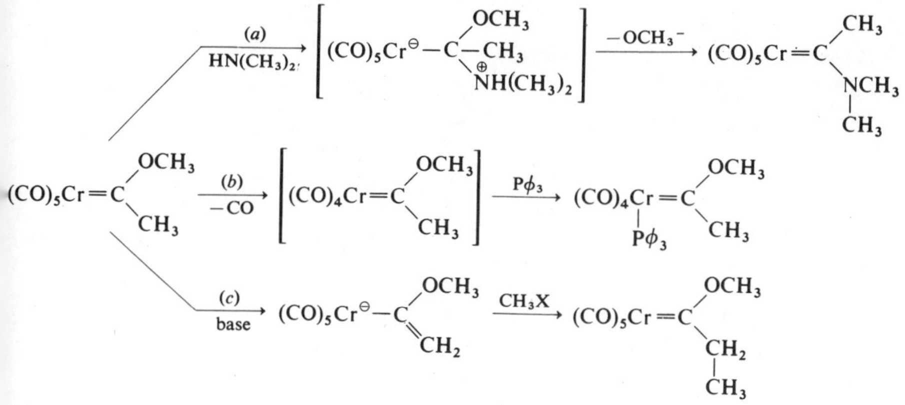

Scheme 4

A. Nucleophilic Attack at the Carbene Carbon Atom

Alkoxy-substituted carbene complexes serve as valuable precursors for a wide variety of metal–carbene complexes as the carbene carbon atom is very susceptible to nucleophilic attack and the alkoxy group is a better leaving group than the entering nitrogen, sulfur, or carbon nucleophiles.

Ammonia reacts with metal–carbene complexes under mild conditions to give amino-substituted carbene complexes (Fischer and Klabunde, 1967). A detailed investigation of the reaction of cyclohexylamine with $(CO)_5$-

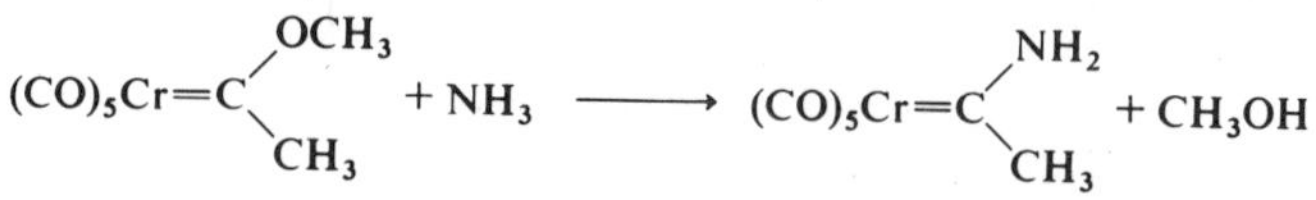

$$(CO)_5Cr{=}C(OCH_3)CH_3 + NH_3 \longrightarrow (CO)_5Cr{=}C(NH_2)CH_3 + CH_3OH$$

$CrC(OCH_3)C_6H_5$ indicates that the reaction is kinetically complex (Heckl *et al.*, 1968; Werner *et al.*, 1971). In hexane, the rate of reaction is first order in carbene complex and third order in amine. In dioxane, the reaction is first order in carbene complex and second order in amine. The high kinetic order in amine is undoubtedly due to the hydrogen bonding requirements of cyclohexylamine in nonpolar solvents. The reaction is thought to proceed by reversible nucleophilic attack of the amine at the carbene carbon atom which produces a tetravalent carbon species. The slow step in the reaction is considered to be the amine-catalyzed elimination of CH_3OH from the tetravalent

$$(CO)_5Cr{=}C(C_6H_5)(OCH_3) \rightleftharpoons (CO)_5Cr^{\ominus}{-}C(OCH_3)(C_6H_5)({}^{\oplus}NH_2{-}C_6H_{11}) \xrightarrow{\text{slow}} (CO)_5Cr{=}C(C_6H_5)(NH{-}C_6H_{11})$$

R—NH; :base (amine or dioxane)

intermediate. The intervention of a tetravalent carbon intermediate is supported by the isolation of an adduct of a tertiary amine (Kreissl *et al.*, 1973a).

$$(CO)_5Cr^{\ominus}{-}C(C_6H_5)(OCH_3){-}{}^{\oplus}N(C_2H_4)_3N$$

The reactions of carbene complexes with most primary and secondary amines give the expected substitution products in high yield. However, the reaction of $(CO)_5CrC(OCH_3)CH_3$ with diisopropylamine unexpectedly gave the monoisopropylaminocarbene complex (Connor and Fischer, 1969). Apparently, elimination of propylene from the very hindered diisopropylaminocarbene complex can occur readily.

$$(CO)_5Cr{=}C(OCH_3)(CH_3) \longrightarrow (CO)_5Cr^{\ominus}{-}C(CH_3){=}{}^{\oplus}N(CH(CH_3)_2)(CH(CH_3)CH_2{-}H) \quad (:\text{base})$$

$$\downarrow$$

$$(CO)_5Cr^{\ominus}{-}C(CH_3){=}N{-}CH(CH_3)_2 \longrightarrow (CO)_5Cr{=}C(CH_3)(NH{-}CH(CH_3)_2)$$

Thiols react with metal–carbene complexes in a manner similar to amines (Fischer and Kiener, 1967; Fischer *et al.*, 1972a).

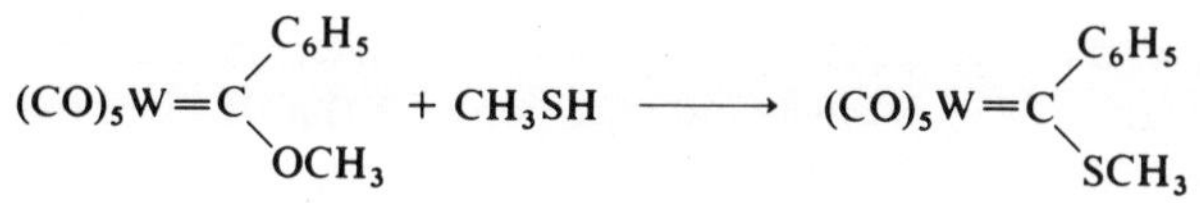

$$(CO)_5W{=}C(C_6H_5)(OCH_3) + CH_3SH \longrightarrow (CO)_5W{=}C(C_6H_5)(SCH_3)$$

The reaction of alkyl selenols occurs normally to give selenium-substituted carbene complexes in low yield (Fischer *et al.*, 1973c). However, reaction of the more acidic selenophenol with carbene complexes leads to cleavage of the carbene ligand. Apparently, the carbon–metal bond in the tetravalent intermediate is protonated by C_6H_5SeH.

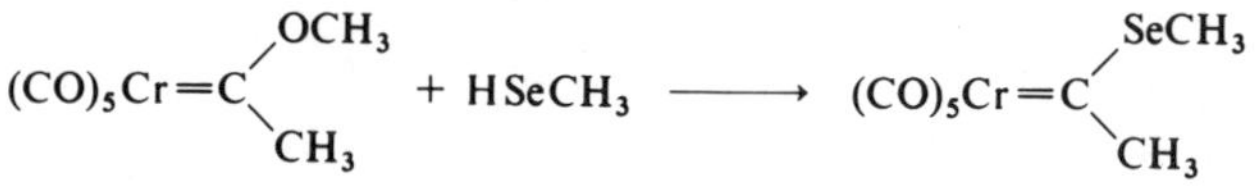

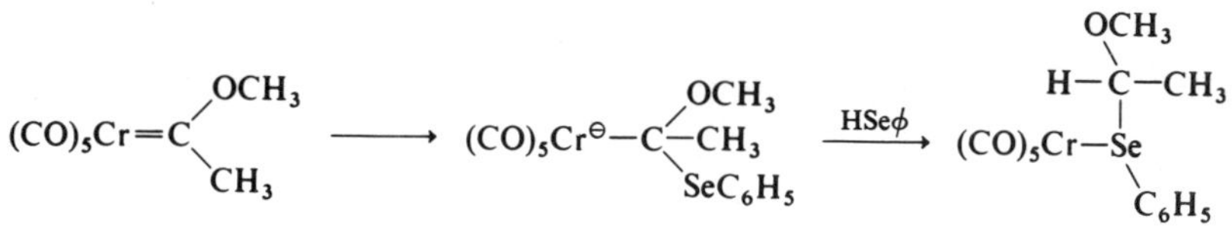

$$(CO)_5Cr{=}C(OCH_3)(CH_3) + HSeCH_3 \longrightarrow (CO)_5Cr{=}C(SeCH_3)(CH_3)$$

$$(CO)_5Cr{=}C(OCH_3)(CH_3) \longrightarrow (CO)_5Cr^{\ominus}{-}C(OCH_3)(CH_3)(SeC_6H_5) \xrightarrow{HSe\phi} (CO)_5Cr{-}Se(C_6H_5){-}CH(OCH_3)(CH_3)$$

Organolithium reagents attack metal–carbene complexes at the carbene carbon atom to give adducts which are stable at low temperature, but decompose at room temperature to give products characteristic of free radicals. Phenyllithium reacts with $(CO)_5WC(OCH_3)C_6H_5$ at $-78°$ to give a tetravalent intermediate which does not undergo spontaneous loss of methoxide, but decomposes on warming to room temperature (see Scheme 5). Treatment of the intermediate with HCl at $-78°$ induces the loss of methanol and produces a 50% yield of a diphenylcarbene complex (Burkhardt and Casey, 1973). An

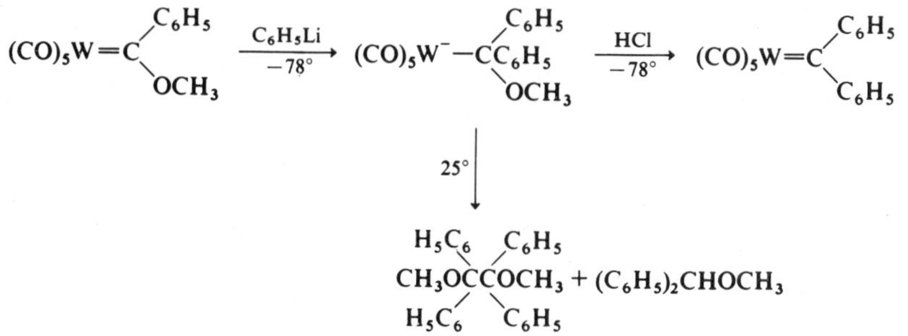

$$(CO)_5W{=}C(C_6H_5)(OCH_3) \xrightarrow[-78°]{C_6H_5Li} (CO)_5W^{-}{-}C(C_6H_5)(C_6H_5)(OCH_3) \xrightarrow[-78°]{HCl} (CO)_5W{=}C(C_6H_5)(C_6H_5)$$

$$(CO)_5W^{-}{-}C(C_6H_5)_2(OCH_3) \xrightarrow{25°} CH_3OC(C_6H_5)_2C(C_6H_5)_2OCH_3 + (C_6H_5)_2CHOCH_3$$

Scheme 5

attempt to prepare a phenylvinylcarbene complex by a similar procedure led to monomeric and dimeric vinyl ethers (Brunsvold and Casey, 1974).

$$(CO)_5Cr{=}C(C_6H_5)(OCH_3) \xrightarrow{CH_2{=}CHLi,\ -78^\circ} (CO)_5Cr^{\ominus}{-}C(C_6H_5)(OCH_3)(CH{=}CH_2) \xrightarrow{HCl} (H_5C_6)(CH_3O)C{=}CHCH_3$$

An attempt to prepare a phenylmethylcarbene complex led to products attributable to an unstable carbene complex (L. D. Albin and C. P. Casey, unpublished results, 1975).

$$(CO)_5W{=}C(C_6H_5)(OCH_3) \xrightarrow[(2)\ HCl]{(1)\ CH_3Li} \left[(CO)_5W{=}C(C_6H_5)(CH_3)\right] \longrightarrow CH_2{=}C(CH_3)C_6H_5 + \text{1-}CH_3\text{-1,2-}(C_6H_5)_2\text{cyclopropane}$$

A novel ring expansion results from attack of nitrogen ylide (**XIV**) on carbene complex **XIII** (Rees and von Angerer, 1972).

$$\underset{\textbf{(XIII)}}{(C_6H_5)_2C_3{=}Mo(CO)_5} + \underset{\textbf{(XIV)}}{C_6H_5\overset{O}{\overset{\|}{C}}{-}C^{\ominus}H{-}N^{\oplus}C_5H_5} \longrightarrow (H_5C_6)(C_6H_5)(C_6H_5)C_5H_2O{=}Mo(CO)_5$$

The conjugate addition of nucleophiles to α,β-unsaturated carbene complexes has been observed in several cases. Dimethylamine was found to add to acetylenic carbene complexes (Fischer and Kreissl, 1972). Phenyllithium

$$(CO)_5W{=}C(OCH_2CH_3)(C{\equiv}C{-}C_6H_5) \xrightarrow{HN(CH_3)_2} (CO)_5W{=}C(N(CH_3)_2)(CH{=}C(C_6H_5)N(CH_3)_2)$$

adds to the carbon–carbon double bond of a styrylcarbene complex to give a low yield of the conjugate addition product; higher yields were obtained from reaction of lithium diphenylcuprate (Brunsvold and Casey, 1974).

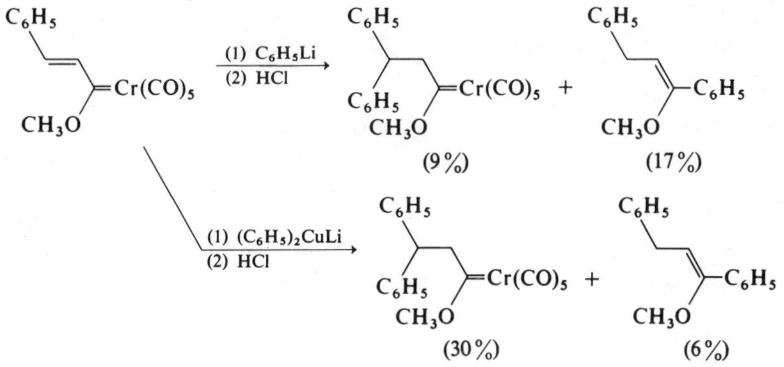

Other interesting reactions of nucleophiles with metal–carbene complexes include the reaction of 1-aminoethanol which leads to imino-substituted carbene complexes (Fischer and Knauss, 1971a) and the reaction of isonitriles which leads to the unusual complex **XV** (Aumann and Fischer, 1968a).

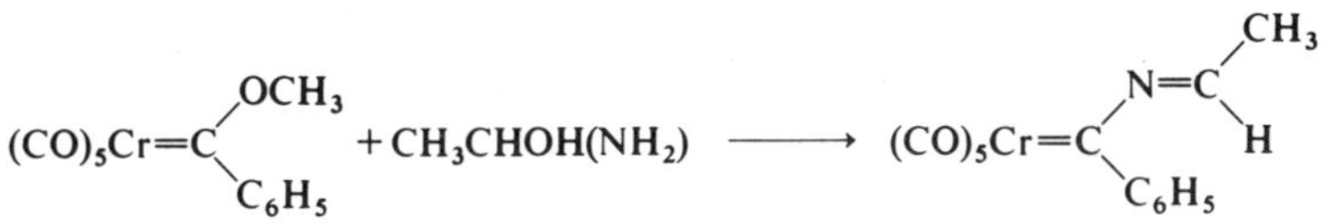

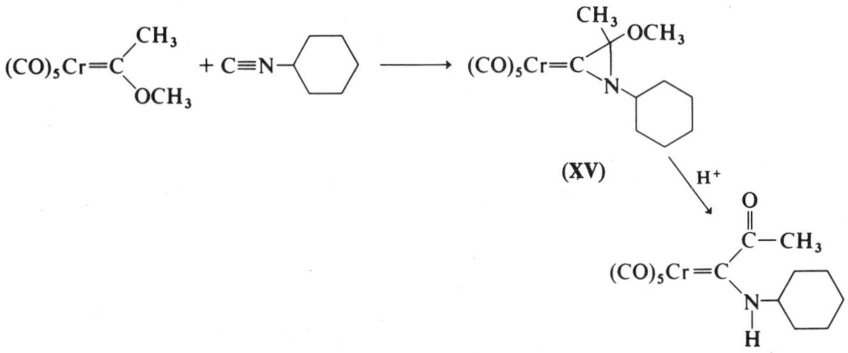

(XV)

The reaction of hydroxy-substituted carbene complexes with the dehydrating agent dicyclohexylcarbodiimide (DCC) gives the adduct **XVI** (Weiss *et al.*, 1974). The reaction is thought to proceed via dehydration to give a vinylidene-carbene complex which then undergoes a [2 + 2] cycloaddition with DCC.

$$(CO)_5Cr{=}C(OH)CH_3 \xrightarrow{DCC} [(CO)_5Cr{=}C{=}CH_2] \xrightarrow{DCC} (CO)_5Cr{=}C\text{(cyclic, N-cyclohexyl, =N-cyclohexyl)}$$

(XVI)

B. Ligand Substitution Reactions

The ligand substitution reactions of carbene complexes such as $(CO)_5$-$CrC(OCH_3)CH_3$ allow the synthesis of many phosphine- and phosphite-substituted carbene complexes. It is expected that these complexes will have modified reactivity and will provide a means of fine tuning reactions of carbene complexes. More importantly, the substitution reactions of carbene complexes proceed by a dissociative mechanism involving coordinatively unsaturated intermediates. Study of the ligand substitution reactions can give valuable information about these coordinatively unsaturated intermediates which are also involved in the important cyclopropanation, alkene scission, and thermolysis reactions of metal–carbene complexes.

The carbene ligand activates the metal complex toward ligand substitution more than CO does. The activation energy for CO loss from $Cr(CO)_6$ is 39 kcal $mole^{-1}$, whereas that for CO loss from $(CO)_5CrC(OCH_3)C_6H_5$ is 27.4 kcal $mole^{-1}$ (Werner, 1968). ^{13}CO exchange of **XVII** has a half-life of about

$$(CO)_5Cr{=}\text{(2-oxacyclopentylidene)} + {}^{13}CO \longrightarrow (CO)_4({}^{13}CO)Cr{=}\text{(2-oxacyclopentylidene)}$$

(XVII)

4.6 min at 140° in decalin (Anderson and Casey, 1975). ^{13}C nmr of labeled **XVII** indicated that labeled ^{13}CO was incorporated statistically into both the cis and trans positions. The ^{13}CO exchange of $(CO)_5WC(C_6H_5)_2$ is much more rapid: 30% exchange was observed in 10 hr at 30° in ether (T. J. Burkhardt and C. P. Casey, unpublished results, 1974).

Metal–carbene complexes undergo facile substitution of phosphines for CO (Rascher and Werner, 1968a; Fischer and Fischer, 1974a). An equilib-

$$\phi_3P + (CO)_5Cr{=}C(C_6H_5)(OCH_3) \xrightarrow[2\ hr]{60^\circ} \textit{cis-}(CO)_4(\phi_3P)CrC(OCH_3)C_6H_5 + \textit{trans-}(CO)_4(\phi_3P)CrC(OCH_3)C_6H_5$$

brium mixture of cis and trans isomers is obtained. The isomers have been separated and shown to equilibrate within minutes under the reaction conditions (Fischer and Fischer, 1974b). The equilibrium ratio of cis: trans isomers varies between 0.35 and 5.1 depending on the nature of the phosphine and the solvent. The equilibration of cis and trans isomers has been proposed to proceed via an intramolecular isomerization (Fischer *et al.*, 1972d, 1974c); however, the observation that *cis*-$(CO)_4[(C_6H_{11})_3P]CrC(OCH_3)CH_3$ undergoes exchange with PEt_3 at a rate similar to its cis–trans isomerization provides suggestive evidence that cis–trans isomerization proceeds via dissociation of phosphine.

The kinetics of phosphine substitution of carbene complexes have been found to depend on the nature of the phosphine (Rascher and Werner, 1968b). The rate of reaction of $(CO)_5CrC(OCH_3)C_6H_5$ with $(C_6H_{11})_3P$ is independent of phosphine concentration; the rate of reaction with $(C_4H_9)_3P$ has a second-order term dependent on phosphine in addition to the phosphine-independent term. Although the phosphine-independent term in the rate expression was

$$\text{Rate}^{58.8°}_{P(C_6H_{11})_3} = 0.87 \times 10^{-4}[(CO)_5CrC(OCH_3)C_6H_5]$$

$$\text{Rate}^{58.8°}_{PBu_3} = 0.89 \times 10^{-4}[(CO)_5CrC(OCH_3)C_6H_5] + 3.06 \times 10^{-4}[(CO_5CrC(OCH_3)C_6H_5][PBu_3]$$

readily assigned to the rate-limiting dissociation of CO from the carbene complex, the process giving rise to the second-order term is complex and still not completely understood. The reaction of secondary and tertiary phosphines with chromium–carbene and tungsten–carbene complexes at low temperature gives isolable adducts (Kreissl *et al.*, 1973b). These adducts have

$$(CO)_5Cr{=}C\begin{matrix}C_6H_5\\ OCH_3\end{matrix} + PEt_3 \rightleftharpoons (CO)_5Cr^{\ominus}{-}C\begin{matrix}C_6H_5\\ {-}OCH_3\\ P^{\oplus}Et_3\end{matrix}$$

been shown to be in equilibrium with starting materials at room temperature (Fischer *et al.*, 1974b). The phosphine-dependent second-order rate term for the substitution of carbene complexes has been ascribed to the reaction of these adducts (Fischer *et al.*, 1974a). The reaction probably involves dissociation of CO from the adduct followed either by capture of the coordinatively unsaturated intermediate by external phosphine or by migration of phosphine from the carbene carbon atom to the metal (see Scheme 6).

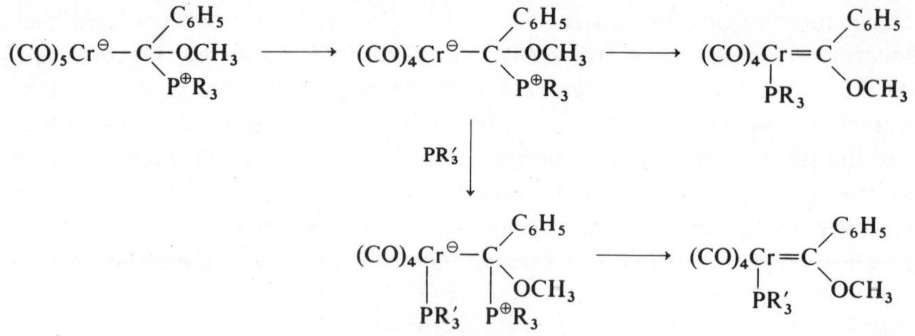

$$(CO)_5Cr^{\ominus}-C(C_6H_5)(OCH_3)(P^{\oplus}R_3) \longrightarrow (CO)_4Cr^{\ominus}-C(C_6H_5)(OCH_3)(P^{\oplus}R_3) \longrightarrow (CO)_4Cr(PR_3)=C(C_6H_5)(OCH_3)$$

$$\downarrow PR'_3$$

$$(CO)_4Cr^{\ominus}(PR'_3)-C(C_6H_5)(OCH_3)(P^{\oplus}R_3) \longrightarrow (CO)_4Cr(PR'_3)=C(C_6H_5)(OCH_3)$$

Scheme 6

C. Reactions of Carbene Anions

Protons attached to the α-carbon atom in metal–carbene complexes undergo rapid base-catalyzed hydrogen–deuterium exchange with hydroxylic solvents (Kreiter, 1968). For example, the half-life for exchange of the α protons in **VII** in acetone–D_2O with no added base is 23 min at 40° (Casey, 1970). This rapid exchange implies the existence of an intermediate carbanion. These

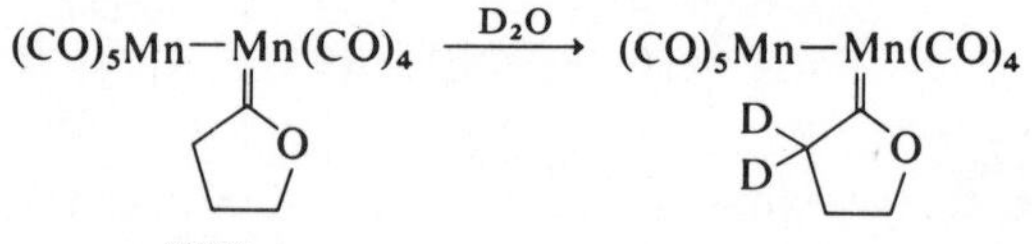

$$(CO)_5Mn-Mn(CO)_4(=\text{C-cyclic oxacarbene}) \xrightarrow{D_2O} (CO)_5Mn-Mn(CO)_4(=\text{C-cyclic oxacarbene}, \alpha\text{-}D_2)$$

(VII)

carbene anion intermediates can be stoichiometrically generated by treatment of carbene complexes with organolithium reagents at low temperature (Casey *et al.*, 1972). Addition of DCl to carbene anion **XVIII** allows the recovery of

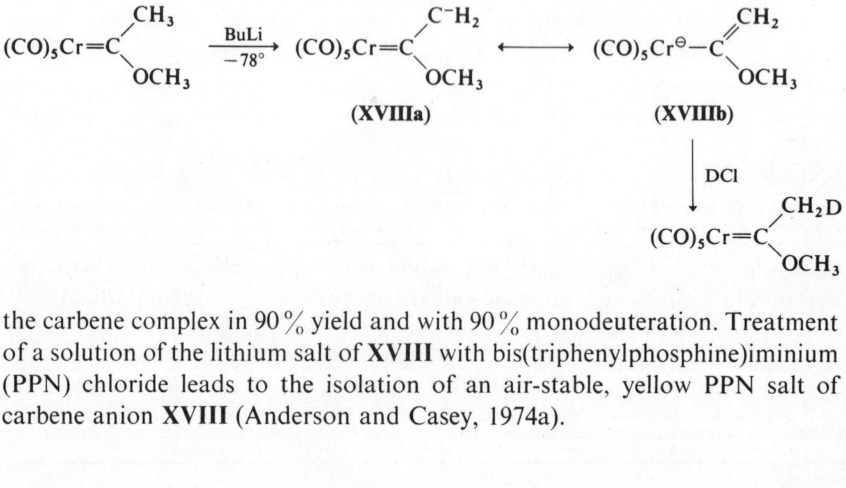

$$(CO)_5Cr=C(CH_3)(OCH_3) \xrightarrow[-78^\circ]{BuLi} (CO)_5Cr=C(C^{-}H_2)(OCH_3) \longleftrightarrow (CO)_5Cr^{\ominus}-C(=CH_2)(OCH_3)$$

(XVIIIa) **(XVIIIb)**

$$\downarrow DCl$$

$$(CO)_5Cr=C(CH_2D)(OCH_3)$$

the carbene complex in 90% yield and with 90% monodeuteration. Treatment of a solution of the lithium salt of **XVIII** with bis(triphenylphosphine)iminium (PPN) chloride leads to the isolation of an air-stable, yellow PPN salt of carbene anion **XVIII** (Anderson and Casey, 1974a).

The nmr and ir spectra of **XVIII** indicate that the structure is best described as the vinylpentacarbonylchromium anion **XVIIIb**. The nmr spectrum of **XVIII** in tetrahydrofuran-d_8 contains two one-proton singlets at δ3.78 and 4.52 due to two nonequivalent vinyl hydrogens. The nmr spectrum of **XVIII** remains unchanged up to 120° where rapid decomposition ensues and indicates a substantial rotational barrier about the carbon–carbon bond of **XVIII**. The carbonyl-stretching frequencies of anion **XVIII** are all shifted to substantially lower frequency than their counterparts in the carbene complex and indicate that substantial negative charge exists on the $Cr(CO)_5$ fragment.

Infrared studies of solutions of carbene anion **XVIII** in the presence of protic acids indicate that $(CO)_5CrC(OCH_3)CH_3$ is a remarkably acidic compound. Compound **XVIII** is not measurably protonated by methanol and is only 55% protonated by one equivalent of *p*-cyanophenol in THF (Anderson and Casey, 1974a). Thus, $(CO)_5CrC(OCH_3)CH_3$ is one of the most acidic neutral carbon acids known. Its high acidity can be understood in terms of the zwitterionic resonance form which draws attention to the analogy between alkylalkoxycarbene complexes and positively charged O-alkylated ketones. The high acidity of carbene complexes can also be ascribed in part to the delocalization of negative charge onto the electronegative oxygen atoms of the pentacarbonylchromium moiety in **XVIII**.

$$(CO)_5Cr{=}C(OCH_3)(CH_3) \longleftrightarrow (CO)_5Cr^{\ominus}{-}C(=O^{\oplus}{-}CH_3)(CH_3)$$

In spite of their great thermodynamic stability, carbene anions are reactive toward a variety of carbon electrophiles. Allylic and benzylic halides react with carbene anions at 0° to give alkylated carbene complexes (W. R. Brunsvold and C. P. Casey, unpublished results, 1975). Methyl fluorosulfonate acts as an efficient methylating agent of carbene anions (Casey *et al.*, 1972).

(CO)₅Cr⁻–[2-(3-methyl-4,5-dihydrofuryl)] anion reacts with:

- CH_3COCl → $(CO)_5Cr{=}$[3-methyl-3-acetyl-2-oxacyclopentylidene] (H_3C, $C{=}O$, CH_3) (66%)
- $C_6H_5CH_2I$ → $(CO)_5Cr{=}$[3-methyl-3-($CH_2C_6H_5$)-2-oxacyclopentylidene] (45%)
- prenyl bromide → $(CO)_5Cr{=}$[3-methyl-3-($CH_2{-}CH{=}C(CH_3)_2$)-2-oxacyclopentylidene] (67%)

α-Bromo esters react with carbene anions to give a mixture of mono- and di-alkylated products (Anderson and Casey, 1974b). Primary alkyl halides

$$(CO)_5Cr{=}C_4H_6O \xrightarrow[(2)\ CH_3OSO_2F]{(1)\ BuLi} \xrightarrow[(2)\ CH_3OSO_2F]{(1)\ BuLi} (CO)_5Cr{=}C_4H_4(CH_3)_2O \quad (58\%)$$

$$(CO)_5Cr{=}C(CH_3)(OCH_3) \xrightarrow[(2)\ BrCH_2CO_2CH_3]{(1)\ BuLi} (CO)_5Cr{=}C(OCH_3)(CH_2CH_2CO_2CH_3) \quad (37\%) + (CO)_5Cr{=}C(OCH_3)CH(CH_2CO_2CH_3)_2 \quad (20\%)$$

are much less reactive toward carbene anions than are allylic halides; refluxing THF is required to give even low yields of alkylated material.

$$(CO)_5Cr{=}C_4H_6O \xrightarrow[(2)\ CH_3CH_2Br,\ 67^\circ]{(1)\ BuLi} (CO)_5Cr{=}C_4H_5(CH_2CH_3)O \quad (15\%)$$

Acid chlorides react readily with carbene anions to give acylated carbene complexes. When the initial acylated product contains an enolizable proton, the corresponding enol ester is isolated (Casey *et al.*, 1972).

$$(CO)_5W{=}C(CH_3)(OCH_3) \xrightarrow[(2)\ CH_3COCl]{(1)\ BuLi} (CO)_5W{=}C(OCH_3)CH{=}C(CH_3)OC(O)CH_3 \quad (23\%)$$

Aldehydes react with carbene anions to produce vinylcarbene complexes. Reaction of carbene anions with formaldehyde gives dimeric products that apparently arise by conjugate addition of a carbene anion to an initially

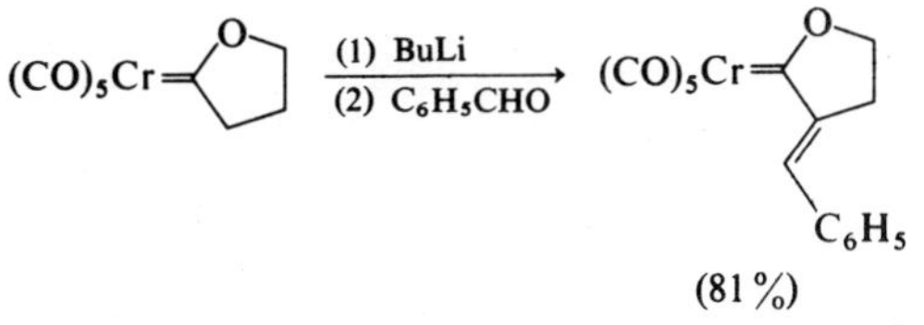

$(CO)_5Cr$ (1) BuLi, (2) C_6H_5CHO → $(CO)_5Cr$ ... C_6H_5 (81%)

formed *exo*-methylenecarbene complex (Brunsvold and Casey, 1975). No reaction of carbene anions with ketones, esters, or alcohols was observed. Thus, these functional groups should not require protection during synthetic sequences involving carbene anions.

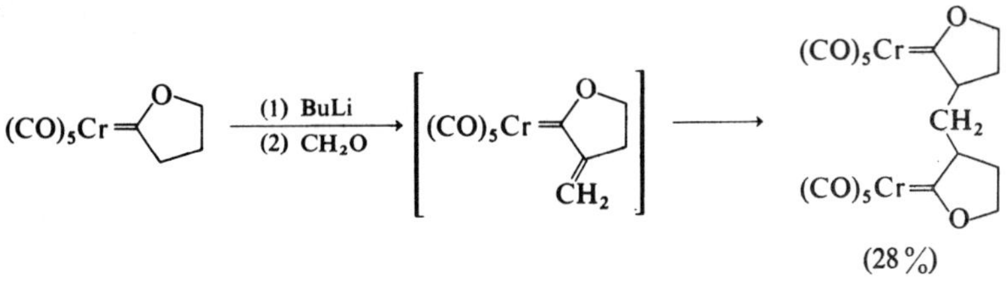

$(CO)_5Cr$ (1) BuLi, (2) CH_2O → [$(CO)_5Cr$ ... CH_2] → $(CO)_5Cr$ ... CH_2 ... $(CO)_5Cr$ (28%)

Carbene anions react with ethylene oxide to give 2-oxacyclopentylidene complexes (Anderson and Casey, 1974b). The reaction can be envisioned as nucleophilic attack of the carbene anion on the epoxide, followed by intramolecular displacement of methoxide from the intermediate adduct. As expected for a nucleophilic attack, propylene oxide is attacked by carbene anions at the least hindered carbon atom.

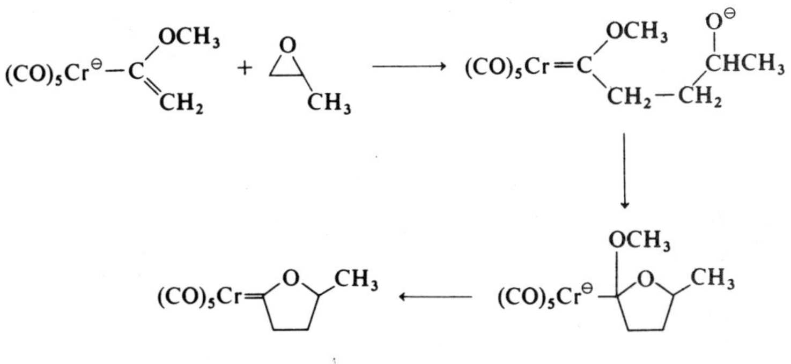

$(CO)_5Cr^{\ominus}-C(OCH_3)=CH_2$ + O, CH_3 → $(CO)_5Cr=C(OCH_3)CH_2-CH_2CH(O^{\ominus})CH_3$

↓

$(CO)_5Cr$ O CH_3 ← $(CO)_5Cr^{\ominus}$ OCH_3 O CH_3

Dialkylation is often an important side reaction in the alkylation of carbene anions. The dependence of the acidity and of the reactivity of carbene anions

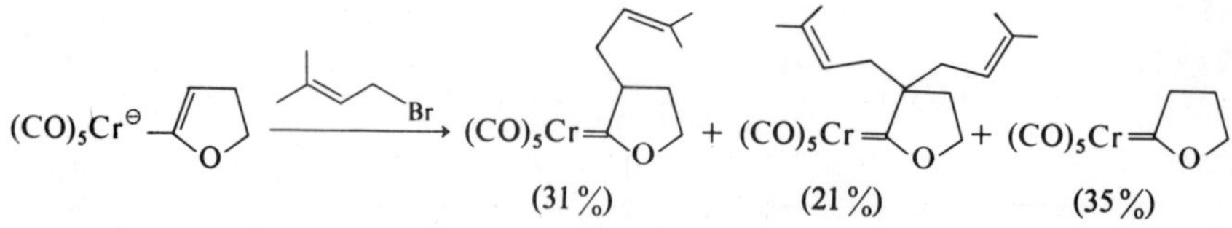

$(CO)_5Cr^{\ominus}$ — Br → $(CO)_5Cr$= + $(CO)_5Cr$= + $(CO)_5Cr$=

(31%) (21%) (35%)

on the substitution of the α-carbon atom has been investigated to gain insight into the dialkylation problem (W. R. Brunsvold and C. P. Casey, unpublished results, 1975). It was found that secondary anion **XIX** and tertiary anion **XX** were of equal basicity. However, the more substituted anion **XX** was three

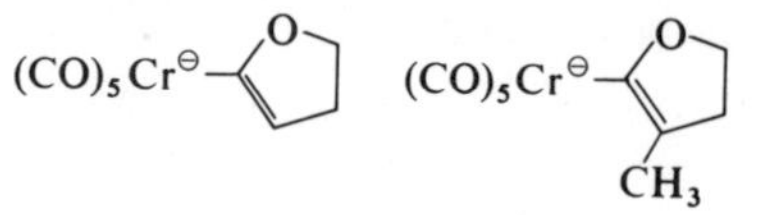

$(CO)_5Cr^{\ominus}$ $(CO)_5Cr^{\ominus}$ CH_3

(XIX) **(XX)**

times more reactive than **XIX** toward benzyl halides. The reasons for the higher reactivity of the more hindered anion are not understood. Nevertheless, the higher reactivity of the more substituted anion accounts for the tendency of carbene anions to undergo dialkylation. Dialkylation can be somewhat suppressed by adding the carbene anion to an excess of a reactive alkylating agent.

The usual preparative route to metal–carbene complexes involves the attack of an organolithium reagent on a coordinated carbonyl group. This synthetic method is limited by the availability of organolithium reagents; only functional groups which are compatible with lithium reagents may be introduced into the carbene complex by such routes. The reactions of carbene anions with carbon electrophiles provide a convenient method for the synthesis of a wide variety of carbene complexes containing functional groups such as esters and ketones which are not compatible with lithium reagents. The carbene anions are very weak bases and as such are mild reagents for the introduction of new carbon–carbon bonds.

V. RELEASE OF THE CARBENE LIGAND FROM METAL COMPLEXES

The ultimate utility of metal–carbene complexes in organic synthesis hinges on the discovery of synthetically useful ways of releasing the carbene ligand from its metal complexes. Several useful means of releasing the carbene ligand have already been found and future research will undoubtedly uncover additional transformations. When carbene complexes were first discovered, it was hoped that they might serve as reagents for generating free carbenes. However, to date, no reaction of carbene complexes has been shown to proceed via a free carbene. The thermal decomposition reactions and the alkene addition reactions of carbene complexes (for which the intervention of free carbenes seemed plausible) have been shown not to proceed via free carbenes.

The fact that metal–carbene complexes do not act as ready sources of free carbenes does not detract from their usefulness. Rather, the differences between the complexed and uncomplexed carbenes can be exploited synthetically.

Some of the reactions useful for releasing the carbene ligand from a metal complex are shown in Scheme 7.

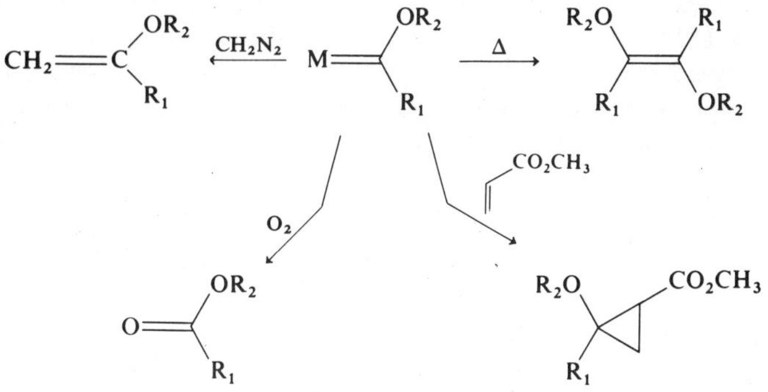

$$CH_2{=}C(OR_2)R_1 \xleftarrow{CH_2N_2} M{=}C(OR_2)R_1 \xrightarrow{\Delta} R_2O(R_1)C{=}C(R_1)OR_2$$

$$M{=}C(OR_2)R_1 \xrightarrow{O_2} O{=}C(OR_2)R_1$$

$$M{=}C(OR_2)R_1 \xrightarrow{CH_2{=}CHCO_2CH_3} \text{1-}R_2O\text{-1-}R_1\text{-2-}CO_2CH_3\text{-cyclopropane}$$

Scheme 7

A. Thermal Decomposition

The thermal decomposition of phenylmethoxycarbene complexes leads predominantly to dimer formation (Fischer *et al.*, 1969). When the decomposition is run in the presence of simple alkenes, no cyclopropanes are formed. Thermolysis of alkylalkoxycarbene complexes leads predominantly

$$(CO)_5Cr{=}C(C_6H_5)OCH_3 \xrightarrow[(CH_3)_2C{=}C(CH_3)_2]{135^\circ} H_5C_6(CH_3O)C{=}C(OCH_3)C_6H_5\ (40\%) + H_5C_6(CH_3O)C{=}C(C_6H_5)OCH_3\ (20\%)$$

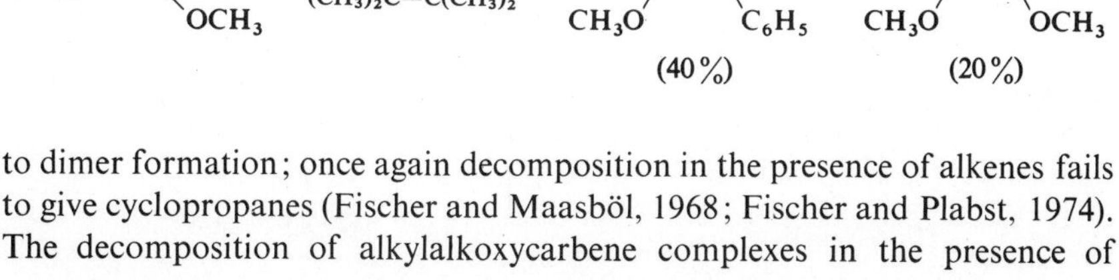

to dimer formation; once again decomposition in the presence of alkenes fails to give cyclopropanes (Fischer and Maasböl, 1968; Fischer and Plabst, 1974). The decomposition of alkylalkoxycarbene complexes in the presence of

$$(CO)_5Cr{=}C(OCH_3)(CH_3) \xrightarrow[\text{decalin}]{150^\circ} H_3C(CH_3O)C{=}C(OCH_3)CH_3 \ (24\%) + H_3C(CH_3O)C{=}C(CH_3)OCH_3 \ (10\%)$$

$$(CO)_5Cr{=}C(OCH_3)(CH_3) \xrightarrow[80^\circ]{\text{pyridine}} CH_3O(H)C{=}CH_2 \ (85\%) + (CO)_5Cr{-}NC_5H_5$$

pyridine occurs under much milder conditions and gives vinyl ethers which are the formal products of insertion of a carbene into an α C–H bond. Fischer and Plabst (1974) have suggested that free carbenes are involved in the pyridine catalyzed reactions. However, the fact that pyridine does not catalyze the decomposition of arylmethoxycarbene complexes (R. L. Anderson and C. P. Casey, unpublished results, 1974) makes an alternate mechanism involving formation of a carbene anion, followed by protonation of the carbon–metal bond, appear more attractive.

$$(CO)_5Cr{=}C(CH_3)(OCH_3) + C_5H_5N \longrightarrow (CO)_5Cr^{\ominus}{-}C(OCH_3){=}CH_2 \quad H{-}\overset{\oplus}{N}C_5H_5$$

$$\downarrow$$

$$H{-}C(OCH_3){=}CH_2$$

To test for the intermediacy of free carbenes in the decomposition of metal–carbene complexes, the thermolysis of the 2-oxacyclopentylidene complex **XVII** was studied (Anderson and Casey, 1975). 2-Oxacyclopentylidene has been generated by pyrolysis of the sodium salt of butyrolactone tosylhydrazone and shown to rearrange to give both dihydrofuran and cyclobutanone (Agosta and Foster, 1972). The thermolysis of **XVII** in decalin at 180° for 15

$$\text{butyrolactone tosylhydrazone anion } (N^{\ominus}{-}Ts) \xrightarrow{310^\circ} [\text{2-oxacyclopentylidene}] \longrightarrow \text{dihydrofuran } (25\%) + \text{cyclobutanone } (20\%)$$

(Ts = Tosyl)

min led only to the formation of dimer and dihydrofuran. The absence of cyclobutanone in the reaction mixture rules out the intervention of free carbenes. Detailed investigation of the decomposition of **XVII** revealed that (1)

$(CO)_5Cr$ (**XVII**) ⟶ (50%) + (5%)

the rate of reaction is second order in **XVII**, (2) CO inhibits decomposition of **XVII**, and (3) ^{13}CO exchanges with **XVII** at a rate more rapid than decomposition. A kinetic mechanism consistent with this data is shown below. The key step involves the formation of a highly reactive coordinatively unsaturated intermediate. An attractive mechanism for dimer formation involves forma-

$(CO)_5Cr$ ⇌ $(CO)_4Cr$ + CO

$(CO)_4Cr$ + $(CO)_5Cr$ ⟶ (slow) $[(CO)_4Cr]$ ⟶ (fast) dimer

tion of a biscarbene intermediate which then rapidly decomposes to dimer. This process requires the transfer of a carbene ligand from one metal to another. Such a transfer from chromium to tungsten was shown to occur under the reaction conditions (see Section III,G).

The pyridine- or triethylamine-catalyzed decomposition of **XVII** gave only dihydrofuran (Anderson and Casey, 1975). The absence of cyclobutanone in the product mixture rules out the intervention of free carbenes in the pyridine-catalyzed decomposition of carbene complexes.

The thermolysis of amino-substituted carbene complexes gives high yields of imines (Fischer and Leupold, 1972; Connor and Rose, 1972).

$$(CO)_5Cr{=}C(NHCH_3)(C_6H_5) \xrightarrow[100\ \text{atm CO}]{170°} Cr(CO)_6 + H_5C_6(H)C{=}N{-}CH_3 \quad (90\%)$$

B. Reactions with Alkenes: Cyclopropane Formation and Alkene Scission

Although no cyclopropane formation was observed in the reactions of heteroatom-stabilized carbene complexes with simple alkenes such as cyclohexene or tetramethylethylene, cyclopropane formation has been observed both with electron-deficient α,β-unsaturated esters and with electron-rich vinyl ethers. The mechanisms involved in cyclopropane formation from these very different classes of alkenes may be substantially different.

Reaction of methyl *trans*-crotonate with metal–carbene complexes occurs at 90°–140° to give mixtures of cyclopropanes in 60% yield (Dötz and Fischer, 1970, 1972a). Both cyclopropanes are formed stereospecifically, but the ratio of isomers depends on the metal. The metal dependence provides evidence that cyclopropane formation does not involve a free carbene. The

$(CO)_5M{=}C(C_6H_5)(OCH_3)$ + $CH_3CH{=}CHCO_2CH_3$ ⟶ cyclopropane (CO_2CH_3, H; H, CH_3; C_6H_5, OCH_3) + cyclopropane (CO_2CH_3, H; H, CH_3; OCH_3, C_6H_5)

M = Cr	70	to	30
M = Mo	85	to	15
M = W	70	to	30

formation of optically active cyclopropanes from the reaction of a chiral metal–carbene complex with diethyl fumarate provides further evidence against the intermediacy of free carbenes in these reactions (Cooke and Fischer, 1973).

$EtO_2CCH{=}CHCO_2Et$ + $(CO)_4Cr{=}C(C_6H_5)(OCH_3)$ [with $P(CH_3)(C_6H_5)(CH_2CH_2CH_3)$ on Cr] ⟶ cyclopropane (CO_2CH_3, H; H, CO_2CH_3; C_6H_5, OCH_3)
optically active

The formation of cyclopropanes from α,β-unsaturated esters occurs under conditions at least as severe as those employed in the phosphine substitution and ^{13}CO exchange reactions of carbene complexes. Coordinatively unsaturated carbene complexes are therefore reasonable intermediates for these reactions. Complexation of an alkene to the metal complex provides a means of bringing the carbene and alkene ligands into close proximity. Formation of a metallocyclobutane and reductive elimination of a cyclopropane complete our suggested mechanism (see Scheme 8).

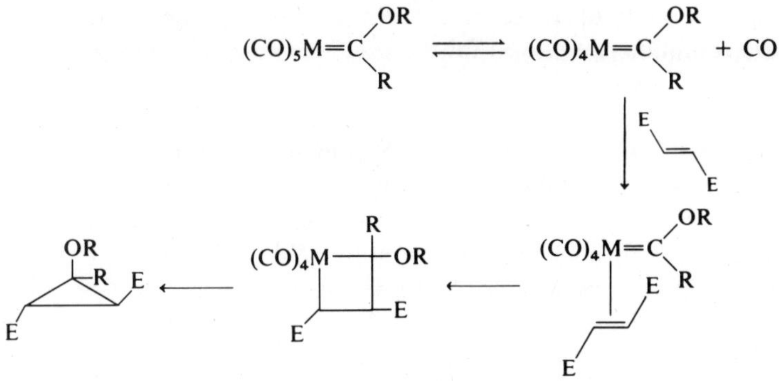

Scheme 8

The reaction of metal–carbene complexes with electron-rich vinyl ethers occurs under milder conditions than the reaction with electron-poor unsaturated esters. The conditions are also milder than those required for ligand substitution of carbene complexes. The reaction products depend strongly on the external CO pressure: with no added CO, alkene scission products predominate; under 100 atm CO pressure, cyclopropanes are formed in 60% yield (Dötz and Fischer, 1972b). The ratio of isomeric cyclopropanes formed

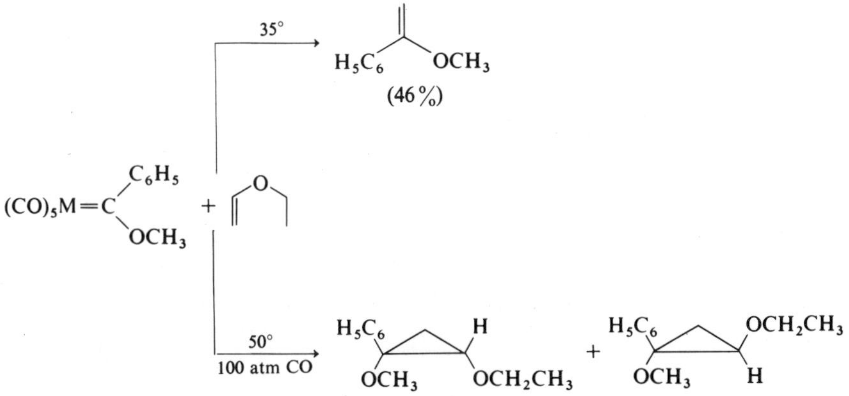

M = Cr	46%	14%
M = Mo	48%	12%
M = W	38%	22%

Scheme 9

varies with the nature of the metal atom of the carbene complex and provides evidence against free carbenes in this reaction (see Scheme 9).

The inaccessibility of a vacant coordination site on the carbene complex under the conditions employed for the reactions of vinyl ethers together with the nucleophilic nature of vinyl ethers makes a reaction mechanism involving initial nucleophilic attack of a vinyl ether on the carbene carbon atom attractive (see Scheme 10). The initial adduct **XXI** might then form a seven-coordinate metallocyclobutane **XXII**. This species could reversibly dissociate CO to form a six-coordinate metallocyclobutane **XXIII**. Seven-coordinate **XXII** would be favored by high CO pressure and could undergo reductive elimination to give cyclopropane and $Cr(CO)_5$. Reductive elimination of cyclopropane from six-coordinate **XXIII** would be a high-energy process because it would produce the 14-electron $Cr(CO)_4$ fragment. Alkene scission could occur by ring opening of the metallocyclobutane from **XXIII** and would be favored at low CO pressure.

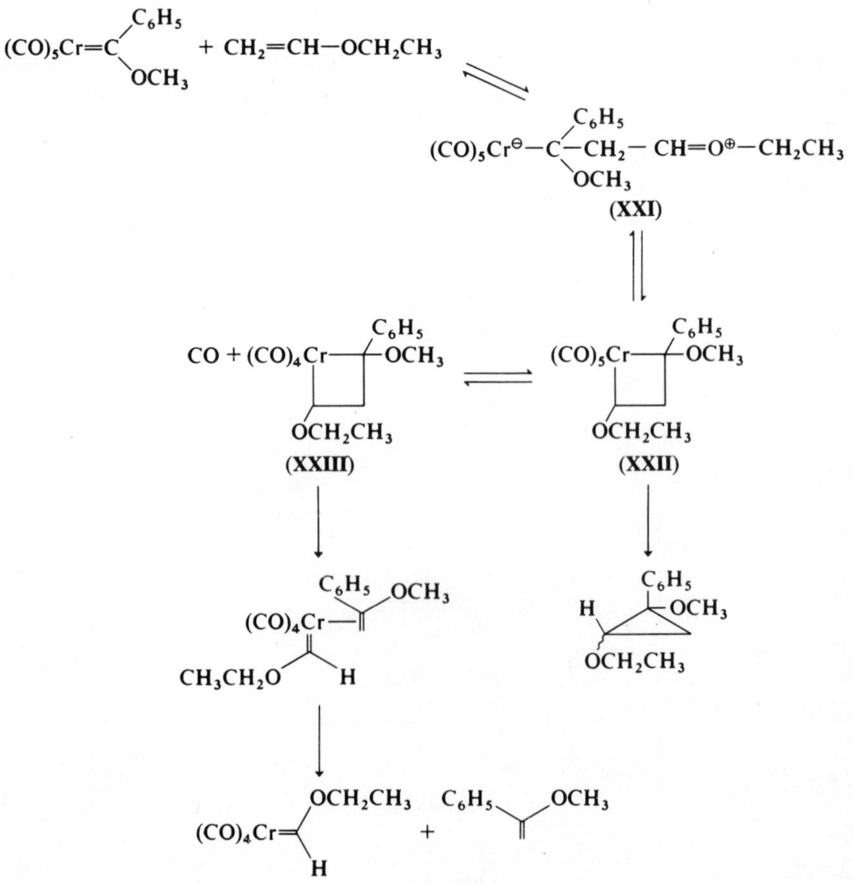

$(CO)_5Cr{=}C(C_6H_5)(OCH_3) + CH_2{=}CH{-}OCH_2CH_3 \rightleftharpoons (CO)_5Cr^{\ominus}{-}C(C_6H_5)(OCH_3){-}CH_2{-}CH{=}O^{\oplus}{-}CH_2CH_3$ **(XXI)**

(XXI) $\rightleftharpoons$ $(CO)_5Cr$ metallocyclobutane bearing C_6H_5, OCH_3 and OCH_2CH_3 **(XXII)**

$CO + (CO)_4Cr$ metallocyclobutane bearing C_6H_5, OCH_3 and OCH_2CH_3 **(XXIII)** $\rightleftharpoons$ **(XXII)**

(XXII) $\rightarrow$ cyclopropane bearing H, C_6H_5, OCH_3, OCH_2CH_3

(XXIII) $\rightarrow$ $(CO)_4Cr$ coordinated to $C_6H_5(OCH_3)C{=}$ and $CH_3CH_2O(H)C{=}$ $\rightarrow$ $(CO)_4Cr{=}C(OCH_2CH_3)(H)$ + $C_6H_5(OCH_3)C{=}$

Scheme 10

The reaction of carbene complexes with enamines has been found to give cyclopropanes in low yield (Dorrer *et al.*, 1974). Reaction of carbene complexes with 1-vinyl-2-pyrrolidone in the absence of added CO pressure gives alkene scission products (Dorrer and Fischer, 1974a), whereas at high CO pressure products possibly derived from ketenes are obtained (Dorrer and Fischer, 1974b) (see Scheme 11).

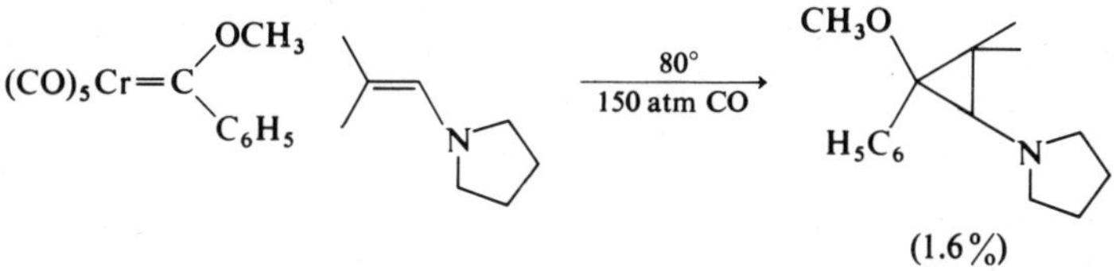

$(CO)_5Cr{=}C(OCH_3)(C_6H_5)$ + enamine $\xrightarrow[150\text{ atm CO}]{80°}$ cyclopropane (CH_3O, H_5C_6, N-pyrrolidinyl) (1.6%)

$(CO)_5WC(C_6H_5)_2$ (**XXIV**), a carbene complex not stabilized by electron-donating heteroatoms attached directly to the carbene carbon, is less stable and more reactive than methoxy-substituted carbene complexes (Burkhardt and Casey, 1973). Compound **XXIV** undergoes exchange with ^{13}CO at 33° and thermally decomposes at 40°–60°. Reaction of **XXIV** with ethyl vinyl ether gave a high yield of a cyclopropane and a small amount of alkene scission product (see Scheme 12). Reaction of **XXIV** with isobutylene gave a

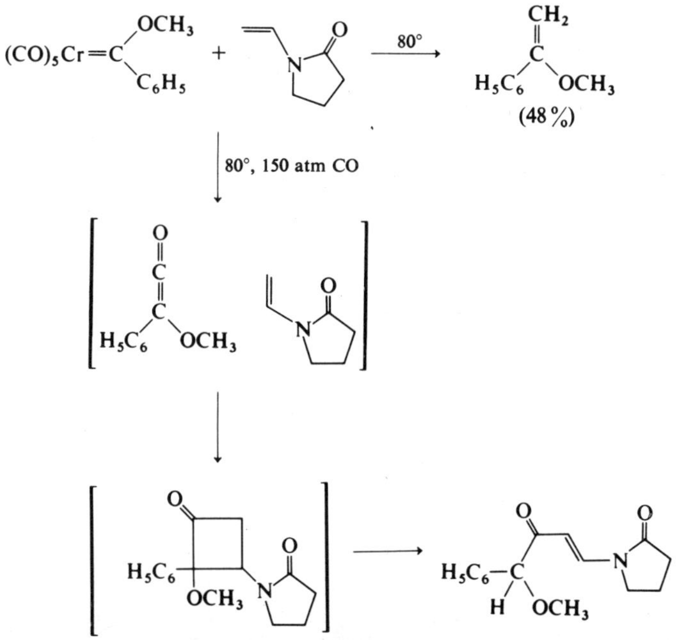

$(CO)_5Cr{=}C(OCH_3)(C_6H_5)$ + 1-vinyl-2-pyrrolidone $\xrightarrow{80°}$ $CH_2{=}C(H_5C_6)(OCH_3)$ (48%)

↓ 80°, 150 atm CO

[$O{=}C{=}C(H_5C_6)(OCH_3)$ + 1-vinyl-2-pyrrolidone]

↓

[cyclobutanone (H_5C_6, OCH_3, N-pyrrolidonyl)] ⟶ $H_5C_6{-}CH(OCH_3){-}CO{-}CH{=}CH{-}N$(pyrrolidone)

Scheme 11

small amount of substituted cyclopropane, the major product, 1,1-diphenylethylene, derived from transfer of the least substituted end of the alkene (Burkhardt and Casey, 1974). Reaction of **XXIV** with 1-methoxy-1-phenylethylene gave alkene scission products and a new carbene complex.

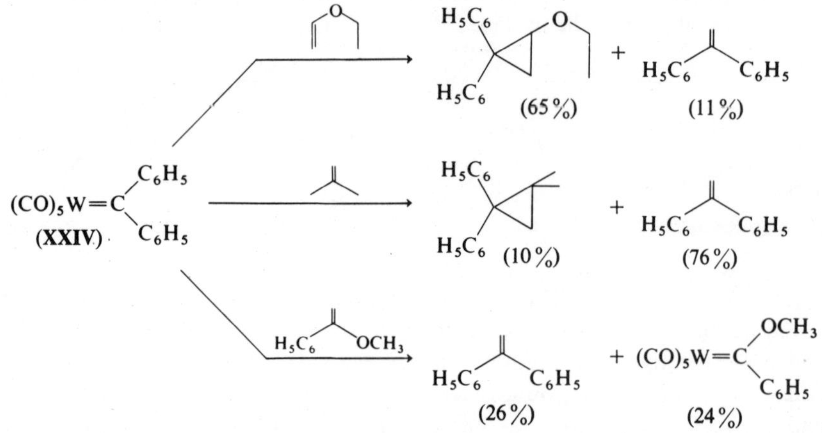

$(CO)_5W{=}C(C_6H_5)_2$ **(XXIV)**

H_5C_6 H_5C_6 (65%) + H_5C_6 C_6H_5 (11%)

H_5C_6 H_5C_6 (10%) + H_5C_6 C_6H_5 (76%)

H_5C_6 OCH_3 → H_5C_6 C_6H_5 (26%) + $(CO)_5W{=}C(OCH_3)C_6H_5$ (24%)

Scheme 12

Both the cyclopropanation and alkene scission reactions of **XXIV** can be explained in terms of the mechanistic scheme (Scheme 13). The metallocyclobutane **XXV** formed by rearrangement of a metal complex containing both a carbene and an alkene ligand is the key intermediate in these reactions. Compound **XXV** can undergo a reductive elimination to give a cyclopropane or it can undergo cleavage to give a metal complex containing both a coordinated 1,1-diphenylethylene and a new carbene ligand.

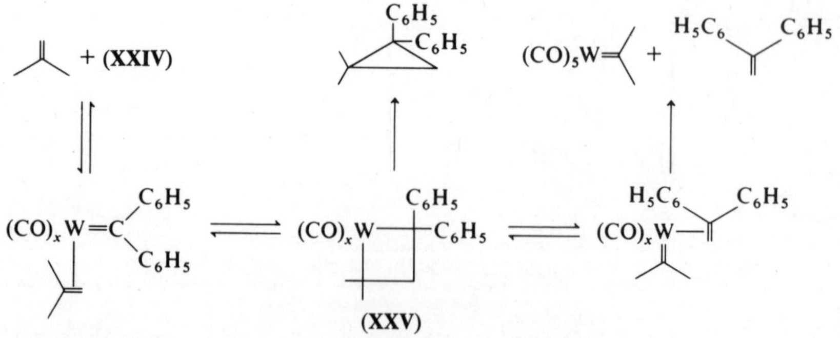

+ (XXIV)

C_6H_5 C_6H_5

$(CO)_5W{=}$ + H_5C_6 C_6H_5

$(CO)_xW{=}C(C_6H_5)_2$ ⇌ $(CO)_xW$ C_6H_5 C_6H_5 ⇌ $(CO)_xW$ H_5C_6 C_6H_5

(XXV)

Scheme 13

It should be noted that the equilibrium between a metallocyclobutane and a metal complex containing both an alkene and a carbene ligand provides a sufficient mechanism for olefin metathesis. Such a scheme has previously been proposed by Chauvin (Chauvin and Herrisson, 1971; Soufflet *et al.*,

1973). Unlike all previous mechanisms for olefin metathesis, the metallocyclobutane mechanism does not involve the pairwise exchange of alkylidene units between two alkenes undergoing dismutation. The nonpairwise exchange of alkenes in the olefin dismutation reaction has now been established (Grubbs *et al.*, 1975; Katz and McGinnis, 1975; Katz and Rothchild, 1976).

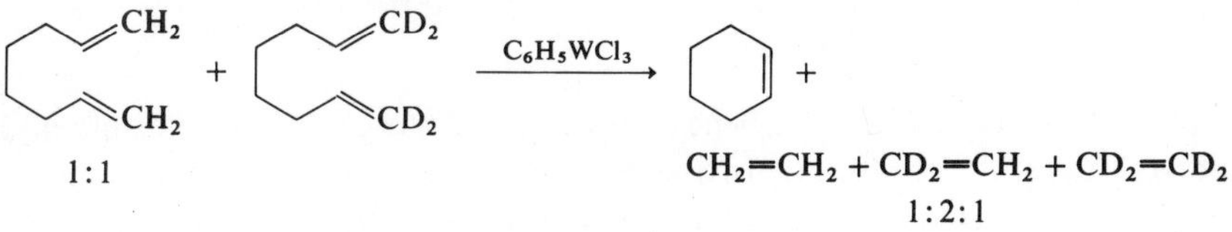

$$\text{(1,7-octadiene, } CH_2 \text{ ends)} + \text{(1,7-octadiene, } CD_2 \text{ ends)} \;(1:1) \xrightarrow{C_6H_5WCl_3} \text{cyclohexene} + CH_2{=}CH_2 + CD_2{=}CH_2 + CD_2{=}CD_2 \;(1:2:1)$$

C. Reactions with Ylides

Phosphorus ylides react with $(CO)_5WC(OCH_3)C_6H_5$ to produce vinyl ethers (Burkhardt and Casey, 1972). High yields of vinyl ethers are obtained

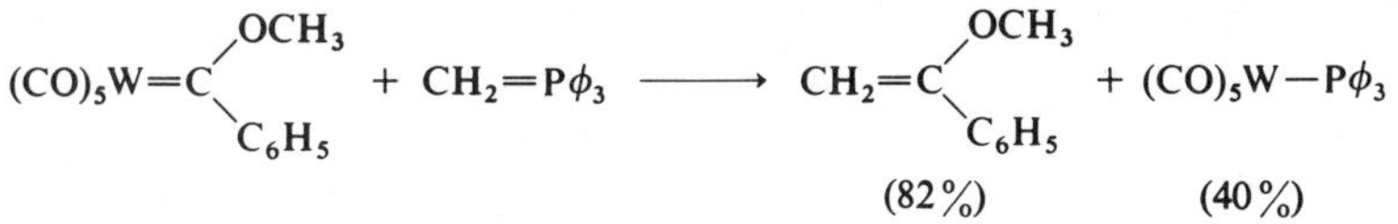

$$(CO)_5W{=}C(OCH_3)(C_6H_5) + CH_2{=}P\phi_3 \longrightarrow \underset{(82\%)}{CH_2{=}C(OCH_3)(C_6H_5)} + \underset{(40\%)}{(CO)_5W{-}P\phi_3}$$

from $\phi_3P{=}CH_2$ and $\phi_3P{=}CHCH_3$ at room temperature, but no vinyl ether could be obtained from the hindered $\phi_3P{=}C(CH_3)_2$. Reaction with the stabilized ylide $\phi_3P{=}CH(C_6H_5)$ required heating to 60° and no reaction of carbonyl-stabilized ylides was observed.

The reaction can be envisioned as proceeding via nucleophilic attack by the phosphorane carbon atom on the electron-deficient carbene carbon atom to form a betaine-like intermediate (**XXVI**) which subsequently fragments to form an enol–ether complex and free triphenylphosphine which then react to give the observed products (see Scheme 14). The observation of $(CO)_5$-$WP(C_6H_4CH_3)_3$ in the product mixture when the reaction was carried out in the presence of $P(C_6H_4CH_3)_3$ supports this mechanism.

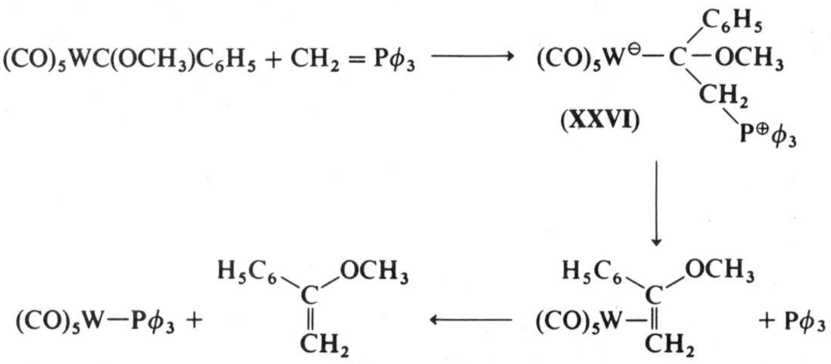

$$(CO)_5WC(OCH_3)C_6H_5 + CH_2 = P\phi_3 \longrightarrow \underset{\textbf{(XXVI)}}{(CO)_5W^{\ominus}{-}C(C_6H_5)(OCH_3){-}CH_2{-}P^{\oplus}\phi_3}$$

$$\downarrow$$

$$(CO)_5W{-}P\phi_3 + H_5C_6(OCH_3)C{=}CH_2 \longleftarrow (CO)_5W{-}\|[H_5C_6(OCH_3)C{=}CH_2] + P\phi_3$$

Scheme 14

The reaction of phosphoranes with alkylmethoxycarbene complexes fails to produce vinyl ethers owing to abstraction of a proton from the carbon alpha to the carbene carbon atom. However, the less basic diazoalkanes react

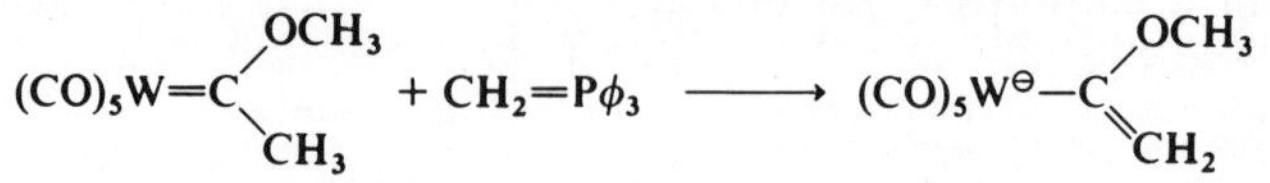

$$(CO)_5W{=}C(OCH_3)(CH_3) + CH_2{=}P\phi_3 \longrightarrow (CO)_5W^{\ominus}{-}C(OCH_3){=}CH_2$$

cleanly with alkylmethoxycarbene complexes at 5° to give high yields of vinyl ethers (Casey *et al.*, 1973a). In the reaction of acetylenic carbene complex **XXVII** with diazomethane, dipolar addition of CH_2N_2 to the acetylene

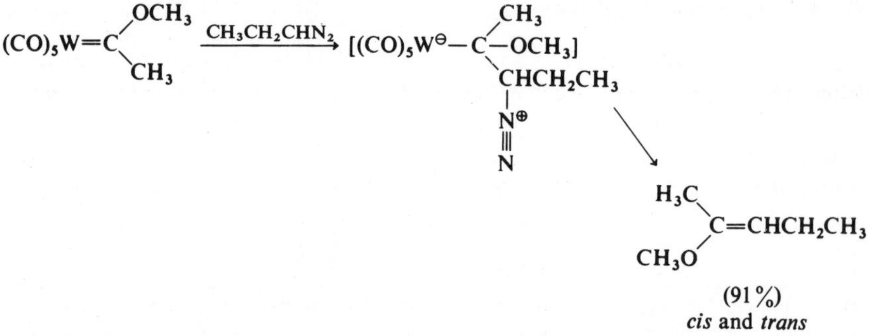

$$(CO)_5W{=}C(OCH_3)(CH_3) \xrightarrow{CH_3CH_2CHN_2} [(CO)_5W^{\ominus}{-}C(CH_3)(OCH_3){-}CH(CH_2CH_3){-}N^{\oplus}{\equiv}N] \longrightarrow (H_3C)(CH_3O)C{=}CHCH_2CH_3$$

(91%) *cis* and *trans*

apparently precedes reaction at the carbene carbon atom (Kreissl *et al.*, 1973c) (see Scheme 15).

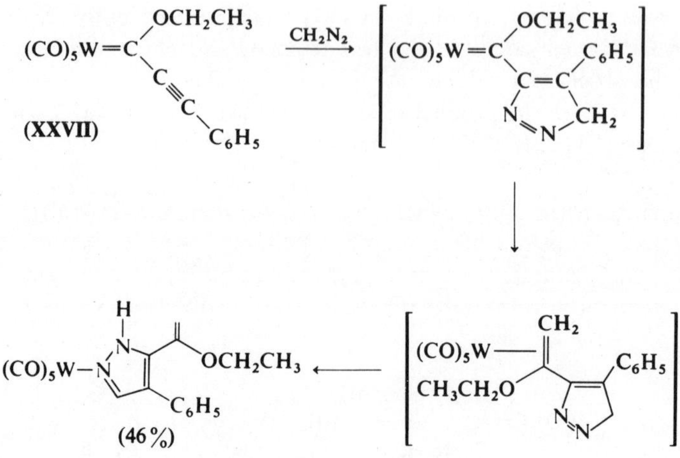

$$(CO)_5W{=}C(OCH_2CH_3){-}C{\equiv}C{-}C_6H_5 \ \textbf{(XXVII)} \xrightarrow{CH_2N_2} \left[(CO)_5W{=}C(OCH_2CH_3){-}C{=}C(C_6H_5){-}CH_2{-}N{=}N \text{ (ring)}\right]$$

$$\downarrow$$

$$(CO)_5W{-}N\text{(pyrazole, N–H)}{-}C({=}CH_2)(OCH_2CH_3),\ C_6H_5 \ (46\%) \longleftarrow \left[(CO)_5W{-}C({=}CH_2)(CH_3CH_2O){-}\text{pyrazoline}(C_6H_5)\right]$$

Scheme 15

D. Oxidative Cleavage

Oxidation replaces the carbon–metal double bond of metal–carbene complexes with a carbon–oxygen double bond. A variety of oxidizing agents including pyridine *N*-oxide (Cotton and Lukehart, 1971), dimethyl sulfoxide (C. P. Casey, R. L. Boggs, and W. R. Brunsvold, unpublished results, 1974), ceric ion (Casey *et al.*, 1972) and oxygen (Fischer and Riedmüller, 1974) have been employed. These oxidations are normally clean, high-yield reactions

$$(CO)_5W{=}C(CH{=}CHC_6H_5)(OCH_3) \xrightarrow{Ce^{4+}} O{=}C(CH{=}CHC_6H_5)(OCH_3) \quad (95\%)$$

which are useful in characterizing the carbene ligand by conversion to known organic compounds.

Reaction of carbene complexes with sulfur or selenium gives moderate yields of thio and seleno esters (Fischer and Riedmüller, 1974).

$$(CO)_5Cr{=}C(OCH_3)(C_6H_4Cl) \xrightarrow[35^\circ]{S} S{=}C(OCH_3)(C_6H_4Cl) \quad (30\%)$$

$$(CO)_5Cr{=}C(OCH_3)(C_6H_4Cl) \xrightarrow[100^\circ]{Se} Se{=}C(OCH_3)(C_6H_4Cl) \quad (12\%)$$

E. Reductive Cleavage

The reductive cleavage of metal–carbene complexes has not been thoroughly investigated. Reduction of an amino-substituted carbene complex gave the corresponding saturated amine in unspecified yield (Connor and Fischer, 1967). Reduction of $(CO)_5CrC(OCH_3)CH_3$ with $LiAlH_4$ gave $<0.1\%$

$$(CO)_5Cr{=}C(CH_3)(NHR) + LiAlH_4 \longrightarrow H_2C(CH_3)(NHR)$$

XXVIII (Fischer and Knauss, 1971b). The reaction of $(CO)_5WC(C_6H_5)_2$ with molecular hydrogen at 50° gave a 33% yield of diphenylmethane (C. P. Casey and S. Neumann, unpublished observations, 1975).

$$(CO)_5Cr{=}C(OCH_3){-}CH{=}CH{-}CH{=}C(OCH_3)CH_3$$

(XXVIII)

F. Acid Cleavage

Metal–carbene complexes are stable in dilute aqueous acid, but strongly basic aminocarbene complexes react with HCl or HBr in ether at −40° to give products derived from protonation of the carbon–metal bond (Fischer *et al.*, 1973b). Alkoxy-substituted carbene complexes react with mixtures of HCl

$$(CO)_5Cr^{\ominus}{-}C(CH_3){=}N^{\oplus}(CH_3)_2 \xrightarrow[-40^\circ]{HCl} (CO)_5Cr^{\ominus}Cl \quad HC(CH_3){=}N^{\oplus}(CH_3)_2$$

and phosphines to give moderate yields of phosphonium salts (Fischer and Schubert, 1973). The reaction has been suggested to proceed via acid cleavage

$$(CO)_5Cr{=}C(C_6H_5)(OCH_3) \xrightarrow{HCl} \left[HC(C_6H_5)(Cl)(OCH_3) \right] \xrightarrow{PR_3} H{-}C(C_6H_5)(OCH_3){-}P^{\oplus}R_3$$

$$(CO)_5Cr{=}C(C_6H_5)(OCH_3) \xrightarrow{PR_3} \left[(CO)_5Cr^{\ominus}{-}C(C_6H_5)(OCH_3){-}P^{\oplus}R_3 \right] \xrightarrow{HCl} H{-}C(C_6H_5)(OCH_3){-}P^{\oplus}R_3$$

of the carbene ligand, followed by reaction of the α-chloro ether with phosphine. However, since the addition of phosphines to the carbene carbon atom is known to be rapid, this reaction may proceed by initial adduct formation, followed by acid cleavage of the carbon–metal bond.

The reaction of sulfur-substituted carbene complexes with HBr at $-30°$ gives sulfide complexes (Fischer and Kreis, 1973).

$$(CO)_5Cr{=}C(SCH_3)(CH_3) \xrightarrow{HBr} (CO)_5Cr{-}S(CH_3){-}CH(Br){-}CH_3$$

G. Miscellaneous Cleavages

The reactions of metal–carbene complexes with silicon hydrides result in addition of the Si–H bond to the carbene ligand (Connor and Rose, 1970; Dötz and Fischer, 1972c). Tin hydrides react similarly (Connor *et al.*, 1973). Trihalomethylmercury compounds react with carbene complexes to give dihaloethylenes in unspecified yield (DeRenzi and Fischer, 1974).

$$(CO)_5Cr{=}C(NC_4H_8)(C_6H_5) + Et_3SiH \xrightarrow{\text{pyridine}} H(NC_4H_8)C(Et_3Si)(C_6H_5) \quad (59\%)$$

$$(CO)_5Cr{=}C(OCH_3)(C_6H_5) \xrightarrow{\phi_2SiH_2} H(OCH_3)C(\phi_2SiH)(C_6H_5) \quad (12\%)$$

$$(CO)_5Cr{=}C(OCH_3)(C_6H_5) \xrightarrow{\phi_3SnH} H(OCH_3)C(\phi_3Sn)(C_6H_5)$$

$$(CO)_5Cr{=}C(C_6H_5)(OCH_3) + C_6H_5HgCCl_3 \longrightarrow Cl_2C{=}C(C_6H_5)(OCH_3)$$

A variety of nitrogen nucleophiles including hydroxylamine (Aumann and Fischer, 1968b), hydrazines (Aumann and Fischer, 1967), oximes (Fischer and

Knauss, 1970), and hydrazoic acid (Conner and Jones, 1971b) react with metal–carbene complexes to give cleavage products.

$$(CO)_5Cr{=}C(CH_3)(OCH_3) + H_2NN(CH_3)_2 \longrightarrow (CO)_5Cr(NCCH_3) + HN(CH_3)_2$$

$$(CO)_5Cr{=}C(OCOCH_3)(C_4H_3O) \xrightarrow{HN_3} \left[(CO)_5Cr{-}C(N_3)(C_4H_3O)\right] \longrightarrow (CO)_5Cr{-}N{\equiv}C{-}C_4H_3O$$

Treatment of alkoxycarbene complexes with strong Lewis acids such as BBr_3 leads to the formation of an interesting new class of organometallic compounds, the carbyne complexes (Fischer *et al.*, 1973d).

$$(CO)_5W{=}C(C_6H_5)(OCH_3) + BBr_3 \longrightarrow Br(CO)_4W{\equiv}C{-}C_6H_5$$

VI. APPLICATION OF METAL–CARBENE COMPLEXES IN ORGANIC SYNTHESIS

The rich and varied chemistry of metal–carbene complexes outlined in this chapter should eventually lead to the wide use of metal–carbene complexes in organic synthesis. To date, carbene complexes have not been used extensively as their chemistry was not widely understood and the reagents are moderately expensive. However, the applications of metal–carbene complexes in peptide synthesis, in the synthesis of an indole alkaloid, and in the synthesis of α-methylenebutyrolactone have been reported. These applications indicate some of the ways in which metal–carbene complexes may prove useful in synthesis.

The use of metal–carbene complexes as an amino-protecting group in peptide synthesis has been reported by Fischer and Weiss (1973). The free amino group of an amino acid ester readily displaces the alkoxy group of an alkoxy-substituted carbene complex. The nitrogen atom of the resulting N-substituted carbene complex is nonbasic and nonnucleophilic because it bears a partial positive charge. The low reactivity of the amino-substituted carbene complex allows a series of reactions to be carried out to construct a peptide chain. Finally, the completed peptide chain can be removed from the carbene complex by treatment with trifluoroacetic acid at 20°. Scheme 16 illustrates the variety of reactions that can be carried out in the presence of the metal–carbene functionality.

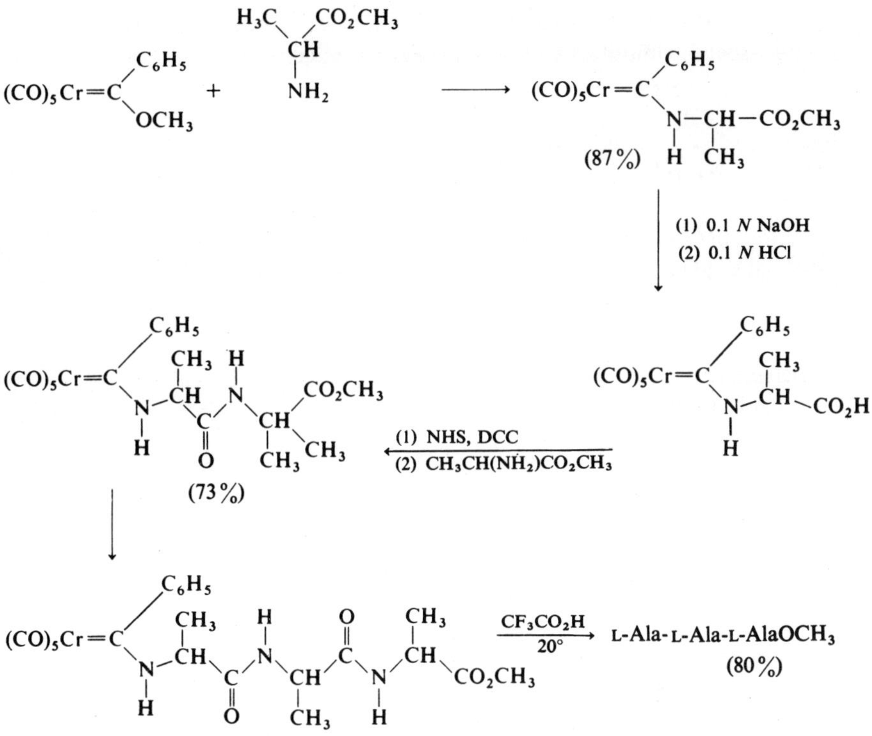

$(CO)_5Cr{=}C(C_6H_5)(OCH_3)$ + $H_3C{-}CH(NH_2){-}CO_2CH_3$ ⟶ $(CO)_5Cr{=}C(C_6H_5){-}N(H){-}CH(CH_3){-}CO_2CH_3$ (87%)

(1) 0.1 *N* NaOH
(2) 0.1 *N* HCl

⟶ $(CO)_5Cr{=}C(C_6H_5){-}N(H){-}CH(CH_3){-}CO_2H$

(1) NHS, DCC
(2) $CH_3CH(NH_2)CO_2CH_3$

⟶ $(CO)_5Cr{=}C(C_6H_5){-}N(H){-}CH(CH_3){-}C(=O){-}N(H){-}CH(CH_3){-}CO_2CH_3$ (73%)

⟶ $(CO)_5Cr{=}C(C_6H_5){-}N(H){-}CH(CH_3){-}C(=O){-}N(H){-}CH(CH_3){-}C(=O){-}N(H){-}CH(CH_3){-}CO_2CH_3$

CF_3CO_2H, 20° ⟶ L-Ala-L-Ala-L-AlaOCH$_3$ (80%)

(NHS=*N*-Hydroxysuccinimide, DCC=dicyclohexylcarbodiimide)

Scheme 16

The conversion of an aminocarbene complex to an imine has been used to synthesize tetrahydroharman (**XXIX**) (Connor and Rose, 1972). This synthesis illustrates the use of metal–carbene complexes in the synthesis of complex organic structures.

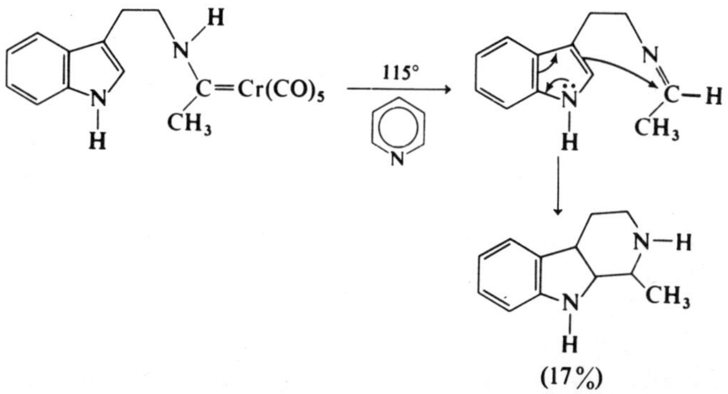

115°, pyridine

(17%)
(XXIX)

The synthesis of α-methylene-γ-butyrolactone (Scheme 17) illustrates the use of carbene anions in organic synthesis (Brunsvold and Casey, 1975). The five-membered ring was constructed by reaction of a carbene anion with an epoxide. The methylene group alpha to the carbene carbon atom in the resulting 2-oxacyclopentylidene complex (**XVII**) can be used as a site of further reaction. Inverse addition of the carbene anion of **XVII** to an excess of $ClCH_2OCH_3$ gave the highest yield of monosubstitution product. Elimination of methanol was achieved on stirring with Al_2O_3. Finally, α-methylene-γ-butyrolactone was obtained on oxidation of the carbene complex with ceric ion.

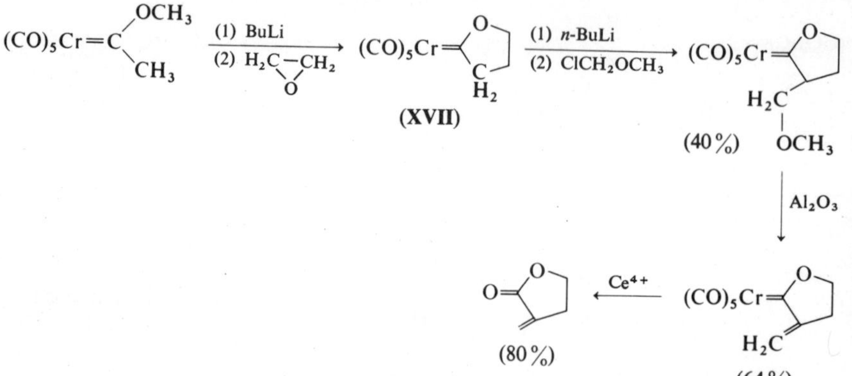

$(CO)_5Cr{=}C(OCH_3)CH_3$ —(1) BuLi; (2) $H_2C{-}CH_2$ (epoxide, O)→ $(CO)_5Cr{=}$(2-oxacyclopentylidene) (**XVII**) —(1) *n*-BuLi; (2) $ClCH_2OCH_3$→ $(CO)_5Cr{=}$(3-CH_2OCH_3-2-oxacyclopentylidene) (40%) —Al_2O_3→ $(CO)_5Cr{=}$(3-methylene-2-oxacyclopentylidene) (64%) —Ce^{4+}→ α-methylene-γ-butyrolactone (80%)

Scheme 17

REFERENCES

Agosta, W. C., and Foster, A. M. (1972). *J. Am. Chem. Soc.* **94**, 5777.
Anderson, R. L., and Casey, C. P. (1971). *J. Am. Chem. Soc.* **93**, 3554.
Anderson, R. L., and Casey, C. P. (1974a). *J. Am. Chem. Soc.* **96**, 1230.
Anderson, R. L., and Casey, C. P. (1974b). *J. Organomet. Chem.* **73**, C28.
Anderson, R. L., and Casey, C. P. (1975). *J. Chem. Soc., Chem. Commun.* p. 895.
Aumann, R., and Fischer, E. O. (1967). *Angew. Chem., Int. Ed. Engl.* **6**, 181.
Aumann, R., and Fischer, E. O. (1968a). *Chem. Ber.* **101**, 954.
Aumann, R., and Fischer, E. O. (1968b). *Chem. Ber.* **101**, 963.
Aumann, R., and Fischer, E. O. (1969). *Chem. Ber.* **102**, 1495.
Badley, E. M., Chatt, J., Richards, R. L., and Sim, G. A. (1969). *Chem. Commun.* p. 1322.
Badley, E. M., Chatt, J., and Richards, R. L. (1971). *J. Chem. Soc. A* p. 21.
Balch, A. L., and Miller, J. (1972). *J. Am. Chem. Soc.* **94**, 417.
Beck, H. J., and Fischer, E. O. (1971). *Chem. Ber.* **104**, 3101.
Beck, H. J., Fischer, E. O., and Kreiter, C. G. (1971). *J. Organomet. Chem.* **26**, C41.
Brunsvold, W. R., and Casey, C. P. (1974). *J. Organomet. Chem.* **77**, 345.
Brunsvold, W. R., and Casey, C. P. (1975). *J. Organomet. Chem.* **102**, 175.
Burkhardt, T. J., and Casey, C. P. (1972). *J. Am. Chem. Soc.* **94**, 6543.

Burkhardt, T. J., and Casey, C. P. (1973). *J. Am. Chem. Soc.* **95**, 5833.
Burkhardt, T. J., and Casey, C. P. (1974). *J. Am. Chem. Soc.* **96**, 7808.
Cardin, D. J., Cetinkaya, B., and Lappert, M. F. (1972a). *Chem. Rev.* **72**, 545.
Cardin, D. J., Cetinkaya, B., Cetinkaya, E., Lappert, M. F., Manojlović-Muir, L. J., and Muir, K. W. (1972b). *J. Organomet. Chem.* **44**, C59.
Cardin, D. J., Doyle, M. J., and Lappert, M. F. (1972c). *J. Chem. Soc., Chem. Commun.* p. 927.
Cardin, D. J., Cetinkaya, B., Doyle, M. J., and Lappert, M. F. (1973). *Chem. Soc. Rev.* **2**, 99.
Casey, C. P. (1970). *Chem. Commun.* p. 1220.
Casey, C. P., Boggs, R. A., and Anderson, R. L. (1972). *J. Am. Chem. Soc.* **94**, 8947.
Casey, C. P., Bertz, S. H., and Burkhardt, T. J. (1973a). *Tetrahedron Lett.* p. 1421.
Casey, C. P., Cyr, C. R., and Boggs, R. A. (1973b). *Synth. Inorg. Met.-Org. Chem.* **3**, 249.
Casey, C. P., Cyr, C. R., Anderson, R. L., and Marten, D. F. (1975). *J. Am. Chem. Soc.* **97**, 3053.
Cetinkaya, B., Lappert, M. F., and Turner, K. (1972). *J. Chem. Soc., Chem. Commun.* p. 851.
Cetinkaya, B., Dixneuf, P., and Lappert, M. F. (1974). *J. Chem. Soc., Dalton Trans.* p. 1827.
Chatt, J., Richards, R. L., and Royston, G. H. D. (1973). *J. Chem. Soc., Dalton Trans.* p. 1433.
Chauvin, Y., and Herrisson, J. L. (1971). *Makromol. Chem.* **141**, 161.
Chisholm, M. H., and Clark, H. C. (1971). *Inorg. Chem.* **10**, 1711.
Chisholm, M. H., Clark, H. C., and Hunter, D. H. (1971). *Chem. Commun.* p. 809.
Connor, J. A., and Fischer, E. O. (1967). *Chem. Commun.* p. 1024.
Connor, J. A., and Fischer, E. O. (1969). *J. Chem. Soc. A* p. 578.
Connor, J. A., and Jones, E. M. (1971a). *J. Chem. Soc. A* p. 3368.
Connor, J. A., and Jones, E. M. (1971b). *Chem. Commun.* p. 570.
Connor, J. A., and Müller, J. (1969). *Chem. Ber.* **102**, 1148.
Connor, J. A., and Rose, P. D. (1970). *J. Organomet. Chem.* **24**, C45.
Connor, J. A., and Rose, P. D. (1972). *J. Organomet. Chem.* **46**, 329.
Connor, J. A., Day, J. P., and Turner, R. M. (1973). *J. Chem. Soc., Chem. Commun.* p. 578.
Cooke, M. D., and Fischer, E. O. (1973). *J. Organomet. Chem.* **56**, 279.
Cooper, H. J., and Green, M. L. H. (1974). *J. Chem. Soc., Chem. Commun.* p. 761.
Cotton, F. A., and Lukehart, C. M. (1971). *J. Am. Chem. Soc.* **93**, 2672.
Cotton, F. A., and Lukehart, C. M. (1972). *Prog. Inorg. Chem.* **16**, 487.
Crociani, B., Boschi, T., Nicolini, M., and Belluco, U. (1972). *Inorg. Chem.* **11**, 1292.
DeRenzi, A., and Fischer, E. O. (1974). *Inorg. Chim. Acta* **8**, 185.
Dorrer, B., and Fischer, E. O. (1974a). *Chem. Ber.* **107**, 1156.
Dorrer, B., and Fischer, E. O. (1974b). *Chem. Ber.* **107**, 2683.
Dorrer, B., Fischer, E. O., and Kalbfus, W. (1974). *J. Organomet. Chem.* **81**, C20.
Dötz, K. H., and Fischer, E. O. (1970). *Chem. Ber.* **103**, 1273.
Dötz, K. H., and Fischer, E. O. (1972a). *Chem. Ber.* **105**, 1356.
Dötz, K. H., and Fischer, E. O. (1972b). *Chem. Ber.* **105**, 3966.
Dötz, K. H., and Fischer, E. O. (1972c). *J. Organomet. Chem.* **36**, C4.
Fischer, E. O. (1970). *Rev. Pure Appl. Chem.* **24**, 407.
Fischer, E. O. (1972). *Rev. Pure Appl. Chem.* **30**, 353.
Fischer, E. O. (1974). *Angew. Chem.* **86**, 651.
Fischer, E. O., and Fischer, H. (1974a). *Chem. Ber.* **107**, 657.
Fischer, E. O., and Fischer, H. (1974b). *Chem. Ber.* **107**, 673.
Fischer, E. O., and Kalbfus, W. (1972). *J. Organomet. Chem.* **46**, C15.
Fischer, E. O., and Kiener, V. (1967). *Angew. Chem., Int. Ed. Engl.* **6**, 961.

Fischer, E. O., and Klabunde, U. (1967). *J. Am. Chem. Soc.* **89**, 7141.
Fischer, E. O., and Knauss, L. (1970). *Chem. Ber.* **103**, 1262.
Fischer, E. O., and Knauss, L. (1971a). *J. Organomet. Chem.* **31**, C68.
Fischer, E. O., and Knauss, L. (1971b). *J. Organomet. Chem.* **31**, C71.
Fischer, E. O., and Kreis, G. (1973). *Chem. Ber.* **106**, 2310.
Fischer, E. O., and Kreissl, F. R. (1972). *J. Organomet. Chem.* **35**, C47.
Fischer, E. O., and Kreiter, C. G. (1969). *Angew. Chem., Int. Ed. Engl.* **8**, 761.
Fischer, E. O., and Leupold, M. (1972). *Chem. Ber.* **105**, 599.
Fischer, E. O., and Maasböl, A. (1964). *Angew. Chem.* **76**, 645.
Fischer, E. O., and Maasböl, A. (1967). *Chem. Ber.* **100**, 2445.
Fischer, E. O., and Maasböl, A. (1968). *J. Organomet. Chem.* **12**, P15.
Fischer, E. O., and Moser, E. (1968a). *J. Organomet. Chem.* **12**, P1.
Fischer, E. O., and Moser, E. (1968b). *J. Organomet. Chem.* **13**, 387.
Fischer, E. O., and Plabst, D. (1974). *Chem. Ber.* **107**, 3326.
Fischer, E. O., and Riedel, A. (1968). *Chem. Ber.* **101**, 156.
Fischer, E. O., and Riedmüller, S. (1974). *Chem. Ber.* **107**, 915.
Fischer, E. O., and Schubert, U. (1973). *Chem. Ber.* **106**, 3882.
Fischer, E. O., and Weiss, K. (1973). *Chem. Ber.* **106**, 1277.
Fischer, E. O., Heckl, B., Dötz, K. H., Müller, J., and Werner, H. (1969). *J. Organomet. Chem.* **16**, P29.
Fischer, E. O., Winkler, E., Kreiter, C. G., Huttner, G., and Krieg, B. (1971). *Angew. Chem., Int. Ed. Engl.* **10**, 922.
Fischer, E. O., Leupold, M., Kreiter, C. G., and Müller, J. (1972a). *Chem. Ber.* **105**, 150.
Fischer, E. O., Kreissl, F. R., Winkler, E., and Kreiter, C. G. (1972b). *Chem. Ber.* **105**, 588.
Fischer, E. O., Kreissl, F. R., Kreiter, C. G., and Meineke, E. W. (1972c). *Chem. Ber.* **105**, 2558.
Fischer, E. O., Fischer, H., and Werner, H. (1972d). *Angew. Chem., Int. Ed. Engl.* **11**, 644.
Fischer, E. O., Kreis, G., and Kreissl, F. R. (1973a). *J. Organomet. Chem.* **56**, C37.
Fischer, E. O., Schmid, K. R., Kalbfus, W., and Kreiter, C. G. (1973b). *Chem. Ber.* **106**, 3893.
Fischer, E. O., Kreis, G., Kreissl, F. R., Kreiter, C. G., and Müller, J. (1973c). *Chem. Ber.* **106**, 3910.
Fischer, E. O., Kreis, G., Kreiter, C. G., Müller, J., Huttner, G., and Lorenz, H. (1973d). *Angew. Chem., Int. Ed. Engl.* **12**, 564.
Fischer, E. O., Fischer, H., and Kreissl, F. R. (1974a). *J. Organomet. Chem.* **64**, C41.
Fischer, E. O., Fischer, H., and Kreiter, C. G. (1974b). *Chem. Ber.* **107**, 2459.
Fischer, E. O., Fischer, H., and Werner, H. (1974c). *J. Organomet. Chem.* **73**, 331.
Formacek, V., and Kreiter, C. G. (1972). *Angew. Chem., Int. Ed. Engl.* **11**, 141.
Gerhart, F., and Schöllkopf, U. (1967). *Angew. Chem., Int. Ed. Engl.* **6**, 560.
Green, M. L. H., Mitchard, L. C., and Swanwick, M. G. (1971). *J. Chem. Soc. A* p. 794.
Grubbs, R. H., Burk, P. L., and Carr, D. D. (1975). *J. Am. Chem. Soc.* **97**, 3265.
Heckl, B., Werner, H., and Fischer, E. O. (1968). *Angew. Chem., Int. Ed. Engl.* **7**, 817.
Herberhold, M., and Öfele, K. (1970). *Angew. Chem., Int. Ed. Engl.* **9**, 739.
Herrmann, W. A. (1974). *Angew. Chem., Int. Ed. Engl.* **13**, 599.
Herrmann, W. A. (1975). *Chem. Ber.* **108**, 486.
Katz, T. J., and McGinnis, J. (1975). *J. Am. Chem. Soc.* **97**, 1592.
Katz, T. J., and Rothchild, R. (1976). *J. Am. Chem. Soc.* **98**, 2519.
King, R. B. (1963). *J. Am. Chem. Soc.* **85**, 1922.
Kreissl, F. R., Fischer, E. O., Kreiter, C. G., and Weiss, K. (1973a). *Angew. Chem., Int. Ed. Engl.* **12**, 563.

Kreissl, F. R., Fischer, E. O., Kreiter, C. G., and Fischer, H. (1973b). *Chem. Ber.* **106**, 1262.
Kreissl, F. R., Fischer, E. O., and Kreiter, C. G. (1973c). *J. Organomet. Chem.* **57**, C9.
Kreiter, C. G. (1968). *Angew. Chem., Int. Ed. Engl.* **7**, 390.
Kreiter, C. G., and Öfele, K. (1972). *Chem. Ber.* **105**, 529.
Lappert, M. F., and Oliver, A. J. (1972). *J. Chem. Soc., Chem. Commun.* p. 274.
Miller, J., Balch, A. L., and Enemark, J. H. (1971). *J. Am. Chem. Soc.* **93**, 4613.
Mills, O. S., and Redhouse, A. D. (1968). *J. Chem. Soc. A* p. 642.
Öfele, K. (1968). *Angew. Chem., Int. Ed. Engl.* **7**, 950.
Öfele, K. (1970). *J. Organomet. Chem.* **22**, C9.
Rascher, H., and Werner, H. (1968a). *Inorg. Chim. Acta* **2**, 181.
Rascher, H., and Werner, H. (1968b). *Helv. Chim. Acta* **51**, 1765.
Rees, C. W., and von Angerer, E. (1972). *J. Chem. Soc., Chem. Commun.* p. 420.
Sanders, A., Cohen, L., Giering, W. P., Kenedy, D., and Magatti, C. V. (1973). *J. Am. Chem. Soc.* **95**, 5430.
Schrock, R. R. (1974). *J. Am. Chem. Soc.* **96**, 6796.
Soufflet, J. P., Commereuc, D., and Chauvin, Y. (1973). *C. R. Hebd. Seances Acad. Sci., Ser. C* **276**, 169.
Sundberg, R. J., Bryan, R. F., Taylor, I. F., and Taube, H. (1974). *J. Am. Chem. Soc.* **96**, 381.
Weiss, K., Fischer, E. O., and Müller, J. (1974). *Chem. Ber.* **107**, 3548.
Werner, H. (1968). *Angew. Chem., Int. Ed. Engl.* **7**, 930.
Werner, H., Fischer, E. O., Heckl, B., and Kreiter, C. G. (1971). *J. Organomet. Chem.* **28**, 367.

AUTHOR INDEX

Numbers in italics refer to the pages on which the complete references are listed.

C

D

G

H

K

L

M

N

R

S

T

U

V

W

Y

Z

SUBJECT INDEX

D

Q

R

S

T

ORGANIC CHEMISTRY

A SERIES OF MONOGRAPHS

1. Wolfgang Kirmse. CARBENE CHEMISTRY, 1964; 2nd Edition, 1971
2. Brandes H. Smith. BRIDGED AROMATIC COMPOUNDS, 1964
3. Michael Hanack. CONFORMATION THEORY, 1965
4. Donald J. Cram. FUNDAMENTALS OF CARBANION CHEMISTRY, 1965
5. Kenneth B. Wiberg (Editor). OXIDATION IN ORGANIC CHEMISTRY, PART A, 1965; Walter S. Trahanovsky (Editor). OXIDATION IN ORGANIC CHEMISTRY, PART B, 1973
6. R. F. Hudson. STRUCTURE AND MECHANISM IN ORGANO-PHOSPHORUS CHEMISTRY, 1965
7. A. William Johnson. YLID CHEMISTRY, 1966
8. Jan Hamer (Editor). 1,4-CYCLOADDITION REACTIONS, 1967
9. Henri Ulrich. CYCLOADDITION REACTIONS OF HETEROCUMULENES, 1967
10. M. P. Cava and M. J. Mitchell. CYCLOBUTADIENE AND RELATED COMPOUNDS, 1967
11. Reinhard W. Hoffman. DEHYDROBENZENE AND CYCLOALKYNES, 1967
12. Stanley R. Sandler and Wolf Karo. ORGANIC FUNCTIONAL GROUP PREPARATIONS, VOLUME I, 1968; VOLUME II, 1971; VOLUME III, 1972
13. Robert J. Cotter and Markus Matzner. RING-FORMING POLYMERIZATIONS, PART A, 1969; PART B, 1; B, 2, 1972
14. R. H. DeWolfe. CARBOXYLIC ORTHO ACID DERIVATIVES, 1970
15. R. Foster. ORGANIC CHARGE-TRANSFER COMPLEXES, 1969
16. James P. Snyder (Editor). NONBENZENOID AROMATICS, VOLUME I, 1969; VOLUME II, 1971

17. C. H. Rochester. ACIDITY FUNCTIONS, 1970

18. Richard J. Sundberg. THE CHEMISTRY OF INDOLES, 1970

19. A. R. Katritzky and J. M. Lagowski. CHEMISTRY OF THE HETEROCYCLIC *N*-OXIDES, 1970

20. Ivar Ugi (Editor). ISONITRILE CHEMISTRY, 1971

21. G. Chiurdoglu (Editor). CONFORMATIONAL ANALYSIS, 1971

22. Gottfried Schill. CATENANES, ROTAXANES, AND KNOTS, 1971

23. M. Liler. REACTION MECHANISMS IN SULPHURIC ACID AND OTHER STRONG ACID SOLUTIONS, 1971

24. J. B. Stothers. CARBON-13 NMR SPECTROSCOPY, 1972

25. Maurice Shamma. THE ISOQUINOLINE ALKALOIDS: CHEMISTRY AND PHARMACOLOGY, 1972

26. Samuel P. McManus (Editor). ORGANIC REACTIVE INTERMEDIATES, 1973

27. H.C. Van der Plas. RING TRANSFORMATIONS OF HETEROCYCLES, VOLUMES 1 AND 2, 1973

28. Paul N. Rylander. ORGANIC SYNTHESES WITH NOBLE METAL CATALYSTS, 1973

29. Stanley R. Sandler and Wolf Karo. POLYMER SYNTHESES, VOLUME I, 1974; VOLUME II, IN PREPARATION

30. Robert T. Blickenstaff, Anil C. Ghosh, and Gordon C. Wolf. TOTAL SYNTHESIS OF STEROIDS, 1974

31. Barry M. Trost and Lawrence S. Melvin, Jr. SULFUR YLIDES: EMERGING SYNTHETIC INTERMEDIATES, 1975

32. Sidney D. Ross, Manuel Finkelstein, and Eric J. Rudd. ANODIC OXIDATION, 1975

33. Howard Alper (Editor). TRANSITION METAL ORGANOMETALLICS IN ORGANIC SYNTHESIS, VOLUME I, 1976

In preparation

Alan P. Marchand and Roland E. Lehr (Editors). PERICYCLIC REACTIONS, VOLUMES I AND II

Pierre Crabbé (Editor). PROSTAGLANDIN RESEARCH